Digital Ethics Lab Yearbook

Series Editors
Luciano Floridi, Oxford Internet Institute, Digital Ethics Lab,
University of Oxford, Oxford, UK
Department of Legal Studies, University of Bologna, Bologna, Italy

Mariarosaria Taddeo, Oxford Internet Institute, Digital Ethics Lab,
University of Oxford, Oxford, UK
The Alan Turing Institute, London, UK

The Digital Ethics Lab Yearbook is an annual publication covering the ethical challenges posed by digital innovation. It provides an overview of the research from the Digital Ethics Lab at the Oxford Internet Institute. Volumes in the series aim to identify the benefits and enhance the positive opportunities of digital innovation as a force for good, and avoid or mitigate its risks and shortcomings. The volumes build on Oxford's world leading expertise in conceptual design, horizon scanning, foresight analysis, and translational research on ethics, governance, and policy making.

Jakob Mökander • Marta Ziosi

Editors

The 2021 Yearbook
of the Digital Ethics Lab

Springer

Editors
Jakob Mökander
Oxford Internet Institute
University of Oxford
Oxford, UK

Center for Information Technology Policy
Princeton University
Princeton, NJ, USA

Marta Ziosi
Oxford Internet Institute
University of Oxfordo
Oxford, UK

ISSN 2524-7719   ISSN 2524-7727   (electronic)
Digital Ethics Lab Yearbook
ISBN 978-3-031-09848-2   ISBN 978-3-031-09846-8   (eBook)
https://doi.org/10.1007/978-3-031-09846-8

This Springer imprint is published by the registered company Springer Nature Switzerland AG
The registered company address is: Gewerbestrasse 11, 6330 Cham, Switzerland

Preface

The field of digital ethics – whether understood as an academic discipline or an area of practice – is maturing. This process has both been propelled and reflected by two long-term trends. First, and most importantly, the focus of the discourse concerning how to design and use digital technologies is increasingly shifting from 'soft ethics' to 'hard governance'. The second trend is an ongoing shift from 'what' to 'how', whereby abstract or ad-hoc approaches to AI governance are giving way to more concrete and systematic solutions. While these trends are neither new nor surprising, the maturing of the field of digital ethics has, as we shall see, been accelerated by a series of recent events.

Consider the shift in focus from soft to hard governance. While the latter is enforced by government institutions, the former relies on mechanisms that allow for some contextual flexibility, such as cultural norms and economic incentives. The plethora of 'AI ethics' guidelines or principles produced by regulators and technology providers alike in recent years, including the *Ethics Guidelines for Trustworthy* (AI HLEG, 2019), the *Montreal Declaration for a Responsible Development of AI* (University of Montreal, 2018), and the *Beijing AI Principles* (Beijing Academy of Artificial Intelligence, 2019), constitute soft governance. In contrast, the *Artificial Intelligence Act* (AIA) published by the European Commission (2021) is an example of hard governance (Mökander et al., 2021).

The AIA is a unique milestone insofar as it is the first attempt to elaborate a general legal framework for AI carried out by any major economy. Yet, the AIA did not come as a surprise. Several recent initiatives and publications have foreshadowed the arrival of hard legislation.[1] Moreover, the need to manage the ethical challenges posed by autonomous and self-learning systems has been pressing and clear for a long time, and the fact that soft and hard mechanisms complement and reinforce each other is well established in the governance literature (Erdelyi & Goldsmith, 2018; Floridi, 2018). The AIA can thus be viewed as one example of how the focus in the field of digital ethics is shifting from soft to hard governance (Floridi, 2021). Following the same logic, the eventual imposition of hard

---

[1] For example, the European Commission's *Whitepaper on AI* (2020).

legislation on the design and use of digital technologies is to be expected outside the EU as well – although the shape such legislation will take is likely to vary between different jurisdictions. A step in that direction was taken with the *Algorithmic Accountability Act* of 2022 (AAA), which was put before the U.S. Senate by the Office of Senator Ron Wyden (2022). The AAA calls on companies to conduct impact assessments for bias, accuracy, and other issues when designing or deploying automated systems that make critical decisions with little or no human intervention (Mökander et al., 2022).

That brings us to the second long-term trend, from theoretical to implemented solutions. While playing an important role in raising awareness of the ethical challenges associated with specific technologies, early works in the field of digital ethics remained largely abstract. Of course, the convergence around a set of high-level ethics principles to guide the design and use of digital technologies was a significant achievement in and of itself. However, researchers quickly established that technology providers lacked both incentives and translational tools to interpret, implement, and demonstrate adherence to abstract ethics principles, namely a link was missing from what to how (Morley et al., 2020; Taddeo & Floridi, 2018). In response to this critical knowledge gap, a rich literature has emerged on how organisations can ensure that the technologies they design or deploy are ethical, legal, and technically robust in practice (see e.g., AIEIG, 2020; Ayling & Chapman, 2021; Mökander & Axente, 2021; Morley et al., 2021). In these attempts to provide more detailed guidance, both policymakers and academic researchers have drawn upon established best practices to provide adequate assurance in adjacent fields, including quality management in systems engineering, auditing in the financial sector, and pre-market testing and approval procedures in safety-sensitive areas such as food and medical devices.

Both trends discussed above are reflected in this volume: the fourth edition of the *Digital Ethics Lab Yearbook*. The shift from soft ethics to hard governance runs like a red thread through the first half of this volume and binds together seven chapters that otherwise cover a wide range of domains and geographic areas. In 'The European Legislation on AI: A Brief Analysis of Its Philosophical Approach', Luciano Floridi highlights some foundational aspects of the AIA and analyses the regulatory approach underpinning it; in 'Informational Privacy with Chinese Characteristics', Huw Roberts discusses the emergence of a new privacy protection regime in China; in 'Lessons Learned from Co-governance Approaches – Developing Effective AI Policy in Europe', Caitlin Corrigan demonstrates that addressing the ethical challenges posed by AI systems will require close collaboration between state and non-state actors; in 'State-Firm Coordination in AI Governance', Noah Schöppl discusses the role of states in digital governance and argues that national governments need to increase and coordinate their regulatory capabilities; in 'The Impact of Australia's News Media Bargaining Code on Journalism, Democracy, and the Battle to Regulate Big Tech', Emmie Hine analyses the new Australian legislation designed to provide financial support to publishers and journalists in terms of its compatibility with the business models of big tech giants; in 'App Store Governance: The Implications and Limitations of Duopolistic Dominance', Josh Cowls and Jessica Morley discuss the challenges and tensions inherent to app store governance; and in 'A Legal Principles-Based Framework for AI Liability Regulation', Massimo Durante and Luciano Floridi review the work of the European Commission's Expert Group on Liability and New Technologies (2019) to show how it has started to lay the basis for a set of legal principles for an AI liability regime.

Similarly, the nine chapters in the latter half of this volume are linked insofar as they concern concrete procedures for implementing digital governance in practice. In 'The New Morality of Debt', Nikita Aggarwal argues for the inadequacy of existing regulatory frameworks governing consumer lending in alleviating harms around privacy, autonomy, and dignity; in 'Site of the Living Dead: Clarifying Our Moral Obligations Towards Digital Remains', Mira Pijselman assesses the absence of a unified roadmap for how digital remains ought to be managed; in 'The Statistics of Interpretable Machine Learning', David Watson provides an in-depth survey of the affordances and constraints in the plethora of existing interpretable machine learning approaches; in 'Formalising Trade-Offs Beyond Algorithmic Fairness: Lessons from Ethical Philosophy and Welfare Economics', Michelle Lee and colleagues introduce the use of Key Ethics Indicators (KEIs) as a way towards understanding whether or not an algorithmic system is aligned to a decision-maker's ethical values; in 'Ethics Auditing Framework for Trustworthy AI: Lessons from the IT Audit Literature', Nathaniel Zinda explores how the emerging field of 'AI auditing' can learn from and build on traditional IT audits; in 'Ethics Auditing: Lessons from Business Ethics for Ethics Auditing of AI', Noah Schöppl and colleagues conduct a similar review of the business ethics literature to establish best practices for how auditing – as a governance mechanism – can help organisations (a) design AI systems in ways that are ethical and (b) make verifiable claims about those systems; in 'AI Ethics and Policies: Why European Journalism Needs More of Both', Guido Romeo and Emanuela Griglié argue that policymakers can help newsrooms manage the ethical issues raised by the use of AI in journalism by supporting the development of tools like checklists and guidance on how to use such tools; in 'Towards Equitable Health Outcomes Using Group Data Rights', Gal Wachtel proposes a framework for practically implementing group data rights in a healthcare setting; and, finally, in 'Ethical Principles for Artificial Intelligence in National Defence', Mariarosaria Taddeo and colleagues propose a framework consisting of five principles and issue-related recommendations to foster ethically sound uses of AI for national defence purposes.

From its very start in 2018, the main purpose of the *Digital Ethics Lab Yearbook* has been to give a non-exhaustive snapshot of the diverse and cutting-edge research agendas being pursued within our research group at the Oxford Internet Institute. However, the 2020/21 Yearbook marks the maturation not only of the field of digital ethics but also of the Digital Ethics Lab itself. As the discipline develops in terms of thematic focus and methodological rigour, so do our ways of working. The Digital Ethics Lab has served its purpose of identifying the opportunities and enhancing the benefits of digital innovation whilst showing how to avoid or mitigate the associated risks. To reflect the shifting focus – from soft to hard governance and from abstract to more concrete solutions – our future efforts will be directed towards supporting the newly formed Centre for Digital Ethics and Governance at the University of Bologna and the Digital Governance Research Group at Exeter College, Oxford.

Oxford, UK   Jakob Mökander
Princeton, NJ, USA
Oxford, UK   Marta Ziosi

References

AI HLEG. (2019). *European Commission's ethics guidelines for trustworthy Artificial Intelligence* (Issue May). https://ec.europa.eu/futurium/en/ai-alliance-consultation/guidelines/1

AIEIG. (2020). *From principles to practice – An interdisciplinary framework to operationalise AI ethics* (pp. 1–56). AI Ethics Impact Group, VDE Association for Electrical Electronic & Information Technologies e.V., Bertelsmann Stiftung. https://doi.org/10.11586/2020013

Ayling, J., & Chapman, A. (2021). Putting AI ethics to work: Are the tools fit for purpose? *AI and Ethics, 0123456789*. https://doi.org/10.1007/s43681-021-00084-x

Beijing Academy of Artificial Intelligence. (2019). *The Beijing AI principles.*

Erdelyi, O. J., & Goldsmith, J. (2018). *Regulating Artificial intelligence proposal for a global solution.* In AAAI/ACM conference on artificial intelligence, ethics and society. http://www.aies-conference.com/wp-content/papers/main/AIES_2018_paper_13.pdf

European Commission. (2020). *White Paper on Artificial Intelligence – A European approach to excellence and trust.* 27.

European Commission. (2021). *Proposal for regulation of the European parliament and of the council – Laying down harmonised rules on artificial intelligence (artificial intelligence act) and amending certain Union legislative acts.*

Floridi, L. (2018). Soft ethics and the governance of the digital. *Philosophy and Technology, 31*(1). https://doi.org/10.1007/s13347-018-0303-9

Floridi, L. (2021). The end of an Era: From self-regulation to hard law for the digital industry. *Philosophy and Technology, 34*(4), 619–622. https://doi.org/10.1007/s13347-021-00493-0

Mökander, J., & Axente, M. (2021). Ethics-based auditing of automated decision-making systems: Intervention points and policy implications. *AI & SOCIETY, 0123456789*. https://doi.org/10.1007/s00146-021-01286-x

Mökander, J., Axente, M., Casolari, F., & Floridi, L. (2021). Conformity assessments and post-market monitoring: A guide to the role of auditing in the proposed European AI regulation. *Minds and Machines, 0123456789*, 1–27. https://doi.org/10.1007/s11023-021-09577-4

Mökander, J., Juneja, P., Watson, D.S. et al. (2022). The US Algorithmic Accountability Act of 2022 vs. The EU Artificial Intelligence Act: what can they learn from each other?. *Minds and Machines.* https://doi.org/10.1007/s11023-022-09612-y

Morley, J., Floridi, L., Kinsey, L., & Elhalal, A. (2020). From what to how: An initial review of publicly available AI ethics tools, methods and research to translate principles into practices [Article]. *Science and Engineering Ethics, 26*(4), 2141. https://doi.org/10.1007/s11948-019-00165-5

Morley, J., Elhalal, A., Garcia, F., Kinsey, L., Mökander, J., & Floridi, L. (2021). Ethics as a service: A pragmatic operationalisation of AI ethics. *Minds and Machines, 31*(2), 239–256. https://doi.org/10.1007/s11023-021-09563-w

Office of U.S. Senator Ron Wyden. (2022). Algorithmic accountability act of 2022. *117th Congress 2D Session.* https://doi.org/10.1016/S0140-6736(02)37657-8

Taddeo, M., & Floridi, L. (2018). How AI can be a force for good. *Science, 361*(6404), 751–752. https://doi.org/10.1126/science.aat5991

The European Commission's Expert Group on Liability and New Technologies-New Technologies Formation. (2019). *Liability for Artificial Intelligence and other emerging digital technologies.* https://doi.org/10.2838/25362

University of Montreal. (2018). *Montréal Declaration responsible AI.* https://www.montrealdeclaration-responsibleai.com/the-declaration

Contents

Contributors

Nikita Aggarwal Faculty of Law, University of Oxford, Oxford, UK
Oxford Internet Institute, University of Oxford, Oxford, UK

Alexander Blanchard The Alan Turing Institute, British Library, London, UK

Caitlin C. Corrigan Institute for Ethics in Artificial Intelligence, Technical University of Munich, Munich, Germany

Josh Cowls Oxford Internet Institute, University of Oxford, Oxford, UK

Massimo Durante Department of Law, University of Turin, Turin, Italy

Elizabeth Edgar Defence Science Technology Laboratory (Dstl), Salisbury, UK

Luciano Floridi Oxford Internet Institute, University of Oxford, Oxford, UK
Department of Legal Studies, University of Bologna, Bologna, Italy

Emanuela Griglié La Stampa, Torino, Italy

Emmie Hine University of Oxford, Oxford, UK

Michelle Seng Ah Lee Department of Computer Science & Technology, Compliant & Accountable Systems Research Group, University of Cambridge, Cambridge, UK

David McNeish Defence Science Technology Laboratory (Dstl), Salisbury, UK

Jakob Mökander Oxford Internet Institute, University of Oxford, Oxford, UK
Center for Information Technology Policy, Princeton University, Princeton, NJ, USA

Jessica Morley Oxford Internet Institute, University of Oxford, Oxford, UK

Mira Pijselman Oxford Internet Institute, University of Oxford, Oxford, UK

Huw Roberts Oxford Internet Institute, University of Oxford, Oxford, UK

Guido Romeo Facta Journalism Centre, Bologna, Italy

Noah Schöppl Oxford Internet Institute, University of Oxford, Oxford, UK

Jatinder Singh Department of Computer Science & Technology, Compliant & Accountable Systems Research Group, University of Cambridge, Cambridge, UK

Mariarosaria Taddeo Oxford Internet Institute, University of Oxford, Oxford, UK
The Alan Turing Institute, British Library, London, UK

Gal Wachtel Oxford University, Oxford, UK
Harvard University, Cambridge, MA, USA

David S. Watson Department of Statistical Science, University College London, London, UK

Nathaniel Zinda Oxford Internet Institute, University of Oxford, Oxford, UK

Marta Ziosi Oxford Internet Institute, University of Oxford, Oxford, UK

The European Legislation on AI: A Brief Analysis of Its Philosophical Approach

Luciano Floridi

Abstract On 21 April 2021, the European Commission published the proposal of the new EU Artificial Intelligence Act (AIA)—one of the most influential steps taken so far to regulate AI internationally. This chapter highlights some foundational aspects of the Act and analyses the philosophy behind its proposal.

Keywords Artificial Intelligence · European Commission · Governance · Legislation

Some European legislation on Artificial Intelligence (AI) had been expected at least since 16 July 2019. On that date, Ursula von der Leyen had pledged that, within 100 days of her election as President of the European Commission, she would have proposed new legislation on AI.[1] At that time, I remarked that it was a reasonable strategy but an unrealistic timeline. The High-Level Expert Group on AI (HLEG, of which I was a member),[2] organised by the European Commission, had only recently published its *Ethics Guidelines for Trustworthy AI* (HLEGAI 8 April 2019a) and its *Policy and Investment Recommendations for Trustworthy AI* (HLEGAI 26 June 2019b). It seemed evident that the next step would have been the translation of those guidelines and recommendations into a legal framework (Floridi, 2019a). However, the work carried out by the HLEG had also shown that the road ahead was going to be long and laborious. I figured it would have taken at least a year, not 3 months. I was optimistic. On 19 February 2020, the Commission published the *White Paper on AI - A European Approach to Excellence and Trust* (European Commission 19

---

[1] https://ec.europa.eu/commission/presscorner/detail/en/ip_20_403

[2] https://digital-strategy.ec.europa.eu/en/policies/expert-group-ai

---

L. Floridi (✉)
Oxford Internet Institute, University of Oxford, Oxford, UK

Department of Legal Studies, University of Bologna, Bologna, Italy
e-mail: luciano.floridi@oii.ox.ac.uk

© The Author(s), under exclusive license to Springer Nature   1
Switzerland AG 2022
J. Mökander, M. Ziosi (eds.), *The 2021 Yearbook of the Digital Ethics Lab*,
Digital Ethics Lab Yearbook, https://doi.org/10.1007/978-3-031-09846-8_1

February 2020). The document outlined a risk-based approach to AI and policies to promote the uptake of such technology. But, meanwhile, the COVID-19 pandemic had begun to spread, with its deadly effects and immense disruptions.[3] Despite this, on 21 April 2021, the European Commission published the proposal of the new EU *Artificial Intelligence Act* (henceforth AIA), or, to use its full name, the *Proposal for a regulation of the European Parliament and the Council laying down harmonised rules on Artificial Intelligence (Artificial Intelligence Act) and amending certain Union legislative acts* (Artificial Intelligence Act 21 April 2021).

According to the European Data Protection Supervisor website, the AIA is "the first initiative, worldwide, that provides a legal framework for Artificial Intelligence (AI)".[4] Regardless of whether this may be true (see, for example, the US National AI Initiative Act, which became law on 1 January 2021), the AIA is one of the most influential regulatory steps taken so far internationally. On the whole, it is a good starting point to ensure that the development of AI in the EU is ethically sound, legally acceptable, socially equitable, and environmentally sustainable, with a vision of AI that seeks to support the economy, society and the environment. This is no small ambition, and it will take time and effort to reach a final text that can come close to fulfil it. Yet the ambition, like von der Leyen's pledge, remains substantially reasonable because the EU is ideally placed to deliver such a normative framework.

Of course, the technical and legal aspects of the AIA will evolve as the proposal goes to the European Parliament and the Council of the European Union, that is, the EU legislator, for further consideration and debate. This is a proposal, and it may take a couple of years before it will be finalised and become binding (the process took 4 years for the GDPR (General Data Protection Regulation 27 April 2016), followed by an implementation period of 2 years). However, the underlying philosophy is commendable, despite some limitations. In what follows, I shall highlight some foundational aspects that seem to be more significant. Just a final note of clarification: there are already plenty of short summaries of the proposed legislation, and I shall not provide another one here. Nor shall I comment on some obvious shortcomings already identified by many, from the definition of AI (always a problem for anyone approaching the subject), to the more or less complete and loop-holy list of AI uses or technologies that should be banned. What I am interested in analysing is the philosophy behind the proposal.

From a general perspective, the AIA is a "regulation", not a "directive", so, like the GDPR, it will enter into force on a set date in all 27 Member States, and it will have binding legal force throughout the EU (a directive only indicates goals that each Member State must achieve but does not indicate how to transpose the goals into national laws). Moreover, AIA provisions conferring rights upon individuals and having a sufficiently clear, precise, and unconditional content will enjoy a direct effect, meaning that natural or legal persons will have the possibility to invoke those

---

[3] https://www.ajmc.com/view/a-timeline-of-covid19-developments-in-2020

[4] The quote comes from https://edps.europa.eu/press-publications/press-news/press-releases/2021/artificial-intelligence-act-welcomed-initiative_en

rights before national courts and tribunals. Also, like the GDPR, the AIA is "extra-territorial", in the more technical sense that it has *territorial extension* (Scott, 2014), or, as I would prefer to put it, it is *aterritorial*, a concept much more coherent with the other three concepts (see below) of *Brussels effect*, *digital constitutionalism*, and *digital sovereignty*. It assumes a post-Westphalian world in which the territoriality of the law no longer applies automatically and may be irrelevant (Floridi, 2014). What counts is whether an AI system or service—for example, a loan management program based on machine learning—has an impact on European citizens, not where the company that provides it or uses it is located, whether physically or legally. This unified, post-Westphalian approach is likely to have several positive effects. The EU presents itself as a single interlocutor, not only in the management of personal data (GDPR), but now also for AI applications. AI companies and vendors[5] will have to deal with the EU, not with individual Member States, when they will have to prove that they comply with the new legislation. Each Member States will appoint a national authority responsible for supervising AI. However, a new *European Artificial Intelligence Board* (EAIB, akin to the GDPR's European Data Protection Board[6])—consisting of representatives from every Member State, the European Data Protection Supervisor and the Commission—will assist national supervisory authorities and EU lawmakers to ensure the consistent application of the AI Regulation, e.g., regarding the list of prohibited AI practices and high-risk systems. Companies offering AI services are unlikely to be able to ignore a market of 450 million people that accounts for about 1/6 of the world economy, so they will comply (there are steep administrative fines for different kinds of infringements, from up to €10 M or 2% of worldwide annual turnover whichever is higher, to €30 M or 6% of annual worldwide turnover, whichever is higher). This will further extend the so-called *Brussel effect* (Bradford, 2020), whereby companies end up complying with EU regulations even in other countries because it is more practical to have a single approach globally, enabling the EU to extend *de facto* (though not *de jure*) its laws internationally, through market mechanisms. Companies will also find it more difficult to explain why they do not adopt standards just as high when operating in other countries (Floridi, 2019c). Furthermore, the AIA will place the EU in a position of "leadership by example" for the good governance of AI technologies, especially when interacting with other countries at the forefront in the field of AI research and development (R&D), like Canada,[7] Israel, Japan, Singapore, South Korea, and the UK. In this case too, one may expect a "harmonising" effect similar to that caused by the GDPR. Collaboration between the EU and other countries will be easier thanks to an explicit and reliable legislative framework. As for China and the US, they will not adopt the same EU approach, but they will need to find ways of collaborating with the EU. China may take inspiration from the AIA to

---

[5] https://www.datamation.com/artificial-intelligence/ai-companies/

[6] The Commission may have more influnce on the EAIB than it has on the EDPB.

[7] A Regulatory Framework for AI: Recommendations for PIPEDA Reform https://www.priv.gc.ca/en/about-the-opc/what-we-do/consultations/completed-consultations/consultation-ai/reg-fw_202011/

develop its legislation tailored to its approach to AI (Roberts et al., 2021a, b). The US is more likely to adopt an antitrust approach[8] (Cath et al., 2018; Roberts et al., forthcoming), but the AIA may influence state-level legislation, as it already happened with the California Consumer Privacy Act (Barrett, 2019). A future EU-US Trade and Technology Council[9] may also provide a shared platform. More generally, AI could become a subject of negotiation within the framework of the contractual relations that the EU is establishing with third countries.

From an ethical perspective, the AIA inherits the same foundational approach seen in the GDPR: it is based on protecting human dignity and fundamental rights. This is a very positive feature, even if the current proposal of the AIA is more top-down, less flexible and less focused on the protection of citizens and their rights than the GDPR. Unfortunately, the AIA uses an anachronistic terminology to define this approach as "human-centric", that is, as an approach that places humanity at the centre of technological development. Yet this is both trivially true and dangerously ambiguous. On the one hand, it is obvious that any technology, AI included, must be at the service of humanity, its values, and needs. On the other hand, one must also consider the environment as crucially important, yet "human-centric" seems to be synonymous with "anthropocentric", and we know how much the planet has suffered from humanity's obsession with its importance and centrality, as if everything must always be at its service, including every aspect of the natural world, no matter at what costs and losses. Fortunately, despite the unfortunate and obsolete terminology, the underlying vision is sound: the AIA emphasises the value of AI as a technology that can be very "green" and provide extraordinary support against pollution and climate change and for the sustainable development of information societies (Cowls et al., 2021). In this, the AIA's approach strengthens the idea that protecting the environment must be a cross-cutting issue at the EU level. It is what I have called the *Green and the Blue human project* (Floridi, 2019b; Floridi & Nobre, 2020), which the EU can and should promote in the world.

Still in terms of ethical approach, the AIA explicitly adopts the ethical guidelines proposed by the HLEG and seeks to eliminate or mitigate the risks of AI, support public trust in these innovative technologies, and further the development and adoption of AI in the EU. This risk-based approach seems convincing (it is a common approach for internal market-based legislation) and aligned with the view that ethics benefits the market, not vice versa. But precisely for this reason, one may argue that the AIA could do much more to protect consumers' rights and be much more incisive about providing measures to redress the possible harms or losses that AI systems may cause. This is the part where one may expect and welcome more improvements in the proposal. It was one of the main recommendations made by the

---

[8] Aiming for truth, fairness, and equity in your company's use of AI: https://www.ftc.gov/news-events/blogs/business-blog/2021/04/aiming-truth-fairness-equity-your-companys-use-ai

[9] Joint communication to the European Parliament, the European Council, and the Council—A new EU-US agenda for global change: https://ec.europa.eu/info/sites/default/files/joint-communication-eu-us-agenda_en.pdf

AI4People project: "7. Develop a redress process or mechanism to remedy or compensate for a wrong" (Floridi et al., 2018).

From a technological perspective, one must also praise what is *not* in the proposal. Following an already established approach,[10] the AIA avoids any sci-fi speculations about AI. Unfortunately, scaremongering stories have irresponsibly distracted not only the general public (Floridi, 2016), but initially also the work of the European Parliament (European Parliament 16 February 2017), which experts had criticised in an open letter to the European Commission,[11] and even of the HLEG, where I was among those who strongly insisted about removing non-scientific statements in the first draft to fanciful things such as "artificial consciousness" or AI with "subjective experience" (HLEGAI 18 December 2018). Correctly, the AIA treats AI as a technology for solving problems and performing tasks, not as some kind of Frankenstein's monster. Therefore, the proposal excludes the possibility of assigning to AI systems any status as a legal person, with rights and duties, such as the possibility of owning property, entering into contracts, suing and being sued, and so forth (Floridi & Taddeo, 2018). The responsibility of any AI system rests entirely with the people who design, manufacture, market, and use it. Coherently, the proposal stresses the importance of human oversight throughout the text.

I mentioned above some limits. This is a legislative proposal, and it is too early to indicate how it will be revised, but some conceptual improvements (I am not talking about legal and technical issues) may be in order.

At times, the text is ambiguous. For example, the definition of high-risk AI systems, a vital concept of the AIA, could be improved. Some AI systems are discussed as low- or zero-risk, and as such, they seem to fall outside the scope of strict compliance and be subject only to voluntary codes of conduct, yet it is unclear how this taxonomy will work in practice, leaving too much room for uncertainty and loopholes. And in the case of high-risk, the proposal explicitly combines two senses that it would be preferable to distinguish. On the one hand, there are AI systems that are high-risk because vital issues depend on their proper functioning. Think of an autonomous driving system: it is a "good thing" that should not fail to work. On the other hand, there are AI systems that are high-risk because, if they are used unethically, they can cause significant troubles, think of the abuse of remote, real-time biometric identification for law enforcement purposes, a kind of technological surveillance banned by the proposal. This is a "bad thing" that should not be put into operation. If one does not distinguish between these two senses of high-risk system—something is high risk if it fails to work vs. something is high risk if it is put to work—then one may end confusing the *resilience* that "good" AI systems must have, with the *resistance* that must be exerted towards the "bad" AI systems. Note that conceptual confusion and uncertainty about the specific nature of the risks involved in the design, development and deployment of AI systems will also affect

---

[10] https://www.euractiv.com/section/digital/opinion/the-eu-is-right-to-refuse-legal-personality-for-artificial-intelligence/

[11] https://www.politico.eu/wp-content/uploads/2018/04/RoboticsOpenLetter.pdf

the feasibility of any conformity assessment (auditing, see (Floridi et al., 2018; Mökander & Floridi, 2021; Mökander et al., 2021)) a crucial element in the AIA and the certification system it proposes (Mökander et al., 2022).

In other cases, the proposal is vague, such as when it comes to banning the use of AI systems intended to distort human behaviour, with probable physical or psychological harm. The intent is commendable, but it might risk banning even unproblematic AI systems if this approach were applied in a Draconian way.

Finally, some expectations in the proposal seem too idealistic. For example, consider the properties that the databases used for training machine learning models should satisfy. "Training, validation and testing data sets should be sufficiently relevant, representative and free of errors and complete in view of the intended purpose of the system." (Recital 44). These are characteristics highly desirable but rarely met in full. Think, for instance, about the incompleteness and incorrectness of any public database. Therefore, it seems preferable to speak of thresholds below which failure to satisfy these characteristics would be unacceptable (note that I am assuming that "sufficiently" in the Recital modifies only "relevant" and not all the following properties—as in "sufficiently representative" and so forth—because if this is not the case, then the whole text is too vague).

The AIA will be added to the GDPR and, over time, to the Digital Services Act (Digital Services Act 15 December 2020) and the Digital Markets Act (Digital Markets Act 15 December 2020), which, once adopted, will regulate online platforms and services. When this "legislative square" will be complete—make it a pentagon if you add the Data Governance Act (Data Governance Act 25 November 2020), or a hexagram, with the announced European Health Data Space legislative proposal[12]—the EU will have developed a "digital constitutionalism" (Celeste 2019, De Gregorio, 2021) for an infosphere where its citizens may live and work better and more sustainably. As I have been arguing for a while, it is clear that the challenge is no longer digital innovation but the governance of the digital, and hence the new morphology of power (Floridi, 2015) and the shaping of digital sovereignty (Floridi, 2020). In tackling these normative challenges, the EU is not simply ahead; it has no competition. There remains an indirect risk to be stressed. The new legislation may not improve but merely push out of the EU some risky AI R&D and its related ethical-legal problems, inviting companies to develop their products and services in other countries where legislation is absent, or less stringent, or not enforced, while the EU turns a wilful blind eye—or just inadequately enforces its legislation—and imports services or products obtained elsewhere, checking their current compliance but not their problematic origin (think of an AI-based medical device originally trained on personal data without respecting the GDPR or the AIA). Not respecting this *atemporality* (it should not matter when unethical steps were taken) would contradict the *aterritoriality* of the legislation. In this case, the recommendation is obvious: the EU must keep both eyes open and apply its ethical and legal requirements consistently and without hypocrisy, not just to the *status quo*, but also to the *history* of what comes from other places (note that the proposal already

---

[12] https://ec.europa.eu/health/ehealth/dataspace_en

goes in this direction). After all, the EU founding Treaties state that, in its relations with the broader world, the EU should uphold and promote its values, contributing to the protection of human rights (Art. 3(5) TEU). It may be technically challenging, expensive and even internationally problematic, but nobody ever said that doing the right thing was going to be cheap and easy.[13]

References

Artificial Intelligence Act. (2021, April 21). Proposal for a regulation of the European Parliament and the Council laying down harmonised rules on Artificial Intelligence (Artificial Intelligence Act) and amending certain Union legislative acts. *EUR-Lex—52021PC0206*. https://eur-lex.europa.eu/legal-content/EN/TXT/?uri=CELLAR:e0649735-a372-11eb-9585-01aa75ed71a1

Barrett, C. (2019). Are the EU GDPR and the California CCPA becoming the de facto global standards for data privacy and protection? *Scitech Lawyer, 15*(3), 24–29.

Bradford, A. (2020). *The Brussels effect: How the European Union rules the world*. Oxford University Press.

Cath, C., Wachter, S., Mittelstadt, B., Taddeo, M., & Floridi, L. (2018). Artificial intelligence and the 'Good society': The US, EU, and UK approach. *Science and Engineering Ethics, 24*(2), 505–528.

Celeste, E. (2019). Digital constitutionalism: A new systematic theorisation' (2019). *International Review of Law, Computers & Technology, 33*, 76–99.

Cowls, J., Tsamados, A., Taddeo, M., & Floridi, L. (2021). *The AI gambit—Leveraging artificial intelligence to combat climate change: Opportunities, challenges, and recommendations*. SSRN.

Data Governance Act. (2020, November 25). Proposal for a regulation of the European Parliament and the Council on European data governance (Data Governance Act). *EUR-Lex—52020PC0767*. https://eur-lex.europa.eu/legal-content/EN/TXT/?uri=CELEX:52020PC0767

De Gregorio, G. (2021). The rise of digital constitutionalism in the European Union. *International Journal of Constitutional Law*. https://doi.org/10.1093/icon/moab001

Digital Markets Act. (2020, December 15). Proposal for a regulation of the European Parliament and the Council on contestable and fair markets in the digital sector (Digital Markets Act). *EUR-Lex—52020PC0842*. https://eur-lex.europa.eu/legal-content/en/ALL/?uri=COM:2020:842:FIN

Digital Services Act. (2020, December 15). Proposal for a regulation of the European Parliament and the Council on a Single Market For Digital Services (Digital Services Act) and amending Directive 2000/31/EC. *EUR-Lex—52020PC0825*. https://eur-lex.europa.eu/legal-content/EN/ALL/?uri=CELEX:52020PC0825

European Commission. (2020, February 19). *White Paper on AI—A European approach to excellence and trust*. https://ec.europa.eu/info/sites/default/files/commission-white-paper-artificial-intelligence-feb2020_en.pdf

European Parliament. (2017, February 16). *European Parliament resolution of 16 February 2017 with recommendations to the Commission on Civil Law Rules on Robotics* (2015/2103(INL)). https://www.europarl.europa.eu/doceo/document/TA-8-2017-0051_EN.html

---

[13] I wish to thank Federico Casolari, Joshua Jaffe, Francesca Mazzi, Oreste Pollicino, Huw Roberts, and Paul Timmers for their very helpful feedback on a previous version of this chapter. The chapter is really much better thanks to them and any remaining shortcomings are only mine.

Floridi, L. (2014). *The fourth revolution—How the infosphere is reshaping human reality*. Oxford University Press.

Floridi, L. (2015). The new grey power. *Philosophy & Technology, 28*(3), 329–332.

Floridi, L. (2016). Should we be afraid of AI? *Aeon Essays*. https://aeon.co/essays/true-ai-is-both-logically-possible-and-utterly-implausible

Floridi, L. (2019a). Establishing the rules for building trustworthy AI. *Nature Machine Intelligence, 1*(6), 261–262.

Floridi, L. (2019b). The green and the blue: Naïve ideas to improve politics in a mature information society. In *The 2018 yearbook of the digital ethics lab* (pp. 183–221). Springer.

Floridi, L. (2019c). Translating principles into practices of digital ethics: Five risks of being unethical. *Philosophy & Technology, 32*(2), 185–193.

Floridi, L. (2020). The fight for digital sovereignty: What it is, and why it matters, especially for the EU. *Philosophy & Technology, 33*(3), 369–378.

Floridi, L., & Nobre, K. (2020). *The green and the blue: How AI may be a force for good*. OECD.

Floridi, L., & Taddeo, M. (2018). Romans would have denied robots legal personhood. *Nature, 557*(7705), 309–309.

Floridi, L., Cowls, J., Beltrametti, M., Chatila, R., Chazerand, P., Dignum, V., Luetge, C., Madelin, R., Pagallo, U., Rossi, F., Schafer, B., Valcke, P., & Vayena, E. (2018). AI4People—An ethical framework for a good AI society: Opportunities, risks, principles, and recommendations. *Minds and Machines, 28*(4), 689–707.

General Data Protection Regulation. (2016, April 27). Regulation (EU) 2016/679 of the European Parliament and of the Council of 27 April 2016 on the protection of natural persons with regard to the processing of personal data and on the free movement of such data, and repealing Directive 95/46/EC (General Data Protection Regulation) (Text with EEA relevance). *EUR-Lex—32016R0679*. https://eur-lex.europa.eu/eli/reg/2016/679/oj

HLEGAI. (2018, December 18). *High-Level Expert Group on Artificial Intelligence, EU—Draft ethics guidelines for trustworthy AI*. https://digital-strategy.ec.europa.eu/en/library/draft-ethics-guidelines-trustworthy-ai

HLEGAI. (2019a, 8 April). *High-Level Expert Group on Artificial Intelligence, EU—Ethics guidelines for trustworthy AI*. https://ec.europa.eu/digital-single-market/en/news/ethics-guidelines-trustworthy-ai

HLEGAI. (2019b, 26 June). *High-Level Expert Group on Artificial Intelligence, EU—Policy and investment recommendations for trustworthy Artificial Intelligence*. https://digital-strategy.ec.europa.eu/en/library/policy-and-investment-recommendations-trustworthy-artificial-intelligence

Mökander, J., & Floridi, L. (2021). Ethics-based auditing to develop trustworthy AI. *Minds and Machines, 31*, 323–327. https://doi.org/10.1007/s11023-021-09557-8

Mökander, J., Morley, J., Taddeo, M., & Floridi, L. (2021). Ethics-based auditing of automated decision-making systems: Nature, scope, and limitations. *Science and Engineering Ethics, 27*(4), 44. https://doi.org/10.1007/s11948-021-00319-4

Mökander, J., Axente, M., Casolari, F., & Floridi, L. (2022). Conformity assessments and post-market monitoring: A guide to the role of auditing in the proposed European AI regulation. *Minds and Machines, 32*, 241–268. https://doi.org/10.1007/s11023-021-09577-4

Roberts, H., Cowls, J., Hine, E., Morley, J., Taddeo, M., Wang, V., & Floridi, L. (2021a). *China's artificial intelligence strategy: Lessons from the European Union's 'ethics-first' approach*. Available at: SSRN 3811034.

Roberts, H., Cowls, J., Morley, J., Taddeo, M., Wang, V., & Floridi, L. (2021b). The Chinese approach to artificial intelligence: An analysis of policy, ethics, and regulation. *AI & Society, 36*(1), 59–77.

Roberts, H., Cowls, J., Hine, E., Mazzi, F., Tsamados, A., Taddeo, M., & Floridi, L. (forthcoming). *Achieving a 'Good AI Society': Comparing the aims and progress of the EU and the US*.

Scott, J. (2014). Extraterritoriality and territorial extension in EU law. *The American Journal of Comparative Law, 62*(1), 87–126.

Informational Privacy with Chinese Characteristics

Huw Roberts

Abstract In recent years there is said to have been a 'privacy awakening' in China, with a demand from citizens for improved data protections being met by comprehensive regulatory measures, including legislation passed by the Standing Committee of the National People's Congress. The measures introduced provide individuals with strong consumer privacy protections but do little to meaningfully curb state mass surveillance and censorship. In this chapter, I explore the dual nature of China's emerging data protection regime and argue that the absence of citizen protections is more than just an example of authoritarian policymaking that is unwilling to curtail its own power. Strong consumer and weak citizen protections are reflective of a socio-political conception of informational privacy rights in China. To make this argument, I develop a conception of informational privacy as mutual obligations between the state and citizens.

Keywords Privacy · China · Data governance · Surveillance · Confucianism

1 Introduction

When responding to a question about the informational privacy concerns of Chinese citizens at a panel discussion in 2018, Baidu CEO, Robin Li stated:

> Chinese people are more open or are not that sensitive about privacy. If they are able to exchange privacy for safety, convenience, or efficiency, in many cases they are willing to do that. Then we can make more use of that data (Shen, 2018).

Li's comments are not unique, with it commonly argued that Chinese citizens do not care about privacy (Cai, 2007; Huang et al., 2020; Lee, 2018; Tam, 2018). However, this view is increasingly being challenged by citizens' actions. Li's remarks sparked fierce backlash on Chinese social media, where users vehemently rejected his claim.

H. Roberts (✉)
Oxford Internet Institute, University of Oxford, Oxford, UK

© The Author(s), under exclusive license to Springer Nature
Switzerland AG 2022
J. Mökander, M. Ziosi (eds.), *The 2021 Yearbook of the Digital Ethics Lab*,
Digital Ethics Lab Yearbook, https://doi.org/10.1007/978-3-031-09846-8_2

Likewise, there have been a number of recent cases of consumers voicing their dissatisfaction over data collection and use (Porter, 2019), including a high-profile lawsuit against a safari park that mandated the use of facial recognition technology for visitor entry (Shen, 2021). This consumer pushback against private sector data practices is being responded to by the government, who have introduced a flurry of data protection measures. Most notable is the Personal Information Protection Law, the first comprehensive consumer data protection legislation, which was passed by the Standing Committee of China's National People's Congress (NPC) in August 2021.[1] These developments have led some commentators to proclaim that China is experiencing a 'privacy awakening' (Ma, 2019a; Sacks & Laskai, 2019; Yang, 2018).

This claim appears at odds with China's mass surveillance regime (Leibold, 2020; Qiang, 2019). The country contains many of the world's most surveilled cities (Bischoff, 2020); there is extensive monitoring and censorship on social media (Bartow, 2013; MacKinnon, 2011)[2]; and the fledgling Social Credit System—a policy effort by the Chinese government that seeks to collect and leverage personal data to regulate economic, social and moral aspects of society through assessing trustworthiness (Chorzempa et al., 2018)—is indicative of the government's continued efforts to leverage data to enact control. Indeed, a closer analysis of the data protection measures proposed or introduced, as well as the wider legal systems in China, reveals that few meaningful privacy rights are provided from the government's mass surveillance regime (Roberts et al., 2021b).

It could reasonably be assumed that the 'privacy awakening' amongst consumers is matched with a concern over growing surveillance capabilities of the state, and that the government is merely unwilling to curtail its own power. It is unlikely that the Communist Party of China (CCP) would meaningfully curtail its surveillance capacities, suggesting that there is worth to this claim. In this sense, the introduction of comprehensive *consumer* but not *citizen* data protections could be considered an example of responsive authoritarianism: the government listening to citizens' concerns and enacting policy changes insofar as it does not threaten the state's capacity to rule (Heurlin, 2016). However, it is also possible that there is a lack of desire for comprehensive rights against government surveillance and that the data protection measures introduced are reflective of citizens' concerns.

This chapter explores whether the way informational privacy is conceptualised within China means that the duality of protections provided by government is permissible. I argue that many people are increasingly willing to demand rights as consumers, yet their privacy expectations as citizens are often conceptualised in terms of mutual obligations. As such, rather than demanding absolute privacy rights from government, there is an expectation that the state and citizens should act in line with their respective obligations.

---

[1] This is China's top law-making body.

[2] Note: censorship on social media is enacted through internet companies and not directly by the state.

To make this argument, I first examine the claim that there has been a 'privacy awakening' within China, and emphasise that although privacy concerns are nothing new, government substantively meeting a demand for informational privacy rights is. I then outline the evolution of the state-individual relationship within China and discuss this in regard to informational privacy expectations. From this foundation, I develop the idea of informational privacy as a form of relationality that is conceptualised in terms of mutual obligations between the state and citizens. I conclude the chapter by arguing that the dual nature of China's data protection regime is permissible to many, as it reflects many citizens' priorities for informational privacy rights. Considering informational privacy as a conception of mutual obligations helps to dispel the myth that Chinese citizens do not care about privacy, whilst also offering a framework for other scholars to use when assessing Chinese government data practices, including mass surveillance.

2 Exploring China's 'Privacy Awakening'

An effective right to informational privacy, understood here as the legal protections in place which facilitate individual control over collection, use and disclosure of data (Solove & Schwartz, 2020), has been historically absent within China (Fry, 2015).[3] Since 2016, a number of measures have been introduced which enact consumer protections. The 2016 Cybersecurity Law codified consent requirements for data collection and processing, as well as requirements of legality, propriety and necessity for network operators. These measures were further refined in the 2017 Personal Information Security Specification, a voluntary standard that provided guidance on how to implement some high-level provisions in the Cybersecurity Law (Pernot-Leplay, 2020).[4] The 2021 Personal Information Protection Law (PIPL), passed by the Standing Committee of the NPC, specifically focuses on data protection and defines key concepts such as personal information, as well as enshrining a data minimisation principle into law. Alongside these national-level regulatory efforts, policies have also been introduced at a local level, with the city of Tianjin, for instance, restricting the private sector's collection and use of biometric data ('Pushback against Surveillance in China Grows', 2021). These measures have led some commentators to tout China as 'Asia's surprise leader in data protection' (Lucas, 2018).

---

[3] A right to privacy is enshrined within the Chinese Constitution yet no broad, enforceable privacy protections were present prior to 2016. Some adjacent rights have been present such as the General Principles of Civil Law have also recognised a right to reputation since 1988, which has treated the disclosure of certain types of personal information in specific contexts as damaging to reputation (Hert & Papakonstantinou, 2016). There were attempts to draft a data protection law in the early 2000s, yet these efforts did not come to fruition.

[4] Although this standard is voluntary, it holds substantive clout as they act as guidance for the implementation of the Cybersecurity Law (Sacks, 2018; Sacks & Li, 2018).

Beyond these comprehensive consumer protections, the extent to which China is experiencing a regulatory 'privacy awakening' can be called into question. Key exemptions exist in the protections, including for consent and notification measures when these will "impede state organs' fulfilment of their statutory duties and responsibility" (Art 35).[5] Moreover, there is significant overlap between the privacy protections afforded in the Personal Information Security Law and the surveillance powers provided to the government in the existing Cybersecurity Law and Data Security Law (Roberts et al., 2021a). When these exemptions and overlaps are considered alongside China's legal system, which is subordinate to the CCP, it can be concluded that informational privacy protections are being introduced for some government practices (Horsley, 2021), but that this will not meaningfully curtail the extensive surveillance infrastructures in place.

The informational privacy landscape in China is thus characterised by a duality: private companies are generally constrained by the regulations in place, whilst the central government and those companies acting on its behalf face few substantive checks. As a result, there is a right to *consumer privacy* that does not extend to *citizen privacy* within China (Roberts et al., 2021b; Sacks & Laskai, 2019). The dual nature of China's data protection regime raises the question of whether these regulatory measures are reflective of individuals' privacy concerns.

Some commentators have argued that there has been a general 'privacy awakening' amongst individuals in China, due to an increasing consciousness over potential harms and a growing desire for privacy rights (Sacks & Laskai, 2019). Putting aside the question of whether 'awakening' is the appropriate term for describing a seeming growth in informational privacy concerns,[6] evidence for this claim is mixed. As highlighted in the introduction, there is much anecdotal evidence of citizens voicing dissatisfaction about extractive private sector practices and a stated desire for stronger consumer protections and rights (Face Swap App Accused of Excessive Data Collection, 2019; Feng, 2020). Likewise, survey data highlights burgeoning concerns amongst individuals about data security (Wang & Yu, 2015; Yang & Liu, 2019), location tracking (Li, 2020), targeted advertisement (Wang et al., 2016), the use of facial recognition technology in commercial areas (Wang, 2021), and AI technologies more generally (Soo, 2018).

The extent to which this can be considered a 'privacy awakening', however, can be called into question due to pre-existing concerns over other types of privacy. There were an increasing number of legal cases concerning the right to reputation in the 1990s (Cao, 2005), and studies from the turn of the century discussed a growing respect for interpersonal privacy (Mcdougall, 2001; Mcdougall & Hansson, 2002; Yan, 2003). Given that privacy concerns have been well-documented in China for at

---

[5] A valid observation would be that other data privacy regulations around the world, such as the GDPR, include similar vaguely defined exemptions to those present in the Specification (Stevens, 2017). Yet the absence of meaningful rule of law in China prevents Government interpretation of these clauses being checked.

[6] 'Awakening' carries teleological undertones and is suggestive of the emergence of a desire for specific type of privacy being a natural process.

least the last 30 years, it can be seen that individuals have not 'awoken' to privacy. Instead, it is more accurate to conclude that there is now a greater wealth of evidence showing contemporary consumer consciousness and concerns about informational privacy.

Evidence of informational privacy concerns related to government is more limited and contested. There have been isolated cases of individuals pushing back against government data practices, such as in Hangzhou after the local government announced its intention to make a Covid-19 contact tracing application permanent (Davidson, 2020). Similarly, Wang and Yu (2015) found in their survey of Chinese citizens that 37.6% of respondents lacked trust in the government's handling of data because it restricts individual freedom, thought and speech. Whilst their study did not specify how many respondents stated that they trusted the government in handling data, leaving it possible that those who lacked trust are in the minority, the results are indicative of at least some individuals being concerned with government data practices.

In general, however, pushback against central government data practices has been limited and there have been few calls for explicit privacy rights from government (Mozur, 2018). Indeed, available survey data shows high support for government surveillance (Kostka, 2019; Su et al., 2020; Zhang et al., 2019). Although support for surveillance should not necessarily be conflated with an absence of desire for privacy rights, Mcdougall (2001) found that privacy concerns relating to government were rarely mentioned when compared to other types, such as consumer and interpersonal. This finding was mirrored in my own fieldwork, where interviewees largely focused on the importance of protections against private sector data misuse (Roberts, 2020). It is possible that these results could in part be explained by a fear of speaking out against government, yet in studies where participants were students studying abroad or individuals circumventing censorship, there were still high levels of support for authoritarian policies (Conflicted Hearts and Minds, 2020; Kou et al., 2017).

This discussion suggests that there has been an increasing consumer consciousness over informational privacy and an associated desire for rights, which is being met by the government. Whether the absence of citizen protections from mass surveillance measures is reflective of a lack of concern about informational privacy protections from government is less certain. Concerns over government data practices appear secondary to other areas, such as interpersonal and consumer privacy; there is general support for state mass surveillance measures; and there have not been widespread calls for informational privacy rights from central government. Nonetheless, there are clearly specific cases where concerns are present over government data practices. This disjointed 'awakening' warrants further investigation, particularly in relation to what seems to be a limited desire for privacy rights from government.

3   The State-Individual Relationship in Confucian Thought

To understand why there is a seemingly high acceptance of government intrusion, one can look at how the state-individual relationship within China has evolved, and how it currently manifests. A useful starting point is to consider the state-individual relationship in the country's philosophical traditions, of which Confucianism is the most influential, with certain elements still relevant in contemporary China (Bell, 2010a; Li et al., 2017; Ma, 2019b).

Confucianism considers the individual as a morally important agent who should focus on 'self-cultivation', with the ultimate aim of becoming an 'exemplary person' (*junzi*) (Ames & Rosemont, 2010, p. 122). Confucian ethics preferences a single way in which human beings should live to achieve a good life, with other ways of living deemed morally unequal (Wong, 2020). To follow this morally correct path, individuals abide by rules of conduct (*li*) that place specific obligations on a person based on their role within given relationships (Tu, 1998). Mencius, one of the most influential early Confucian thinkers, theorised the individual as being embedded within five core relationships (*wu lun*): (1) father and son, (2) ruler and subordinate, (3) husband and wife, (4) brother and brother and (5) friend and friend (Liu, 2004). The individual is considered the locus of cultivation (Brindley, 2010), yet this cultivation is inherently relational in that it is achieved through interacting with others (Yao, 1996). Through fulfilling obligations that are appropriate to specific relationships, the individual undertakes moral self-cultivation whilst also bringing about good for society.

This is not to say that individuals do not have self-interest. In conceptualising the individual in Confucian thought, it is appropriate to consider the 'small self' and the 'great self' (Yang, 2006, p. 330). The 'small self' represents an individual's self-interest which is not considered to be a moral self-proper, instead being viewed as something that should be controlled or constrained so one's true self, in which public interests are given priority over selfish interests, can emerge (Yao, 1996). The 'great self' is brought about by moral cultivation through rules of conduct in interpersonal relationships that brings the self into maturity. Thus, Confucian thought does not consider the individual as possessing unfettered autonomy but understands existence in relational terms (Brindley, 2010).

In terms of the relationship between the ruler and the subordinate, the ruler has to show benevolence whilst the subject should show obedience to achieve harmony. A benevolent government is a paternalistic one that seeks to ensure a 'harmonious community in which people can live in happiness' (Shin et al., 2012, p. 115). Foremost, this requires a government to be trustworthy; without trust, social order breaks down and poses the biggest threat to achieving harmony (Yao, 1999). To build trust within society, the primary aim of government is to cater for individuals' basic needs, such as economic prosperity and physical safety. To that end, an individual's obedience to government is contingent on benevolence being shown by the ruler (Shin et al., 2012).

Relating this understanding of the individual back to informational privacy, two conclusions can be reached. First, it suggests that an atomistic view of the individual, upon which much of Western privacy studies is built, is alien to Confucian thought (Whitman, 1985).[7] In Western thought, privacy rights are typically premised on clear divides between the individual and the group, and between the private and public spheres. No such differences have historically existed in the relational organisation of social life in China, where these divides are relative (Bell, 2010b).

Second, Confucianism focuses on relational duties to provide to one another rather than a notion of absolute rights (Peerenboom, 1993; Shin et al., 2012). The duties of government are conceived in terms of positive material rights that are provided to the individual, such as ensuring economic welfare and social stability, rather than negative rights that protect the individual, such as liberty or freedom (Chan, 2013). Confucian thought would consider an egoistic claim of negative rights a failure (Peerenboom, 1993), in that it would undermine relational duties and harmony (Lee, 1992). A Confucian theory of privacy would thus place little emphasis on the importance of absolute informational privacy rights from the state.

4   Individualisation and the 'Great Self'

This Confucian grounding for informational privacy in China provides a good starting point; however, it is inadequate alone for explaining contemporary empirical phenomena given the many ideologies that shape modern day China. Of particular note is how Maoist ideology and China's subsequent marketisation have impacted the individual and their relationship to the state.

There has been a well-documented growth of individualism in China since the Maoist period (Cao, 2009; Ess, 2005). Under Mao, the Chinese government denounced the traditional *wu lun* conception of interpersonal relationships as feudalistic, instead promoting a comradeship that was based on a universal morality to other citizens under the state (Gold, 1985). The ultimate aim of this process was to replace notions of filial piety and family loyalty with allegiance to the party state above all else. In this drive for modernity, Mao enacted a state-led process of individualisation that enhanced societal collectivism, but also liberated traditional family structures, particularly in relation to increased gender equality, marriage freedom, and independence (Yan, 2009, 2010).

This individualisation process was furthered by a state-led marketisation of the economy since the 1970s, which saw rural collectives dismantled and private enterprises established, ending state-led resource allocation that had characterised the

---

[7] The Confucian understanding of autonomy might appear similar to more recent privacy studies literature that is embedded within notions of relational autonomy (Mokrosinska, 2018). Nonetheless, as has been pointed out be Ma (2019b), Western relational autonomy still focuses on controlling access to different social contexts, which is different to the mutual obligations present within Confucian thought.

Maoist period. An individualistic capitalist drive for profit eroded ideas of comradeship, whilst mass rural migration to the cities for employment led individuals to become further disembedded from their traditional family bonds (Yan, 2010).[8] Traditional Chinese *wu lun* obligations remain in society (Qi, 2015), with many commentators highlighting their revival on account of a resurgence of neo-Confucianism (Bell, 2010a; Biao, 2011), yet the individualisation process has led the individual to become an important unit of discourse in Chinese society.

For privacy, individualisation has led to a greater societal acceptance of more individualistic notions of privacy (Wang, 2016; Yan, 2003). Nonetheless, many scholars emphasise the importance of collective interests for understanding informational privacy in China (Li et al., 2017; Lü, 2005; Wang, 2012). Considering relationality between different actors, particularly in regard to rights consciousness, helps to demystify the mixture of individual and collective considerations that appear in informational privacy preferences.

Individualisation within China has led to the emergence of a rights consciousness amongst citizens (Li, 2010), but one that is distinct from how rights are typically understood in the West (Perry, 2009; Yan, 2009). Throughout China's socio-political changes, the central government has maintained its position at the apex of society, holding the same type of relational authority that was present within the traditional Confucian model (Tong, 2011; Yan, 2010). Contemporary rights are not granted at birth or considered universal but rather are an earned privilege that is granted by the Chinese government for the performance of duties (Hooper, 2005; Li & Wu, 1999). Because of this, many individuals do not conceive of themselves as holding inalienable rights from central government (Perry, 2009; Yan, 2010). The state is perceived as the provider, protector and mediator of rights, which has left individuals as 'consumer citizens', who assert rights not vis-à-vis the state, but vis-à-vis the market, with the endorsement and encouragement of the state (Hooper, 2005, p. 2).

Yan (2010, p. 494) argues that the language of the 'small self' and the 'great self' is appropriate for conceptualising the state-individual relationship in contemporary China. In regard to interpersonal relationships and the private sector, there has been a growth in the legitimacy of the 'small self', with the contemporary individual able to claim self-interests beyond the duties-based model that is present within Confucian thought. However, this 'small self' is subordinate to the 'great self', in a similar fashion to in the Confucian model, meaning that an individual's autonomy is still subordinate to the interests of society, as defined by the state. In this sense, the normatively correct position is one of subsuming self-interests under that of wider society, but self-interests are permissible in interpersonal contexts and in regard to companies. That being said, the legitimacy of the state and its presentation of what is beneficial for society is still underpinned by the state fulfilling its duties (Eaton & Hasmath, 2020; Li, 2010).

---

[8] The household registration (hukou) system that is still present in contemporary China means that citizens do not have access social welfare benefits outside of their local jurisdiction, meaning migrant works often enter new cities alone.

5 Informational Privacy as Relational Obligations

Significant shifts have taken place since the *wu lun* conception of relationships, with the individual gaining rights consciousness in relation to private sector actors, which has led to a strong desire to control access to information about oneself as a consumer. In contrast, relationships with the state are still based on a model of obligations, which means that concerns might be present but that an absolute right to informational privacy is not considered the appropriate mechanism for resolving these.

Although shifts in dominant ideologies have altered the exact framing of obligations,[9] government legitimacy throughout has been based on a benevolent paternalism (Cheng et al., 2004; Tong, 2011). This entails providing citizens with positive rights, whilst also being morally upright in its own actions (Chen, 2013; Peerenboom, 1993; Tong, 2011).[10] In return, citizens are expected to trust and comply with government rules, insofar as it fulfils its duties. This framing indicates that a model of autonomous agents who have an absolute right to control access from government is inappropriate, as individuals are relationally bound to government based on a notion of mutual obligations. The government ensures societal stability and harmony, provides positive rights, and does not abuse its position in exchange for compliance. Extensive data collection and use are within the remit of the government's duty, as long as it ensures societal stability and harmony. When considered through this lens, a desire for, and enactment of, consumer informational privacy protections without accompanying citizen protections becomes more understandable.

Considering informational privacy as an idea of mutual obligations also helps to explain the pushback against specific government decisions, such as that by Hangzhou's local government to extend the use of the city's contact tracing application. In an obligations model of informational privacy, the government can only enact data practices insofar as it fulfils its duties to society. Acting in its own self-interest of maintaining and enhancing power rather than for the benefit of society, as was perceived to be the case in Hangzhou, is unacceptable. My own fieldwork supports this conclusions, with individuals reporting being unhappy about government surveillance and censorship on social media when it was seen to be excessive and stifle daily activities (Roberts, 2020). Accordingly, concerns about certain government practices can still be present, without an associated demand for explicit privacy rights or structural changes relating to mass surveillance.

Before turning to the implications of this new understanding of informational privacy in China, it is important to acknowledge how the actions of the state are

---

[9] For instance, contemporary duties that would not have been present in the Confucian period are the aforementioned ensuring of consumer rights and also to sustain economic growth for material wellbeing (Chen, 2013).

[10] This understanding helps to explain the widespread support that Xi Jinping's government receives (Cunningham et al., 2020), despite cracking down on rights (Pils, 2018).

influential in shaping the subjectivities of individuals in terms of the remit provided for government intrusions. In particular, it has been recognised that the Chinese government has securitised the issue of domestic terrorism (Guan & Liu, 2020), creating a need and justification for measures of control. This does not alter the understanding of mutual obligations, but it could mould the subjectivities of individuals into conceiving safety issues to be substantial problems, thus justifying greater government control than may have been considered legitimate if there was a freer media landscape.

6   Conclusion

The duality of privacy protections proposed and introduced by the Chinese government in recent years appears to be a typical example of responsive authoritarian policymaking that is responsive to citizens' concerns, only insofar as it does not threaten the controlling capacity of the state. The conception of informational privacy introduced in this chapter suggests that the reality is more nuanced. Whilst there has been a strong desire to enact consumer privacy protections, particularly from data misuse, the model of mutual obligations that is present between citizens and the state mean a demand for informational privacy rights from government has been limited.

This conception of informational privacy is consequential. It suggests that an absence of citizen informational privacy protections might be acceptable as long as the government is perceived to act appropriately, which contrasts to the heavy emphasis on rights-based protections from government in much of Western privacy studies (Posner, 1977; Thomson, 1975; Warren & Brandeis, 1890). The idea of informational privacy as mutual obligations allows Chinese government data practices, such as mass surveillance, to be critiqued on account of failing to live up to the obligations that citizens expect rather than based on a failure to introduce absolute rights.

It should be stressed that the conception of informational privacy as mutual obligations does not provide a single authoritative view of privacy in China. Concerns over informational privacy within China, like anywhere else, are diverse with a multitude of relevant conceptions likely present. Accordingly, further qualitative research on conceptions of informational privacy within China would complement the theoretical arguments put forward in this chapter, and would provide for a more refined understanding of the privacies present. Continued scholarly effort to understand the ways in which privacy manifests within China will help to further move discussions beyond clumsily applying imported conceptions of informational privacy, which has so often led to misplaced conclusions.

References

Ames, R. T., & Rosemont, H. (2010). *The analects of Confucius: A philosophical translation.* Random House Publishing Group.

Bartow, A. (2013). Privacy Laws and Privacy levers: Online surveillance versus economic development in the People's republic of China symposium: The second wave of global privacy protection. *Ohio State Law Journal, 74*(6), 853–896.

Bell, D. A. (2010a). *China's new Confucianism: Politics and everyday life in a changing society.* Princeton University Press.

Bell, D. A. (2010b). *Confucian political ethics.* Princeton University Press.

Biao, X. (2011). The individualization of Chinese society. By Yunxiang Yan. pp. 384. (Berg, Oxford, 2009.) £17.99, ISBN 978-1-84788-378-0, paperback. *Journal of Biosocial Science, 43*(1), 126–127. https://doi.org/10.1017/S0021932010000362

Bischoff, P. (2020, July 20). Surveillance camera statistics: Which City has the Most CCTV cameras? *Comparitech.* https://www.comparitech.com/vpn-privacy/the-worlds-most-surveilled-cities/

Brindley, E. (2010). *Individualism in early China: Human agency and the self in thought and politics.* University of Hawaii Press.

Cai, F. (2007). 中西方隐私观探析 Analysis of Chinese and Western views on privacy. 江苏工业学院学报 *Journal of Jiangsu Polytecnic, 8* (2). https://xueshu.baidu.com/usercenter/paper/show?paperid=bcc5e6a6dd02169f9b8d3b1510a30bd3&site=xueshu_se

Cao, J. (2005). Protecting the right to privacy in China. *Victoria University of Wellington Law Review, 36*(3), 645–664.

Cao, J. (2009). *The analysis of tendency of transition from collectivism to individualism in China* (p. 9). Canadian Academy of Oriental and Occidental Culture.

Chan, J. (2013). *Confucian perfectionism: A political philosophy for modern times.* Princeton University Press.

Chen, J. (2013). *A middle class without democracy: Economic growth and the prospects for democratization in China.* Oxford University Press.

Cheng, B.-S., Chou, L.-F., Wu, T.-Y., Huang, M.-P., & Farh, J.-L. (2004). Paternalistic leadership and subordinate responses: Establishing a leadership model in Chinese organizations. *Asian Journal of Social Psychology, 7*(1), 89–117. https://doi.org/10.1111/j.1467-839X.2004.00137.x

Chorzempa, M., Triolo, P., & Sacks, S. (2018). *Policy brief 18–14: China's social credit system: A mark of Progress or a threat to privacy?* p. 11.

Conflicted hearts and minds. (2020, March). [2020–03]. Merics. https://merics.org/en/report/conflicted-hearts-and-minds

Cunningham, E. A., Saich, T., & Turiel, J. (2020). *Understanding CCP resilience: Surveying Chinese public opinion through time* (p. 18). Ash Center for Democratic Governance and Innovation.

Davidson, H. (2020, May 26). Chinese city plans to turn coronavirus app into permanent health tracker. *The Guardian.* http://www.theguardian.com/world/2020/may/26/chinese-city-plans-to-turn-coronavirus-app-into-permanent-health-tracker

Hert, P. de, & Papakonstantinou, V. (2016). *The data protection regime in China: In-depth analysis.* http://op.europa.eu/en/publication-detail/-/publication/82ca9bb2-9ad0-11e6-868c-01aa75ed71a1/language-en/format-PDF

Eaton, S., & Hasmath, R. (2020 C.E.). *Economic legitimation in a new era: Public attitudes about state ownership and market regulation* (SSRN scholarly paper ID 2984141). Social Science Research Network. https://doi.org/10.2139/ssrn.2984141.

Ess, C. (2005). "Lost in translation"?: Intercultural dialogues on privacy and information ethics (Introduction to special issue on privacy and data privacy protection in Asia). *Ethics and Information Technology, 7*(1), 1–6. https://doi.org/10.1007/s10676-005-0454-0

Face swap app accused of excessive data collection. (2019, September). ECNS.Cn. http://www.ecns.cn/news/sci-tech/2019-09-02/detail-ifznpqie3826518.shtml

Feng, E. (2020, January). In China, A New Call To Protect Data Privacy. NPR. https://text.npr.org/793014617

Fry, J. (2015). Privacy, predictability and internet surveillance in the U.S. and China: Better the devil you know? *University of Pennsylvania Journal of International Law, 37*(2), 419.

Gold, T. B. (1985). After comradeship: Personal relations in China since the cultural revolution. *The China Quarterly, 104*, 657–675. JSTOR.

Guan, T., & Liu, T. (2020). Polarised security: How do Chinese netizens respond to the securitisation of terrorism? *Asian Studies Review, 44*(2), 335–354. https://doi.org/10.1080/1035782 3.2019.1697205

Heurlin, C. (2016). *Responsive authoritarianism in China.* Cambridge University Press.

Hooper, B. (2005). *The consumer citizen in contemporary China* (Working paper no. 12) (p. 24). Lund University.

Horsley, J. P. (2021, January 29). *How will China's privacy law apply to the Chinese state?* Brookings. https://www.brookings.edu/articles/how-will-chinas-privacy-law-apply-to-the-chinese-state/

Huang, Y., Sun, M., & Sui, Y. (2020, April 15). *How digital contact tracing slowed Covid-19 in East Asia.* Harvard Business Review. https://hbr.org/2020/04/how-digital-contact-tracing-slowed-covid-19-in-east-asia

Kostka, G. (2019). China's social credit systems and public opinion: Explaining high levels of approval. *New Media & Society, 1461444819826402.* https://doi.org/10.1177/1461444819826402

Kou, Y., Semaan, B., & Nardi, B. (2017). A Confucian look at internet censorship in China. In R. Bernhaupt, G. Dalvi, A. Joshi, D. K. Balkrishan, J. O'Neill, & M. Winckler (Eds.), *Human-computer interaction—INTERACT 2017* (pp. 377–398). Springer International Publishing. https://doi.org/10.1007/978-3-319-67744-6_25

Lee, S.-H. (1992). Was there a concept of rights in Confucian virtue-based morality? *Journal of Chinese Philosophy, 19*(3), 241–261. https://doi.org/10.1111/j.1540-6253.1992.tb00358.x

Lee, K.-F. (2018). *AI superpowers: China, Silicon Valley, and the New World order.* Houghton Mifflin Harcourt.

Leibold, J. (2020). Surveillance in China's Xinjiang region: Ethnic sorting, coercion, and inducement. *Journal of Contemporary China, 29*(121), 46–60. https://doi.org/10.1080/1067056 4.2019.1621529

Li, L. (2010). Rights consciousness and rules consciousness in contemporary China. *The China Journal, 64*, 47–68. https://doi.org/10.1086/tcj.64.20749246

Li, H. (2020). Negotiating privacy and mobile socializing: Chinese University Students' concerns and strategies for using geosocial networking applications. *Social Media + Society, 6, 2056305120913887*(1). https://doi.org/10.1177/2056305120913887

Li, B., & Wu, Y. (1999). The concept of citizenship in the People's republic of China. In A. Davidson & K. Weekley (Eds.), *Globalization and citizenship in the Asia-Pacific.* Springer.

Li, T. C., Bronfman, J., & Zhou, Z. (2017). *Saving face: Unfolding the screen of Chinese privacy law* (SSRN scholarly paper ID 2826087). Social Science Research Network. https://papers.ssrn.com/abstract=2826087

Liu, Y. (2004). The self and Li in Confucianism. *Journal of Chinese Philosophy, 31*(3), 363–376. https://doi.org/10.1111/j.1540-6253.2004.00159.x

Lü, Y.-H. (2005). Privacy and data privacy issues in contemporary China. *Ethics and Information Technology, 7*(1), 7–15. https://doi.org/10.1007/s10676-005-0456-y

Lucas, L. (2018, May 30). *China emerges as Asia's surprise leader on data protection.* https://www.ft.com/content/e07849b6-59b3-11e8-b8b2-d6ceb45fa9d0.

Ma, W. (2019a, November 12). *China is waking up to data protection and privacy. Here's why that matters.* World Economic Forum. https://www.weforum.org/agenda/2019/11/china-data-privacy-laws-guideline/

Ma, Y. (2019b). Relational privacy: Where the east and the west could meet. *Proceedings of the Association for Information Science and Technology, 56*(1), 196–205. https://doi.org/10.1002/pra2.65

MacKinnon, R. (2011). Liberation technology: China's 'networked authoritarianism'. *Journal of Democracy, 22*(2), 32–46. https://doi.org/10.1353/jod.2011.0033

Mcdougall, B. S. (2001). Privacy in contemporary China: A survey of student opinion, June 2000. *China Information, 15*(2), 140–152. https://doi.org/10.1177/0920203X0101500206

Mcdougall, B. S., & Hansson, A. (2002). *Chinese concepts of privacy*. BRILL. http://ebookcentral. proquest.com/lib/oxford/detail.action?docID=253522

Mokrosinska, D. (2018). Privacy and autonomy: On some misconceptions concerning the political dimensions of privacy. *Law and Philosophy, 37*(2), 117–143. https://doi.org/10.1007/ s10982-017-9307-3

Mozur, P. (2018, July 8). Inside China's dystopian dreams: A.I., shame and lots of cameras. *The New York Times*. https://www.nytimes.com/2018/07/08/business/china-surveillance-technology.html

Peerenboom, R. P. (1993). What's wrong with Chinese rights: Toward a theory of rights with Chinese characteristics. *Harvard Human Rights Journal, 6*, 29–58.

Pernot-Leplay, E. (2020). China's approach on data privacy law: A third way between the U.S. and the E.U.? *Penn State Journal of Law & International Affairs, 8*(1), 49.

Perry, E. J. (2009). China since Tiananmen: A new rights consciousness? *Journal of Democracy, 20*(3), 17–20. https://doi.org/10.1353/jod.0.0111

Pils, E. (2018). The Party's turn to public repression: An analysis of the '709' crackdown on human rights lawyers in China. *China Law and Society Review, 3*(1), 1–48. https://doi. org/10.1163/25427466-00301001

Porter, J. (2019, September 2). *Another convincing deepfake app goes viral prompting immediate privacy backlash*. The Verge. https://www.theverge.com/2019/9/2/20844338/ zao-deepfake-app-movie-tv-show-face-replace-privacy-policy-concerns

Posner, R. A. (1977). The right of privacy. *Georgia Law Review, 12*, 393.

Pushback against surveillance in China grows. (2021). *Biometric Technology Today, 2021*(1), 3. https://doi.org/10.1016/S0969-4765(21)00003-5

Qi, X. (2015). Filial obligation in contemporary China: Evolution of the culture-system. *Journal for the Theory of Social Behaviour, 45*(1), 141–161. https://doi.org/10.1111/jtsb.12052

Qiang, X. (2019). The road to digital Unfreedom: President Xi's surveillance state. *Journal of Democracy, 30*(1), 53–67. https://doi.org/10.1353/jod.2019.0004

Roberts, H. (2020). *Informational privacy with Confucian characteristics: Conceptualising government data collection and use from the perspective of Chinese students*. Oxford Internet Institute Masters Thesis.

Roberts, H., Cowls, J., Hine, E., Morley, J., Taddeo, M., Wang, V., & Floridi, L. (2021a). *Governing artificial intelligence in China and the European Union: Comparing aims and promoting ethical outcomes* (SSRN scholarly paper ID 3811034). Social Science Research Network. https:// papers.ssrn.com/abstract=3811034

Roberts, H., Cowls, J., Morley, J., Taddeo, M., Wang, V., & Floridi, L. (2021b). The Chinese approach to artificial intelligence: An analysis of policy, ethics, and regulation. *AI & SOCIETY, 36*(1), 59–77. https://doi.org/10.1007/s00146-020-00992-2

Sacks, S. (2018, January 29). *New China data privacy standard looks more far-reaching than GDPR*. https://www.csis.org/analysis/new-china-data-privacy-standard-looks-more-far-reaching-gdpr

Sacks, S., & Laskai, L. (2019, February 7). *China is having an unexpected privacy awakening*. Slate Magazine. https://slate.com/technology/2019/02/china-consumer-data-protection-privacy-surveillance.html

Sacks, S., & Li, M. (2018, August 2). *How Chinese cybersecurity standards impact doing business in China*. https://www.csis.org/analysis/ how-chinese-cybersecurity-standards-impact-doing-business-china

Shen, X. (2018, March 28). *Chinese internet users criticize Baidu CEO for saying people in China are willing to give up data privacy for convenience*. South China Morning Post. https://www.scmp.com/ abacus/tech/article/3028402/chinese-internet-users-criticize-baidu-ceo-saying-people-china-are

Shen, X. (2021, April 12). *China's first facial recognition case raises more questions than it answers.* South China Morning Post. https://www.scmp.com/tech/policy/article/3129226/chinas-first-facial-recognition-lawsuit-comes-end-new-ruling-and-new

Shin, D. C., Sin, T., & Sin, T. (2012). *Confucianism and democratization in East Asia.* Cambridge University Press.

Solove, D. J., & Schwartz, P. M. (2020). *ALI data privacy: Overview and black letter text* (SSRN scholarly paper ID 3457563). Social Science Research Network. https://doi.org/10.2139/ssrn.3457563.

Soo, Z. (2018, March 5). *Who says the Chinese don't care about privacy?* South China Morning Post. https://www.scmp.com/business/companies/article/2135713/increasing-use-artificial-intelligence-stoking-privacy-concerns

Stevens, L. A. (2017). *Public interest approach to data protection law: The meaning, value and utility of the public interest for research uses of data.* https://era.ed.ac.uk/handle/1842/25772

Su, Z., Xu, X., & Cao, X. (2020). *What explains popular support for government surveillance in China?* Penn State Preprint. https://sites.psu.edu/xuncao/files/2020/09/Support-for-Government-Surveillance-in-China_April_19_2020_Manuscript.pdf

Tam, L. (2018, April 2). *Why privacy is an alien concept in Chinese culture.* South China Morning Post. https://www.scmp.com/news/hong-kong/article/2139946/why-privacy-alien-concept-chinese-culture

Thomson, J. J. (1975). The right to privacy. *Philosophy & Public Affairs, 4*(4), 295–314. JSTOR.

Tong, Y. (2011). Morality, benevolence, and responsibility: Regime legitimacy in China from past to the present. *Journal of Chinese Political Science, 16*(2), 141–159. https://doi.org/10.1007/s11366-011-9141-7

Tu, W. (1998). Probing the 'three relationships' and 'five bonds' in Confucian harmony. In W. H. Slote & G. A. D. Vos (Eds.), *Confucianism and the family: A study of indo-Tibetan scholasticism.* SUNY Press.

Wang, H. (2012). The conceptual basis of privacy standards in China and its implications for China's privacy law. *Frontiers of Law in China, 7*(1), 134–160. https://doi.org/10.3868/s050-001-012-0007-4

Wang, X. (2016). *Social media in industrial China.* UCL Press. https://doi.org/10.14324/111.9781910634646

Wang, C. (2021, January). 超八成受访者反对公共消费场所使用人脸识别 *over 80% of respondents opposed the use of facial recognition in public consumption places.* 新京报 Beijing News. http://www.bjnews.com.cn/detail/161162356815701.html

Wang, Z., & Yu, Q. (2015). Privacy trust crisis of personal data in China in the era of big data: The survey and countermeasures. *Computer Law & Security Review, 31*(6), 782–792. https://doi.org/10.1016/j.clsr.2015.08.006

Wang, Y., Xia, H., & Huang, Y. (2016). Examining American and Chinese internet users' contextual privacy preferences of Behavioral advertising. In *Proceedings of the 19th ACM conference on computer-supported cooperative work & social computing* (pp. 539–552). ACM. https://doi.org/10.1145/2818048.2819941

Warren, S. D., & Brandeis, L. D. (1890). The right to privacy. *Harvard Law Review, 4*(5). https://www.cs.cornell.edu/~shmat/courses/cs5436/warren-brandeis.pdf

Whitman, C. (1985). Privacy in Confucian and Taoist thought. In D. J. Munro (Ed.), *Individualism and holism: Studies in Confucian and Taoist values.* Centre for Chinese Studies Publications. https://repository.law.umich.edu/book_chapters/21

Wong, D. (2020). Chinese ethics. In E. N. Zalta (Ed.), *The Stanford Encyclopedia of philosophy.* Metaphysics Research Lab, Stanford University. https://plato.stanford.edu/archives/sum2020/entries/ethics-chinese/

Yan, Y. (2003). *Private life under socialism: Love, intimacy, and family change in a Chinese Village, 1949–1999.* Stanford University Press.

Yan, Y. (2009). *The individualization of Chinese society.* Berg Publishers.

Yan, Y. (2010). The Chinese path to individualization. *The British Journal of Sociology, 61*(3), 489–512. https://doi.org/10.1111/j.1468-4446.2010.01323.x

Yang, C.-F. (2006). The Chinese conception of the self towards a person-making perspective. In U. Kim, K.-S. Yang, & K.-K. Hwang (Eds.), *Indigenous and cultural psychology: Understanding people in context.* Springer.

Yang, Y. (2018, October 2). China's data privacy outcry fuels case for tighter rules. *Financial Times.* https://www.ft.com/content/fdeaf22a-c09a-11e8-95b1-d36dfef1b89a

Yang, Y., & Liu, N. (2019, December 5). *China survey shows high concern over facial recognition abuse.* https://www.ft.com/content/7c32c7a8-172e-11ea-9ee4-11f260415385

Yao, X. (1996). Self-construction and identity: The Confucian self in relation to some western perceptions. *Asian Philosophy, 6*(3), 179–195. https://doi.org/10.1080/09552369608575442

Yao, X. (1999). Confucianism and its modern values: Confucian moral, educational and spiritual heritages revisited. *Journal of Beliefs & Values, 20*(1), 30–40. https://doi.org/10.1080/1361767990200103

Zhang, H., Guo, J., Deng, C., Fan, Y., & Gu, F. (2019). Can video surveillance systems promote the perception of safety? Evidence from surveys on residents in Beijing, China. *Sustainability, 11*(6), 1595. https://doi.org/10.3390/su11061595

Lessons Learned from Co-governance Approaches – Developing Effective AI Policy in Europe

Caitlin C. Corrigan

Abstract Advancements in artificial intelligence (AI) are moving faster than the State's ability to fully govern it, resulting in a need for innovative approaches that also involve non-state actors, or "co-governance" mechanisms. But the question remains as to exactly how co-governance mechanisms can be incorporated into AI governance in the EU. While governing the use of AI presents certain challenges to traditional modes of governance, these challenges are not unique to AI. This chapter provides an in-depth case study of co-governance mechanisms used in relation to services and supply chains that have similar governance challenges to AI in order to map a spectrum of co-governance mechanisms that may be useful for governing AI. After analyzing the applicability of various co-governance mechanisms to the AI governing space, the chapter offers recommendations for how and where these approaches may be most worthwhile. Effective AI governance will require a combination of approaches. Key to success will be State support, cooperation or buy-in in order to maintain legitimacy, while capitalizing upon non-state capacity and distributed implementation. Specifically, first, States, educational institutions and companies should make a major push to integrate responsible AI practices in education curriculums, setting the foundation for effective accreditation and professionalization mechanisms that employers recognize and value. Second, for lower-risk applications, EU Governments should support or compliment multi-stakeholder initiatives to assess and label products. Finally, for higher-risk AI systems, legislative points of control that include the use of co-governance mechanisms for implementation should be put in place.

Keywords Alternative governance · Artificial intelligence · AI governance · Co-governance · Governance challenges

C. C. Corrigan (✉)
Institute for Ethics in Artificial Intelligence, Technical University of Munich, Munich, Germany
e-mail: c.corrigan@tum.de

© The Author(s), under exclusive license to Springer Nature
Switzerland AG 2022
J. Mökander, M. Ziosi (eds.), *The 2021 Yearbook of the Digital Ethics Lab*,
Digital Ethics Lab Yearbook, https://doi.org/10.1007/978-3-031-09846-8_3

1 Introduction

Technological advancement, particularly in artificial intelligence (AI), often develops faster than the State's ability to fully regulate or govern it. While EU and national governments have made several notable moves to govern this space, most recently in the form of the Artificial Intelligence Act (2021), there is often a real or perceived inability to keep up with technological developments and maintain the needed expertise to design effective rules and monitor compliance. This can lead to gaps in the regulatory sphere or a rush to implement policies that may have limited effectiveness.

These circumstances can be seen as similar to other contexts where "governance gaps" (Ruggie, 2004) exist, such as in post-conflict situations, shared ecosystems, or developing country settings (Cammett & MacLean, 2014; Corrigan, 2016; Draude et al., 2018; Martin & Webb, 2020). Where the State is not able to govern a system fully, non-state actors from industry, civil society, international organizations or traditional institutions may fill in some of these gaps in order for society to operate effectively. While these "alternative governance" approaches may be able to fill in some gaps, there are also a variety of cases of the negative and political impacts of moving governance away from the State (Cammett & MacLean, 2014; Corrigan, 2016; Hönke & Thomas, 2012; Jones-Luong, 2014), including issues related to equity, democratic stability, and accountability. From these cases, it becomes clear that substantial and effective cooperation *between* State and non-state actors, or effective "co-governance" (Finck, 2017), is vital for the success of these approaches.

In the current move to *implement* suggested guidelines for AI governance, it is also important to examine how these guidelines can be *improved* thanks to lessons learned while governing other industries, products or services confronting similar constraints. Examining the potential for using co-governance actors and approaches from a public policy perspective can provide insights into how EU countries may approach working with other non-governmental actors to provide effective AI governance and fill "governance gaps." Therefore, this chapter addresses the question of *how co-governance mechanisms can be incorporated into AI governance in the EU*. Using lessons learned from industries, products or services with similar regulatory challenges where co-governance has been successfully employed, it is possible to provide meaningful guidance for employing co-governance mechanisms within AI policymaking. By analyzing examples of co-governance in fields outside of AI, the chapter will suggest recommendations for how States could employ or support innovative co-governance mechanisms to improve their approach to governing the use of AI.

The chapter is organized as follows. Section 2 outlines the concept of co-governance, its operational dimensions and its applicability within the complex environment of AI governance. Section 3 describes a multiple case study methodology and the approach used for selecting cases for review. Section 4 dives into the chosen cases related to supply chain management and building reliability in professional services. Section 5 analyzes these findings in terms of when they might apply to successful AI governance. Section 6 outlines the lessons learned from this analysis, providing recommendations for governance mechanisms going forward. Section 7 provides overall conclusions. Before closing this section, it may be worth adding a note on the conceptual vocabulary used in the following pages. Throughout the rest of this chapter, the term "co-governance" will be used inclusively, to refer to the wide range of potential mechanisms that would fall under alternative/co-governance/co-regulation. Co-governance indicates the presence of multiple stake-holders in the governing process, as opposed to alternative governance which merely implies the use of actors outside the state. Co-governance also moves beyond regulatory mechanisms, which co-regulation implies. Therefore, it is the most applicable term for covering all of the examples that will be discussed and the analysis that follows.

2   Co-governance as an Approach

Because of the nature and use of AI-enabled technology, and the complexity of the environment in which it is being developed, governance of this technology requires novel solutions that fall somewhere between classic regulatory protocols and self-regulation. The Data Ethics Commission itself has argued for innovative forms of co-regulation as potential solutions to the highly complex and ever changing field of algorithmic systems (see Daten Ethik Kommission (DEK), 2019 – recommendation 58).[1] More generally, co-governance (in the broad sense specified above, which includes co-regulation) is often the result of dealing with complex systems or processes where traditional approaches to governance, such as prescriptive and universal regulation, are difficult. In such situations, there may be disconnects between the problem and the causal agents, nonlinearity and disagreements on thresholds of safe operation and sudden and unpredictable transformations, or bifurcations, in the operating environment. These characteristics lead to cases where there may be abrupt and often hard to anticipate shifts in a system. Moreover, as is the case with the AI debate, multiple opinions on the types and locations of dangers and their degree of threat to society. To govern such systems, it is important to have mechanisms that can deal with uncertainty, sudden change, and difficulty in identifying causality (Young, 2017).

Co-governance solutions not only fit within the range between pure self-regulation and pure State-led regulation, but also fall into the interval from mechanisms aimed at individual employees and practitioners to those aimed at entire

---

[1] In the EU context, co-regulation has been defined as a 'mechanism whereby an [EU] legislative act entrusts the attainment of the objectives defined by the legislative authority to parties which are recognized in the field (such as economic operators, the social partners, non-governmental organizations, or associations)' (2003 Interinstitutional Agreement on Better Law-Making, (n 43) para 18.) (Finck, 2017).

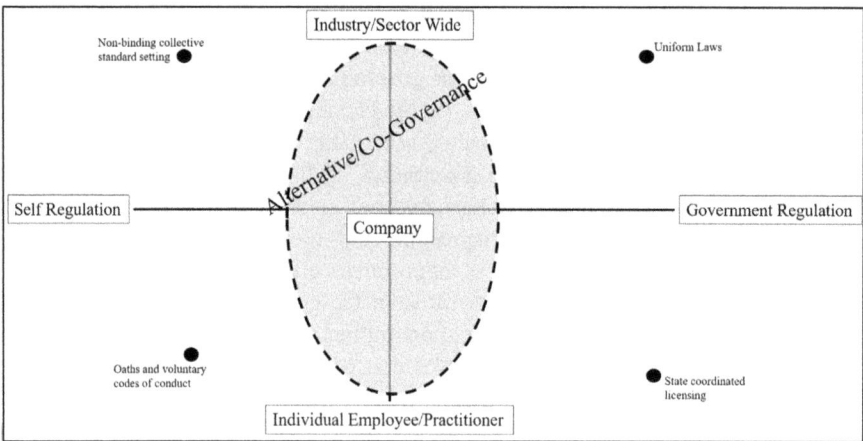

Fig. 1 Locating co-governance

industries or sectors. In the example of AI governance, part of the complication lies in the fact that the development and use of AI is spread over various sectors, spaces, and actors. There is a chain of development and responsibility that may start with individual software developers and extend all the way to an entire sector. While co-governance options fall somewhere in-between purely self- or State-led regulation, there is also a potential to locate these mechanisms anywhere from the level of individual responsibility to international/sector-wide efforts. Thus, this chapter assesses various co-governance approaches with respect to two criteria: the (1) level of State involvement and (2) group or actor the approaches are aimed at governing (see Fig. 1).

Figure 1 depicts these two criteria, outlining the space of co-governance, with examples of the more extreme governance options that exist outside of this space. These include purely self-regulatory options at both the individual level (such as oaths or voluntary codes of conduct) and the industry-wide level (such as collective voluntary standard setting), with company level approaches falling somewhere in between, such as voluntary company codes of conduct or corporate social investment. On the State-led regulation side, we might see mechanisms at the individual level, such as strict licensing of individuals to be allowed to practice a trade, or at the industry wide level with binding national or international laws with strict oversight. Where these approaches fail to govern successfully a space, innovative co-governance approaches, falling somewhere in the shaded circle, may be more effective.

It is important to note that support for co-governance does not mean that there is no room for complimentary State-led or self-regulation. Nor does it imply that the State is not involved in the governance process. In the case of co-governance, government entities may support, structure or monitor an actor or industry in conjunction with other private actors (Martinez et al., 2013), capitalizing on non-governmental expertise, and creating inclusivity, buy-in and legitimacy (Marsden, 2011), while maintaining a functional role in the process. In fact, as was mentioned in Sect. 1, successful alternative or co-governance observed in other contexts relies on buy-in and guidance from the State. Without some State involvement or buy-in, co-governance approaches may not support effectively governmental capacity to govern, potentially undermining the role of states or limiting their capacity to provide accountability and protection to citizens (Milward & Provan, 2000; Peters, 1994). However, as well be displayed in the examples below, the form of State involvement varies widely across examples of co-governance.

3   Approach and Methodology – Challenges to Regulating AI and Identifying Common Themes and Examples

In order to address the research question, we begin with a case study approach (Yin, 2014) to assess how different co-governance mechanisms could work when considering AI governance. The co-governance space is a large one with numerous examples, making case selection crucial to providing appropriate lessons learned (Herron & Quinn, 2016). In order to identify the most relevant mechanisms, it is necessary to identify the main challenges in AI governance that makes co-governance a needed solution. The logic is that examples of co-governance solutions in industries, services or supply chains that confront the same challenges as AI governance would provide the most applicable lessons learned. By looking at these mechanisms in particular then, we are able to draw more relevant recommendations for using co-governance as a form of AI governance going forward. Therefore, this chapter draws on multiple case study examples that are nested within the challenges that confront AI governance in particular (outlined below) (see Fig. 2). This makes possible narrowing down the relevant co-governance mechanisms.

3.1   Challenges in AI Governance

AI governance faces three major challenges:

1. The Transparency Challenge
Algorithmic-based decision-making draws on very large amounts of data to arrive at decisions or provide suggestions. It does so in ways that are often hard for users of the system to understand (see Tsamados et al., 2021 for an overview). Moreover, machine learning algorithms inherently adapt and change as they receive more information, creating a dynamic situation where it can be difficult to provide the needed information consistently to those who should monitor and regulate the use of such systems. Challenges therefore emerge in how governing bodies, or users, can gather the information required to assess the impact and outcomes of such decision-making devices, particularly if this information is constantly changing.

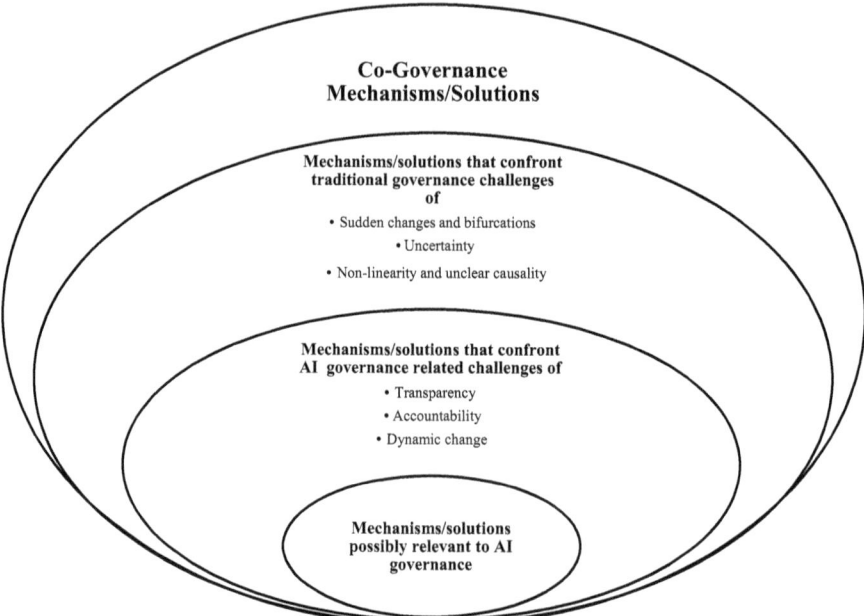

Fig. 2 Nesting of co-governance examples

Furthermore, how can those looking for information not be overwhelmed by super-fluous or out of date information (Ananny & Crawford, 2018; Diakopoulos, 2016; Diakopoulos & Koliska, 2017)? Keeping up with the multitude of algorithms being employed in a variety of forms, over multiple industries and for numerous purposes, and assessing their risk or impact would require significant capacity and expertise for any government agency, along with massive coordination across jurisdictions in order to avoid "governance gaps."

2. The Accountability Challenge

The accountability challenge emerges from two issues in AI governance. First, AI is used across many industries, all of which have their own relevant regulation and various uses for AI-based technology. AI tools are deployed in business to business (b2b) and in business to consumers (b2c) applications. Thus, the chain of custody for the finished products is complex and, with that, the accountability structures for when something goes wrong are hard to decipher. In considering regulation, one would have to consider how regulations would apply to a specific sector or whether the law should be based on the use of a specific tool across sectors. Are the regulations aimed at the developers, consumers or the b2b users and how does this work for companies located outside of the jurisdiction?

The second issue arises when one considers that even if one can monitor a system effectively enough to attribute accountability or liability for its harmful impact due to the misuse of an AI-enabled tool, it remains unclear in complex organization environments (Buhmann et al., 2020) *who* should be held accountable and liable. Is it the company that used the tool or the company that created it? Even if one applies liability at an individual level, AI applications are developed by teams of people trying to find a consensus on an approach to a complex process (Zweig, 2018). Thus, there are often no linear causal connections between the design/development of an AI tool and its outcomes (Seaver, 2014) or little documentation on the data used to back this up (Gebru et al., 2020). This is in stark contrast to other domains, say, medical accountability and liability where you can often trace treatment and decision making back to one hospital or set of doctors.

3. The Dynamic Challenge

As mentioned above, AI-based algorithms inherently change as they learn (Burrell, 2016). Thus, the initial assessment of their use is as necessary as the management of adapting systems (Diakopoulos, 2016). Additionally, the field of AI as a whole is changing rapidly and in non-linear and unpredictable ways. This is why it is important to create the capacity to keep up with the large amounts of often opaque information (the Transparency Challenge), but also why there is a growing need – which is hard to meet – for expertise to monitor AI capabilities (the Dynamic Challenge) as they change (Ananny & Crawford, 2018). AI developers usually move faster and retain higher amounts of talent than governmental regulatory agencies. Therefore, the classic set up of State-led regulation and monitoring runs the risk of being unable to manage both the large amount of information received over a diverse landscape of users, and the risk of focusing on the wrong types of assessments or not understanding the process. In this case, State regulators need to work *with* developers and b2b users to assess, monitor and govern.

From all of these challenges, questions then arise as to how regulators may place some of this responsibility to assess and monitor the use of AI on non-government actors without undermining policy effectiveness or "hollowing out" (Milward & Provan, 2000; Peters, 1994) the role of the State. For instance, by mandating internal assessments or the release of curated information, looking to accreditation bodies, or creating points of control to focus on the highest risk tools. Nonetheless, as the Accountability Challenge makes clear, there are still questions of mandating to whom, how often, with what oversight and which stakeholders.

3.2   Identifying Examples

These challenges seem daunting. However, asymmetric information or expertise, complex chains of command and dynamic environments are not unique to AI-enabled products and services. Drawing on previous examples of how States have dealt with governance in such realms removes the need for the EU and national governments to start from scratch, and instead enables them to rely on findings about co-governance approaches already used under comparable conditions.

As Fig. 2 outlines, the selection of examples for this study is predicated on first considering the attributes that define the complexity in regulating AI noted above. It follows that situations where at least one of these challenges were also observed, and approaches to govern them were designed within a European (or similar) context, would be the most relevant and enlightening cases. This narrows the space of co-governance mechanisms down to the most applicable examples.

In order to get an overview of potential co-governance examples that vary in terms of dimensions outlined in Fig. 1, ranging from self- to governmental regulation *and* from individual to sector wide orientation, a multiple case study methodology is employed examining "diverse examples" (Herron & Quinn, 2016; Seawright & Gerring, 2008). This will yield a mapping of co-governance examples across the co-governance "space." The research then evaluates their applicability to AI governance and reflects on the lessons that can be learned from the examples. The examples explored relate to solutions for (1) complex supply chain management and (2) building reliable and professional services. In these cases, co-governance solutions are essential and have developed over extended periods of time and geographic areas.

4 Co-governance Examples from Outside AI – Ideas, Implementation and Challenges

4.1 Complex Supply Chain Management – The Transparency/ Accountability Challenge

Production and supply chains that cross borders, include various informal or unregulated actors, and have complex information flows offer potentially important insights. Sectoral examples include "conflict" minerals, sustainable forestry, fair trade products and "clean" clothing. A major roadblock to effective governance is that the consumers of the final products (or the States where the products are sold) are often at great physical distance from the locations where human rights, environmental or other violations are occurring. Thus, demanding accountability and transparency using purely State-led approaches becomes difficult as there are complex webs of stakeholders spread across jurisdictions (Transparency and Accountability Challenge). Because of this, many sectors and industries have developed examples of co-governance solutions that, while not perfect, revolve around the use of *multi-stakeholder monitoring, certifying* and *labeling*.

Multi-stakeholder approach, however, vary in their combination and influence of various stakeholders, as well as the level of State involvement. For instance, moves to curb the use of conflict or "blood minerals"[2] in supply chains relies heavily on

---

[2]This includes minerals (mainly tin, tungsten, tantalum and gold) that were mined using child labor, or where the revenue from that mining funds war and unrest.

State-led (regulatory) approaches[3] with the onus for proving compliance placed on listed companies that import these minerals (e.g., smelters) or produce end use products that use the minerals (e.g., technology companies) through annual reports on their assessments and due diligence. The use of the companies themselves to submit reports verifying their activities capitalizes to an extent on the companies' expertise about their own complex supply chains and allows them to use their preferred methods to identify risk, reducing the institutional capacity burden of the regulator.

Similar efforts are also aimed at human rights abuses that exist in supply chains of agricultural product and textiles production. This includes National Action Plans (NAP) on Business and Human Rights, which are largely voluntary compliance support efforts (DIHR, 2017), or more stringent proposed legislation, such as the German Supply Chains Law (BMZ, 2020), which would *require* large-scale companies to participate (Herbert Smith Freehills, 2020).

What the above mentioned approaches to managing supply chains have in common is that they rely on regulation at "points of control" (Finck, 2018). The international nature of these supply chains means that single or regional governments do not have jurisdiction over the whole process. However, States can request information and compliance from the downstream companies that are within their governing jurisdiction and hold them accountable for the entire process. The focus on requesting information from companies located inside the jurisdiction, in addition to placing the information collecting burden on the companies themselves, alleviates the burden on the State. This approach also focuses governance on stakeholders that the State can actually control, while incentivizing companies to work with all those in their supply chain who have the required information.

More voluntary and multi-stakeholder schemes that involve public-private partnerships have also been used to bring government, civil society and companies together to monitor and improve the human rights aspects related to supply chains. The Kimberly Process Certification Scheme (KPCS) is one of the most prominent examples aimed at governing the global diamond industry. It overlaps a voluntary company led-certification scheme, with State driven export/import controls designed to track rough diamonds and ensure that gems from conflict zones do not enter the legitimate world market (Haufler, 2010).[4] With very strong buy-in from all major companies due to concern with consumer outcry over blood diamonds and participation from all major consuming countries and activist groups, it is able to capitalize on participant motivation, public/NGO outcry (Bieri, 2010) and multi-level approaches.

Other examples are seen in food safety and certification. The EU food hygiene legislation, for instance, relies on enforced self-regulation, an arrangement that reduces the burden on the State to collect information and provide expertise, as well as given a sense of ownership to companies/producers, while still including the State as the main oversight actor (Martinez et al., 2013). However, the success of

---

[3] The most prominent examples being the U.S. Dodd-Frank Wall Street Reform and Consumer Protection Act (Dodd Frank) – Section 1502 and the upcoming EU Conflict Minerals Regulation – 2021(European Commission, n.d.; Global Witness, 2017).

[4] It represents 82 countries and 99.8% of the rough diamond market (The Kimberley Process, n.d.).

public-private partnership approaches depends on the institutional setting, the capacity of the relevant firms, the levels of public participation, the levels of transparency and the alignment of interests, which is not a given for every industry.

Other certification mechanisms have emerged in agriculture, forestry and textile industries that are private led, but include state support for the process. Many of these initiatives revolve around labeling of products in order to help guide consumers towards products they can trust and build in levels of governance to help prevent human rights violations along industry supply chains, often using third party independent certification. Key to the discussion of the role of the State, while led by private organizations with various governance structures, these types of efforts are supported in many ways by governments. States can play advisory roles or promote a scheme's legitimacy through lending their official support to an initiative, for instance. These roles not only give States some insight into how initiatives are operating, but also serve as a form of power in terms of their ability to remove support when the initiative is ineffective (Arnold et al., 2020; Auld et al., 2008; E. A. Bennett, 2017). Prominent examples include the Forestry Stewardship Council (FSC)[5] and the plethora of programs that fall under "Fair Trade."

There are major challenges to effectively relying on these largely voluntary schemes. One is fracturing or the potential for stakeholders (particularly big corporations) to split off and create their own schemes and standards when multistakeholder versions become inflexible, or one stakeholder's preferences start to dominate (Arnold et al., 2020; Crouch, 2019), undermining integrity. Overlapping or competing schemes make it difficult for consumers to navigate the options and arrive at well-informed decisions (Castka & Corbett, 2016). Another challenge is motivation. These types of initiatives emerge when companies or producers are pressured to act in response to governmental, consumer or civil society groups' outcry to "do better," such as in the case of Fair Trade schemes (Sen & Majumder, 2011). Therefore, attempts at co-governance of supply and production chains through multi-stakeholder private-led initiatives require aligning and coordinating of interests between governments, society and business, a task which is not always easy or possible (Martinez et al., 2013).

4.2   Building Reliable and Professional Services – Transparency/Accountability/Dynamic Challenge

Professional services, particularly those that can have a dramatic effect on the consumers or beneficiaries of the service, such as in law or medicine, face governing challenges related to transparency, accountability, and dynamic environments.

---

[5] Founded in 1993 by the WWF, environmental NGOs, timber traders, indigenous peoples' groups and forest worker organizations, the FSC sets standards for members and is comprised of environmental, social and economic chambers that balance and represent interests of the various stakeholders (Auld et al., 2008; FSC International, 2020).

When someone, or some institution, is relied upon for their expertise, and it is often unclear how they arrived at an "expert" decision (Transparency Challenge) or at what exact step of the process something went wrong (Accountability Challenge), it can become difficult to assess liability and accountability. Much of this ambiguity emerges from the asymmetric relationship between service provider and beneficiary (Dynamic Challenge). The "expert" or service provider often holds the information and receives the benefit of the doubt that beneficiaries, with less expertise, will trust their opinion. If there is a dispute or a bad outcome (accusations of malpractice for instance), it is often hard to trace the negative outcome to a particular action taken by the service provider. In this case, beneficiaries often are at a disadvantage in terms of information gathering, accountability tracing, and burden of proof.

Moreover, with services that are urgently needed, the beneficiary often is in a position where they have no other choice but to work with a service provider even if they cannot gather the needed information about the service provider's motivations, reliability, and record. In this case, consumers or beneficiaries must have some way to trust that the service provider will personally adhere to a set of standards in procedures and decision making that will best serve their client. From the practitioner's perspective, a lack of professional identifiers lowers the ability to differentiate unqualified competitors from those who have proper training, leading to bad practices in the field that can drive down the value of investing in training and study (Waddington, 1990).

Because of these challenges, professionalization mechanisms, such as membership associations and accreditation have been used as a way to promote this trust externally, and bind individual service providers to a set of standards. Unlike the examples in supply chain management that place accreditation/certification on a product or company, these solutions set the governance of services mainly at the *individual practitioner level* (Cook et al., 2005).

While many examples of professionalization rely heavily on ethics and self-regulation (See Filipović et al., 2018 for a detailed look), this does not mean the approach is absent of institutionalized co-governance mechanisms. Examples around Europe, such as the German Medical Association (Bundesärztekammer), the British Medical Association or the UK Bar Council are institutionalized membership associations that work with the State and help practitioners settle disputes, but also promote accountability and set best practices and standards for the field (British Medical Association, 2021; Bundesärztekammer, 2021; The General Council of the Bar, 2021).

Such professionalization organizations can work in combination with governmental or employer oversight to institutionalize many of the aspects related to professionalization. For instance, admittance to an organization may be predicated on independently or State run licensing.[6] Licensing can rely on completion of recognized educational programs with common standards of educational training, and

---

[6] For example, State-run Approbation in Germany for professions from physicians, to veterinarians, to pharmacists (Regierung von Oberbayern, 2021) or independently run General Medical Council in the UK (General Medical Council, 2021).

may include continuing education and adherence to best practices. This requires cooperation with public or private educational institutes and programs. Employers can also create incentives to join associations by preferring or requiring these affiliation in choosing employees. Through external parties creating incentives for individuals to join and comply with set standards, these professional associations can move beyond self-imposed ethical standards and purely self-interested actions to provide quality assurance to those using the services.

There are certain advantages to this approach. By creating a sense of community, it builds on an ingrained personal sense of responsibility that focuses the practitioner to consider ethical decision making in their daily practice, which would be hard to fully monitor by any other means. By relying on member-led organizational structure, it also creates rules and codes that are derived from the experts themselves, eliminating some of the dynamic challenges linked to asymmetric information and constantly changing environments.

However, challenges or shortcomings of this approach exist as well. Membership associations serve their members and may act in this self-interest or yield wide lobbying power to the detriment of the population they serve (Cook et al., 2005). Moreover, the successful establishment of a membership body depends on creating a whole ecosystem that reinforces their worth and this can be a complex and slow-moving task.[7] Educational institutions have to revolve their programs and curriculums around the goal of the students receiving membership or accreditation upon graduation, in order to justify its importance. Associations also need to be supported and monitored by related governance structures, such as State licensing requirements, medical review boards or institutional review boards. Indeed, State encouragement and recognition of professional or business associations has been argued to be key in promoting their use and reducing fragmentation of sectoral bodies (Bennett, 2000). Moreover, employers and companies must recognize (or incentivize) expertise if it comes from the applicable accreditation body. These are not things that evolve over months, or even years, but have to be instituted and slowly merged with the current system.

5   Co-governance as It Relates to AI

The examples in Sect. 4 illustrate co-governance mechanisms that have been employed under similar challenges to those faced in AI governance. Figure 3 zooms in on the co-governance field identified in Fig. 1 to map out these mechanisms. As Fig. 3 shows, the mechanisms vary widely in terms of the make-up of stakeholders involved and the level of actor at which they are aimed (vertical axis), for example

---

[7]The medical profession for instance, something we now automatically consider as highly governed throughout Europe, existed for centuries without proper oversight. Only in the mid-nineteenth century did medical professionals and educational institutions begin to institutionalize membership and set common training/education and practice standards (Waddington, 1990).

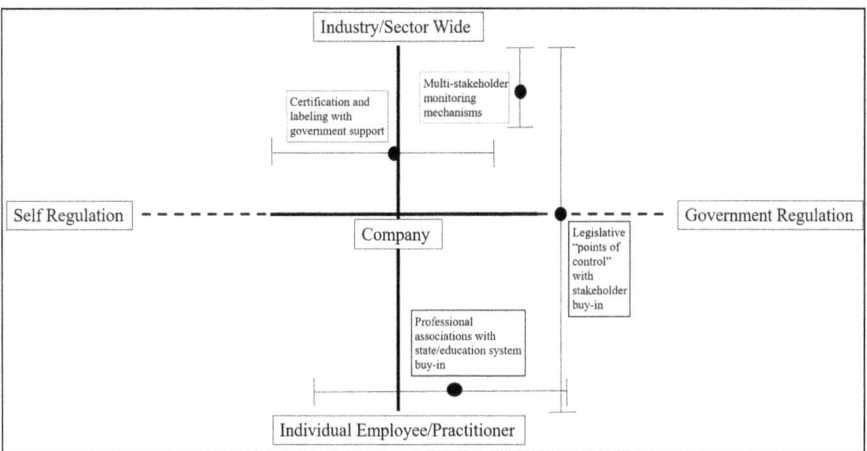

Fig. 3 Locating options for co-governance

does a mechanism focus on governing individuals (bottom) vs. companies (middle) vs. industries (top).

They also vary in terms of the role and power of State actors in the process (horizontal axis), from examples that are more private sector led, to those that are strongly government reinforced. Another way to think about the role of the State in these co-governance regimes is top down vs. bottom up. Top-down approaches assign State issued regulatory tasks to private entities, setting conditions of operation and offering procedures for achieving those conditions. This would be the case with many legislative points of control outlined in the supply chain management examples or State licensing of professions. In bottom up models, governmental stakeholders acknowledge, support and facilitate private sector initiatives (as was the case with many of the food labeling schemes) (Martinez et al., 2013). However, just because the potential, relevant co-governance examples have been created under similar challenges to those faced by AI governance, this does not mean that they will be fully applicable to governing the AI environment. The next section analyses the potential for these mechanisms to be used in AI governance specifically.

5.1 Multi-stakeholder Monitoring, Certification and Labeling of Products

As seen in the examples above, options related to co-governance mechanisms for monitoring, certification and labeling vary in terms of which stakeholders are involved and what power they have. The more voluntary, the higher the likelihood that only the highest motivated actors with the lowest barriers to entry will join or comply. In the case of the diamond industry, where there were just a few major players with very strong economic interest to improve the image of the industry, a largely voluntary process was able to bind the corporate stakeholders. In agricultural products, with a more diversified group of stakeholders, both geographically and in terms of the number of companies, we have seen a fracturing in certification initiatives that reduces the power of all of them. These types of mechanisms also mostly have a common motivation point – State or consumer interest in having more information about the products they are purchasing. Given these observations, multi-stakeholder monitoring, certification, and labeling mechanisms may be most effectively used for certain AI-enabled products where public concern is high and the diversity of corporate stakeholders is low.

Moreover, as the examples illustrated, voluntary schemes can have support of the State or be complimented by governmental oversight. Other research has shown that the presence of some sort of external stakeholder control in these type of governance schemes, such as from independent accreditation bodies or States, improves the perceived quality of governance of the scheme and, thus, is most likely to increase the organization's/label's/certificate's worth (Castka & Corbett, 2016). Sub-sector voluntary schemes for AI-enabled products, therefore, would still benefit from well-placed State recognition and third-party quality assurance to build legitimacy and increase standards.

One already emerging example of a multi-stakeholder certification approach in AI governance is the Institute of Electrical and Electronics Engineers (IEEE) SA's Ethics Certification Program for Autonomous and Intelligent Systems (ECPAIS) (IEEE SA, n.d.). Given the IEEE's standing in the area of standards and certification, this has the potential to provide some legitimacy and obtain a level of external governance or third party buy-in found to be useful in other sectors.

5.2   Professionalization

There are many reasons why it is difficult to apply current professional association examples to the field of AI. Unlike professions where there is a direct patient or client relationship, the technology sector often lacks a direct connection between user and developer. Moreover, at this point there is a lack of clarity about whom would be included under this "profession" (Zweig, 2018). This means incentives between practitioner and beneficiary are less likely to align naturally. Ethical codes that do exist are also vague and do not give the AI practitioner definitive guidelines, such as ones that could also be used to prove malpractice. Furthermore, as mentioned in the previous section, the timelines for establishing such systems of governance are long, in that a whole ecosystem needs to develop around recognizing associations, operationalizing standards and creating accountability mechanisms.

Principles in isolation are insufficient to develop AI ethically, but incentivizing personal ethical conduct through *institutionalization* of such principles (even if not fully regulated) does have a role to play. Filipović et al. (2018), for instance, suggested that institutionalization and sanctioning mechanisms are among the factors necessary for professionalizing ethical behavior for AI developers. Merging education and training with professional accreditation would also be a component albeit a long but necessary process. However, the extended time horizon issue does not preclude this from being a worthwhile path, but it is important to recognize that there will need to be numerous adjustments and reinforcements as the system builds.

Several examples of AI personal codes of conduct or oaths crafted by those in the field already exist. Oren Etzioni, for instance, calls on universities to require graduates to swear an oath for AI practitioners that he adapted from medicine (Etzioni, 2018). Carnegie Mellon University's Software Engineering Institute has created a "checklist and agreement" for developing ethical AI (Smith, 2019). Driven Data has put out a similar checklist, but in a command line tool for ease of adding it directly to data science projects (Driven Data, n.d.). A number of other examples of these "toolkits" exist (Saucedo, 2018).

Moreover, several older organizations, such as the British Computer Society, the German Informatics Society, the IEEE and the Association of Computing Machinery, provide possible initial institutional structures, but their role has to be adapted to accommodate the above mentioned differences of AI developers as a profession and adjust to incentivize buy-in from potential employers. These examples could provide an initial jumping off point for institutionalizing these codes or integrating them into education. Nonetheless, it is possible that professionalization can only help deal with some of the governance challenges of AI, and the efforts would need to be complimented by other governance mechanisms, even in the long run.

5.3 Points of Control (Targeted Regulation with Stakeholder Buy-in)

While veering close to being out of the realm of co-governance, legislation targeted at practical points of control is playing an increasingly important role in supply chain management, as exemplified by their important role in Dodd-Frank Act, EU Conflict Minerals Regulation – 2021 and the debated German Supply Chains Law. What all these mechanisms have in common is that they take a complex web of interactions and provide governmental oversight only at the points at which the State has actual control. In this respect, they are actually a way of providing indirectly governance for the whole system. In these cases, State actors mandate that the companies operating inside of their own jurisdiction provide proof of their due diligence for the whole system (which may be out of the State's jurisdiction), shifting the information collecting burden to the industry actors, and indirectly asserting control over external actors by mandating that the process be put in place. It can be considered co-governance to an extent because it relies on company buy-in to the process in order for it to run effectively. Therefore, the most effective mechanisms would be put in place in consultation with the affected industry stakeholders, and support for smaller stakeholders, in order to increase effectiveness. Examples of this targeted control can be seen in industries more similar to AI, such as in governing data use through the General Data Protection Regulation (GDPR) or in governing e-commerce (Finck, 2018).

Due to the fact that these approaches require considerable governmental capacity and oversight and are most useful where there is some industrial buy-in (in that industry stakeholders also see a need to "level the playing field"), they are more likely to be effective in certain sub-sectors of the AI governance space. For instance, these tactics might be most successful for use with AI applications that are high risk to users or society or for AI applications in the public sector. In low risk applications or those with narrower reach and fewer stakeholders, some of the more voluntary multi-stakeholder options may provide ample solutions with a lower burden to the State.

6   "Lessons Learned" – Recommendations for AI Governance in the EU

In moving from the formulation of AI policy to actual implementation, there will not be just one governance or even co-governance mechanism that will apply to all AI-enabled tools. As the previous examples illustrate, co-governance is a diverse space and AI governance will likely require both sector-wide and sub-sector mechanisms, as well as long and short-term strategies aimed at both companies and individual practitioners.

However, drawing on the above analysis, it is possible to make several recommendations for where and when co-governance can be used to govern the use of AI. Longer-term mechanisms, such as professionalization maybe be daunting and may never be fully solidified in the same way we now see medical or legal professions, but this does not mean they are not worthwhile. *Efforts for co-governance solutions aimed at professionalization should first include a major push by EU governments and educational institutions to integrate responsible AI practices into higher education*[8] *curriculums.* Recognizing and even requiring that the capacity to evaluate AI from an ethical perspective (including content on ethical principles and methods of detecting risks of harm, bias, unfairness, power inequities, transparency etc....) be taught to students who will be developing, but also using, AI in their careers and lives is a first step in creating an environment where these skills are desired and necessary for future employment. This can lay the foundation for professional associations, recognized by States and companies that have responsible AI as a cornerstone of their mission. Additionally, it will build a cadre of informed users of AI who will be able to demand that the developers meet ethical standards and to themselves be informed consumers of AI enabled products.

Current efforts should also have the longer term view of evolving some sort of *accreditation of certain types of degrees that include course work on ethical or responsible AI that potential employers recognize and value.* This could be granted

---

[8] This could also be considered in the ever increasing amount of computer science oriented classes in secondary schools.

by some sort of educational or professionalization body (ideally with State support or recognition[9]), but would also require buy-in from companies to recognize and prefer employees with these accreditations, creating value for students or potential employees who take that extra step to become accredited or obtain membership in associations. Examples of this already exist in some higher education institutions in the form of Digital Badges or Micro Credentialing that are issued for completion of certain milestones, courses or requirements. Ideally, these badges are recognized/valued by employers and signify extra efforts by a student in certain subject or career development areas. They are also often easier to share or seen as more relevant for potential employers than just a traditional transcript (Educause, 2014; Grant, 2014; Markowitz, 2018). In the much longer term, accreditation or certification in the responsible use and development of AI could turn from a preference into a requirement.

More quickly implementable mechanisms, such as certification and monitoring, may provide targeted solutions for certain types of AI-enabled tools. *EU Governments should support or compliment multi-stakeholder initiatives to assess and label products particularly for applications deemed a lower risk to users and society.* Again, given the diversity of where AI-enabled tools are employed, creating schemes for a sub-sector of products will probably be less subject to fracturing. Particularly for low-risk AI applications,[10] governmental support of these types of initiatives, through oversight or recognition, plus third-party accreditation, would provide some form of governance while limiting the burden to the State. State support of independent labeling or certification can help signify to business or consumer (b2b or b2c) users that there is a level of trust in the scheme, also creating value for the members of the scheme and increasing membership. Moreover, if States are involved even in this limited way, they retain power in their ability to remove their support, thus undermining the legitimacy of the scheme. Therefore, even more informal governmental support would have use.

For higher risk AI systems, legislative points of control that include co-governance mechanisms should be put in place. In these cases, companies developing or using such applications categorized as high risk should be required to provide evidence to the State of due diligence in tracing and dealing with potential ethical issues within their applications. These mechanisms can vary depending on the level of risk in terms of whether they are organized as more of a public/private partnership in which the State helps set baseline standards and collects information, but leaves the process and the ability to go beyond the baseline standards up the companies (as in the food safety and organic labeling examples), or whether the State sets strict reporting guidelines with oversight and mandatory compliance for a set of companies (as in the proposed Conflict Mineral and German Supply Chains Laws). For extremely high-risk applications, where there are life or death, human rights or democratic

---

[9]As the examples above displayed, State support for mechanism improves their legitimacy and, therefore, effectiveness. Thus, getting State buy-in to support higher-education requirements is key.

[10]Or what the proposed EU regulation terms "non-high-risk"(Artificial Intelligence Act, 2021, pp. 8, 10), for which they suggests building independent codes of conduct.

implications, it will be worth the capacity burden to the State to invest in more direct oversight assessment mechanisms, adjusting current legislation or administrative capacities, or creating new legislation or agencies. Where the risk is lower, they may be able to divert some of the workload to third parties or the companies themselves.

This recommendation largely lines up with the EU proposed rule on governing "high-risk AI systems." In these cases, systems that pose "significant harmful impact on the health, safety and fundamental rights of persons" need to meet *minimum* standards of transparency and access to information[11] so that that can assessed and certified (Artificial Intelligence Act, 2021, p. 25). Much like the conflict mineral regulation examples, this approach sets State mandated standards, but places the reporting burden on companies and utilizes third party "conformity" assessments, lowering the information gathering burden on the State.

7 Conclusions

As initially argued in this chapter, because of the nature and use of AI-enabled technology, and the complexity of the environment in which it is being developed, governance of these tools requires creative solutions that fall somewhere between classic regulatory protocols and self-regulation. Thus, investigating the potential for mechanisms that fall under the broad categorization of co-governance is a worthwhile endeavor as the EU moves from the development of policies aimed at AI governance to actually defining mechanisms for implementing those broader policies. Through examining examples of mechanisms successfully employed in response to governance challenges similar to those faced by AI, in complex supply chain management and building reliable professions, potential co-governance solutions were mapped with reference to the inclusion of stakeholders, as well as the focus of the mechanisms. Based on this analysis, the potential applicability of these mechanisms for AI governance was evaluated and broad lessons learned were derived.

The analysis developed in this chapter highlights that we can indeed learn from co-governance mechanisms used in other sectors or fields in Europe and beyond and there are ample opportunities for AI governance stakeholders to capitalize on these examples in order to begin to provide effective solutions to this complex governing space, both immediately and in the longer term. It is also clear that AI presents its own and varied governing challenges to which co-governance mechanisms will have to be adapted. This means, as the overarching lesson learned of the paper

---

[11] According to the EU proposed regulation, "this requires keeping records and the availability of a technical documentation, containing information which is necessary to assess the compliance of the AI system with the relevant requirements. Such information should include the general characteristics, capabilities and limitations of the system, algorithms, data, training, testing and validation processes used as well as documentation on the relevant risk management system. The technical documentation should be kept up to date" (Artificial Intelligence Act, 2021, p. 31)

suggests, that there is not going to be just one governance or even co-governance mechanism that will apply to all AI sectors or tools. Co-governance is a large space, with a multitude of mechanisms that can work in conjunction with self-regulation, State-led regulation and complimentary co-governance mechanisms. Policymakers and stakeholders will have to consider how the development, use and challenges to governance differ across AI-based sectors and tools and, therefore, how these different sectors or tools may require their own governance schemes. The level and type of risks that emerge from the use of AI for the consumer, the economy or society may also determine which type of governance schemes are appropriate. Moreover, the type and diversity of stakeholders involved in the process and their level of ability to hold AI developers and users accountable will also play a role in how co-governance mechanisms are effectively organized.

This chapter represents an initial step in analyzing the potential for co-governance to effectively offer an avenue for AI governance and providing indications as to the timeframes and universality in which the different mechanisms might be appropriate. But more steps are needed to begin to implement co-governance mechanisms for AI effectively. Thus, as is the case with much of the current research and rhetoric on AI governance, the next step in this line of research is to understand which mechanisms are best suited for which AI applications and design concrete and tailored co-governance schemes for implementation. Future research would do well to concern itself with questions such as: Do AI applications aimed at healthcare lend themselves more or less to one form of co-governance, and how and when might this have to be overlapped with other co-governance or formal governance solutions as well? What might need to be adapted for the case for AI used in the workplace, or social media or for the military? What aspects of AI development may lend themselves to professionalization mechanisms and for which roles might this not be necessary or not be enough? Effective implementation of these mechanisms will above all require expertise from those with technical backgrounds and those working where the tools are being applied (i.e. healthcare, social media, self driving etc....), but also importantly from those with expertise in implementing policy in complex governing environments.

References

Ananny, M., & Crawford, K. (2018). Seeing without knowing: Limitations of the transparency ideal and its application to algorithmic accountability. *New Media & Society, 20*(3), 973–989. https://doi.org/10.1177/1461444816676645

Arnold, N., Bennett, E., Blendin, M., Brochard, M., Carimentrand, A., Coulibaly, M., De Ferran, F., Durochat, É., Gautrey, G., Geffner, D., Leyssene, C., Lorenz, J., Maisonhaute, J., Paulsen, O., Ripoll, J., Sirdcy, N., & Stoll, J. (2020). *International guide to fair trade labels—Edition 2020*. Commerce Équitable France; Fair World Project; FairNESS; Forum Fairer Handel.

Auld, G., Gulbrandsen, L. H., & McDermott, C. L. (2008). Certification schemes and the impacts on forests and forestry. *Annual Review of Environment and Resources, 33*, 187–211. https://doi.org/10.1146/annurev.environ.33.013007.103754

Bennett, R. J. (2000). The logic of membership of sectoral business associations. *Review of Social Economy, LVIII*(1), 19–42.

Bennett, E. A. (2017). Who governs socially-oriented voluntary sustainability standards? Not the producers of certified products. *World Development, 91*, 53–69. https://doi.org/10.1016/j.worlddev.2016.10.010

Bieri, F. (2010). The roles of NGOs in the Kimberley process. *Global Studies Journal, 20*.

British Medical Association. (2021). *What we do*. https://www.bma.org.uk/what-we-do

Buhmann, A., Paßmann, J., & Fieseler, C. (2020). Managing algorithmic accountability: Balancing reputational concerns, engagement strategies, and the potential of rational discourse. *Journal of Business Ethics, 163*(2), 265–280. https://doi.org/10.1007/s10551-019-04226-4

Bundesärztekammer. (2021). *About the German Medical Association*. https://www.bundesaerztekammer.de/weitere-sprachen/english/german-medical-association/

Burrell, J. (2016). How the machine 'thinks': Understanding opacity in machine learning algorithms. *Big Data & Society, 3*(1), 205395171562251. https://doi.org/10.1177/2053951715622512

Cammett, M. C., & MacLean, L. M. (2014). Introduction. In I. Gough, L. M. MacLean, & M. C. Cammett (Eds.), *The politics of non-state social welfare* (pp. 1–15). Cornell Universtiy Press.

Castka, P., & Corbett, C. J. (2016). Governance of eco-labels: Expert opinion and media coverage. *Journal of Business Ethics, 135*(2), 309–326. https://doi.org/10.1007/s10551-014-2474-3

Cook, K. S., Hardin, R., & Levi, M. (2005). *Cooperation without trust*. Russel Sage Foundation.

Corrigan, C. C. (2016). *The politics of privatizing governance: The political and institutional determinants of corporate social responsibility in Africa*. University of Pittsburgh.

Crouch, D. (2019, January 8). Fair trade food schemes battle to promote better standards. *Financial Times*. https://www.ft.com/content/83247fda-e0f1-11e8-a8a0-99b2e340ffeb

Danish Institute for Human Rights (DIHR). (2017). *National Action Plans on Business and Human Rights—Supply Chains*. https://globalnaps.org/issue/supply-chains/

Daten Ethik Kommission (DEK). (2019). *Opinion of the Data Ethics Commission*. Data Ethics Commission of the Federal Government.

Diakopoulos, N. (2016). Accountability in algorithmic decision-making. *Communications of the ACM, 59*(2), 56–62.

Diakopoulos, N., & Koliska, M. (2017). Algorithmic transparency in the news media. *Digital Journalism, 5*(7), 809–828. https://doi.org/10.1080/21670811.2016.1208053

Draude, A., Börzel, T. A., & Risse, T. (Eds.). (2018). *The Oxford handbook of governance and limited statehood*. Oxford University Press. https://www.oxfordhandbooks.com/view/10.1093/oxfordhb/9780198797203.001.0001/oxfordhb-9780198797203

Driven Data. (n.d.). *An ethics checklist for data scientists*. https://deon.drivendata.org/

Educause. (2014). *7 things you should know about badging for professional development*. 2.

Etzioni, O. (2018, March 14). A Hippocratic Oath for artificial intelligence practitioners. *Tech Crunch*. https://techcrunch.com/2018/03/14/a-hippocratic-oath-for-artificial-intelligence-practitioners

European Commission. (n.d.). *Conflict Minerals—The Regulation Explained*. Retrieved October 30, 2020, from https://ec.europa.eu/trade/policy/in-focus/conflict-minerals-regulation/regulation-explained/

Federal Ministry for Economic Cooperation and Development (BMZ). (2020). *Federal Ministers Heil and Müller: "Now the Coalition Agreement will come into play for a supply chain law. The aim is finalisation before the end of this legislative term."* http://www.bmz.de/en/press/aktuelleMeldungen/2020/juli/200714_pm_21_Federal-Ministers-Heil-and-Mueller-Now-the-Coalition-Agreement-will-come-into-play-for-a-supply-chain-law-The-aim-is-finalisation-before-the-end-of-this-legislative-term/index.html

Filipović, A., Koska, C., & Paganini, C. (2018). Developing a professional ethics for algorithmists: Learning from the examples of established ethics. *Bertelsmann Stiftung Working Paper, 9*, doi:10.11586/2018034.

Finck, M. (2017). Digital regulation: Designing a supranational legal framework for the platform economy. *LSE Law, Society and Economy Working Papers, 15*. https://doi.org/10.2139/ssrn.2990043

Finck, M. (2018). *Blockchain regulation and governance in Europe* (1st ed.). Cambridge University Press. https://doi.org/10.1017/9781108609708

FSC International. (2020). *Forest Stewardship Council.* https://www.fsc.org/

Gebru, T., Morgenstern, J., Vecchione, B., Vaughan, J. W., Wallach, H., Daumé III, H., & Crawford, K. (2020). Datasheets for datasets. *ArXiv:1803.09010 [Cs].* http://arxiv.org/abs/1803.09010

General Medical Council. (2021). *Who we are.* https://www.gmc-uk.org/about/who-we-are

Global Witness. (2017, November 15). *Section 1502 of the US Dodd Frank Act: The landmark US law requiring responsible minerals sourcing.* BRIEFING: US Conflict Minerals Law. https://www.globalwitness.org/en/campaigns/conflict-minerals/dodd-frank-act-section-1502/

Grant, S. L. (2014). *What counts as learning: Open digital badges for new opportunities.* Digitial Media and Learning Research Hub. https://public.ebookcentral.proquest.com/choice/publicfullrecord.aspx?p=5869977

Haufler, V. (2010). The Kimberley process certification scheme: An innovation in global governance and conflict prevention. *Journal of Business Ethics, 89*, 403–416. https://doi.org/10.1007/s10551-010-0401-9

Herbert Smith Freehills. (2020). *Supply chain law in Germany: Current steps towards a mandatory human rights due diligence law* [legal briefing]. https://www.herbertsmithfreehills.com/latest-thinking/supply-chain-law-in-germany-current-steps-towards-a-mandatory-human-rights-due.

Herron, M. C., & Quinn, K. M. (2016). A careful look at modern case selection methods. *Sociological Methods & Research, 45*(3), 458–492.

Hönke, J., & Thomas, E. (2012). Governance for whom?: Capturing the inclusiveness and unintended effects of governance. *SFB-Governance Working Paper Series, 31*(May).

IEEE SA. (n.d.). *The Ethics Certification Program for Autonomous and Intelligent Systems (ECPAIS).* Retrieved November 23, 2020, from https://standards.ieee.org/industry-connections/ecpais.html

Jones-Luong, P. (2014). Empowering local communities and enervating the state?: Foreign oil companies as public goods providers in Azerbaijan and Kazhstan. In I. Gough, L. M. MacLean, & M. C. Cammett (Eds.), *The politics of non-state social welfare* (pp. 57–76). Cornell University Press.

Markowitz, T. (2018, September 16). The seven deadly sins of digital badging in education. *Forbes.* https://www.forbes.com/sites/troymarkowitz/2018/09/16/the-seven-deadly-sins-of-digital-badging-in-education-making-badges-student-centered/

Marsden, C. T. (2011). *Internet co-regulation: European law, regulatory governance and legitimacy in cyberspace.* Cambridge University Press. https://doi.org/10.1017/CBO9780511763410

Martin, M., & Webb, K. (2020). Water quality protection of the Canada-US Great Lakes: Examining the emerging state/nonstate governance approach. *International Journal of Innovation and Sustainable Development, 14*(102–124) https://www.inderscienceonline.com/doi/abs/10.1504/IJISD.2020.104245

Martinez, M. G., Verbruggen, P., & Fearne, A. (2013). Risk-based approaches to food safety regulation: What role for co-regulation? *Journal of Risk Research, 16*(9), 1101–1121. https://doi.org/10.1080/13669877.2012.743157

Milward, H. B., & Provan, K. G. (2000). Governing the hollow state. *Journal of Pulic Adminstration Research and Theory, 10*(2), 359–379.

Peters, B. G. (1994). Managing the hollow state. *International Journal of Public Adminstration, 17*(3–4), 739–756.

Regierung von Oberbayern. (2021). *Arzt/Ärztin; Beantragung einer Approbation.* https://www.regierung.oberbayern.bayern.de/aufgaben/37198/244210/leistung/leistung_12109/index.html

Regulation of the European Parliament and of the Council Laying Down Harmonised Rules on Artificial Intelligence (Artificial Intelligence Act) and Amending Certain Union Legislative Acts, 2021/0106 (COD) (2021).

Ruggie, J. G. (2004). Reconstituting the global public domain—Issues, actors, and practices. *European Journal of International Relations, 10*(4), 499–531. https://doi.org/10.1177/1354066104047847

Saucedo, A. (2018). Awesome AI Guidelines. *GitHub.* https://github.com/EthicalML/awesome-artificial-intelligence-guidelines

Seaver, N. (2014). Knowing algorithms. *Media in Transition, 8.*

Seawright, J., & Gerring, J. (2008). Case selection techniques in case study research: A menu of qualitative and quantitative options. *Political Research Quarterly, 61*(2), 294–308. https://doi.org/10.1177/1065912907313077

Sen, D., & Majumder, S. (2011). Fair trade and fair trade certification of food and agricultural commodities: Promises, pitfalls, and possibilities. *Environment and Society: Advances in Research, 2*(1), 29–47.

Smith, C. (2019). *Designing ethical AI experiences: Checklist and agreement.* Canegie Mellon University - Software Engineering Institute.

The General Council of the Bar. (2021). *About Us and the Bar.* https://www.barcouncil.org.uk/about.html.

The Kimberley Process. (n.d.). *What is the Kimberley Process?* Retrieved October 30, 2020, from https://www.kimberleyprocess.com/en/what-kp

Tsamados, A., Aggarwal, N., Cowls, J., Morley, J., Roberts, H., Taddeo, M., & Floridi, L. (2021). The ethics of algorithms: Key problems and solutions. *AI & SOCIETY.* https://doi.org/10.1007/s00146-021-01154-8

Waddington, I. (1990). The movement towards the professionalisation of medicine. *British Medical Journal, 301*, 3.

Yin, R. K. (2014). *Case study research: Design and methods* (5th ed.). Sage.

Young, O. R. (2017). Beyond regulation: Innovative strategies for governing large complex systems. *Sustainability, 9*(6), 938. https://doi.org/10.3390/su9060938

Zweig, K. A. (2018). Wo Maschinen irren können: Verantwortlichkeiten und Fehlerquellen in Prozessen algorithmischer Entscheidungsfindung. *Impuls Algorithmenethik - Bertelsmann Stiftung, 4.* https://doi.org/10.11586/2018006

State-Firm Coordination in AI Governance

Noah Schöppl

Abstract This chapter addresses two questions: What is the role of states and technology firms in AI governance? And how can an AI governance approach that relies on both states and firms succeed? In response to the first question, it is argued that approaches which rely exclusively on either states or firms are likely to fail and that only a combination of public and private governance regimes fulfils the necessary conditions to govern AI successfully. In exploring the question of how such an approach may succeed, it is argued that states need to increase and coordinate their regulatory capability, while fundamental reform is required for firms, which need to realign their purpose and corporate governance. For this to succeed, technology firms need to learn to see ethics not as a constraint but as a primary objective of their activities. While currently the dominant paradigm for tech firms is to aim to maximise profits (optimisation goal) while being socially acceptable (bounding condition), they need to shift to do what is socially preferable (optimisation goal) while being commercially viable (bounding condition).

Keywords AI governance · State-firm coordination · Public governance · Corporate governance

1 Introduction

There is a consensus that AI needs to be governed, yet, when it comes to who should make the rules for AI, there is a fundamental divide between those who propose private self-governance regimes and those who favour strict regulation by lawmakers (Susskind, 2018; Wagner, 2018; Webb, 2019). This chapter attempts to answer two related questions. The first one is: What is the role of states and technology firms in AI governance? I will argue that approaches which rely exclusively on either states

N. Schöppl (✉)
Oxford Internet Institute, University of Oxford, Oxford, UK

ALLAI, Amsterdam, The Netherlands
e-mail: noah.schoeppl@allai.nl

© The Author(s), under exclusive license to Springer Nature
Switzerland AG 2022
J. Mökander, M. Ziosi (eds.), *The 2021 Yearbook of the Digital Ethics Lab*,
Digital Ethics Lab Yearbook, https://doi.org/10.1007/978-3-031-09846-8_4

47

or firms are likely to fail and that only a combination of public and private governance regimes fulfils the necessary conditions to govern AI successfully. The second and more specific question is: How can an AI governance approach that relies on both, states and firms, succeed? I will argue that states need to increase their regulatory capability; yet more fundamental reform is required for firms: they need to realign their purpose and corporate governance. This chapter draws on and aims to contribute to theories of AI governance and policy, by analysing the state-firm relationship through the lenses of power and legitimacy.

Let me begin by clarifying some terms, as well as the relevance and scope of this argument. Emerging digital environments raise new ethical challenges and in order to resolve them translational ethics is required (Floridi, 2017). The interrelated issues at stake in AI governance are: (a) the protection of fundamental rights, (b) the fairness, accountability, and transparency of institutions, (c) the creation and distribution of economic benefits, (d) security challenges through cybercrime and cyberwar, (e) technologically enabled industrial and geopolitical power shifts, as well as, (f) novel technological possibilities for solving social and environmental problems (Cath, 2018; Dafoe, 2018; Susskind, 2018). For instance, AI is able to perform many tasks that previously were exclusive to humans and which can challenge human responsibility, oversight, and participation (Danaher, 2016). Unintelligible black boxes increasingly determine which opportunities, such as jobs, loans or medical treatments are offered to us. At a more abstract level, AI challenges us to rethink concepts such as power and politics in new contexts as well as fundamental values such liberty, democracy, and justice (Susskind, 2018). Stakeholders—even former supporters—of technology firms have increasingly pointed out irresponsible behaviours, such as designing interfaces that exploit systematic human biases and prioritising profit over the common good (Haugan, 2021; LaJeunesse, 2020; McNamee, 2018). To address these emerging ethical and policy challenges, AI governance is required, which is the attempt to navigate the transition to increasingly advanced AI systems (Cath, 2018; Dafoe, 2018; Maas, 2021).

This chapter focuses on the state-firm relationship since these two types of actors are the most powerful in AI governance (Hare, 2016; Webb, 2019). While other forms of organisations, such as non-governmental organisations (NGOs) or intergovernmental organisations (IGOs) tie into the argument where necessary, they are not the focus of this chapter (Abbott & Snidal, 2009). Specifically, the chapter focuses on liberal democratic states with an acceptable level of the rule of law and protection of fundamental rights (Mounk, 2018). Technology firms will primarily refer to the set of the largest and most dominant AI firms, namely Google, Microsoft, Apple, Facebook, IBM, Amazon, Baidu, Alibaba and Tencent (Webb, 2019). Nevertheless, in this chapter the terms state and firm refer to ideal types rather than specific institutions.

2 Theoretical Framework: Power and Legitimacy

The state-firm relationship will be analysed through the lens of two central concepts: power and legitimacy. Power is required for a governance regime to be feasible and legitimacy is required for it to be acceptable. To conceptualise a term as elusive as power, this chapter combines the theories of Strange (2015), Nye (2009, 2011), as well as Hood and Margetts (2007). Strange provides four interrelated dimensions of structural power: security, production, finance, and knowledge. These roughly correspond to the four tools of government by Hood & Margetts: authority, organisation, treasure and nodality (see Table 1). The dimension of nodality, which is about the controlling of information and beliefs approximates what Nye (2009, 2011) describes as soft power (attraction), while the other three dimensions, fall under hard power (coercion).

While there are a number of differences in what they emphasise, each framework describes the capacity of political agents to influence and shape the options available to others. The concepts by Strange and Nye will be used when emphasising the more abstract, philosophical dimension of power, and the terms by Hood & Margetts will be used when referring to operationalised governance tools.

The second concept of legitimacy, the right to rule, is a normative one. It is the property that any coercive institution which rightfully decides over the lives of others must possess (Danaher, 2016). Legitimacy has three interrelated dimensions. Governance regimes can derive legitimacy from (a) inputs, e.g. democratic elections, (b) procedures, e.g. the rule of law, and (c) outputs, e.g. effectively improved

Table 1 Combined frameworks of governance capacities

Structures: The four resources of structural power by Strange (2015)	**Security**: a monopoly of violence enforces authority	**Production**: organising labour and the means of production	**Finance**: controlling the distribution of credit and funds	**Knowledge**: controlling data, information, and beliefs
Tools: Four instruments of government by Hood & Margetts (2007)	**Authority**: The (legal) power to demand, forbid, or adjudicate	**Organisation**: Arrangement of (skilled) people and capital	**Treasure**: Possession of convertible and valuable stocks	**Nodality**: Centrality in an information or social network
Strategies: Dimensions of relational power by Nye (2011)	**Hard power:** Applying coercive threats of violence (sticks) or economic rewards (carrots) as power-conversion strategies. Hard power is the contextual ability to achieve desired outcomes against the preferences of others.			**Soft power:** Setting agendas and preferences through attraction

living conditions (Danaher, 2016; Mena & Palazzo, 2012).[1] To be sufficiently legitimate, any governance regime needs to satisfy the threshold of an acceptability
requirement in each dimension (Estlund, 2009). For the purpose of this chapter, the
necessary conditions for the successful governance of AI are that a regime needs to
be both feasible and acceptable. Power and legitimacy are both necessary but individually insufficient for successful AI governance. Armed with this theoretical
basis, we may now turn to the analysis of the state-firm relationship in AI governance.

3 The Role of States and Technology Firms in AI Governance

Generally, there are two diametrically opposed camps in the discussion about who
should drive AI governance. The first argues that states should govern AI and the
second argues that technology firms should govern AI. Of course, there is also a lot
of grey area in between, yet, these two ideal types dominate the discussion on which
agents should govern AI. Let us consider each position in turn.

3.1 Evaluating State-Driven AI Governance

Conventionally, governance has primarily been considered to be the domain of
Westphalian states and some claim that AI governance is also best left to states,
because they have the necessary power and legitimacy to set and enforce general
norms (Leung, 2019; Nemitz, 2018; Wagner, 2018). However, governance trends
such as globalisation and the rise of AI challenge both of these features (Maas,
2021). States only rule national jurisdictions and are largely unable to control the
data flows, technological developments and companies that operate across national
boundaries, which undermines their authority and nodality. These limits of conven-
tional governance mechanisms such as the rule of law and the protection of fundamental rights create a "governance gap" (Ebert, 2019) or "regulatory void" (Abbott
& Snidal, 2009).

States do not have a monopoly of violence in cyberspace, thus as the relevance
of cyberspace increases, the authority of states decreases and they struggle to ensure
security for their citizens (Owen, 2016). Technology firms find ways to effectively
lobby against regulation or circumvent tax regimes, which undermines the authority
and financial power of states (The British Academy, 2018). Moreover, states

---

[1] These three categories are ideal types and not always mutually exclusive. For instance, protection
of rights or timely decisions may be regarded as procedural virtues but also as requirements for
effective outputs (Danaher, 2016; Estlund, 2009; Mena & Palazzo, 2012). Moreover, it should be
noted that power and legitimacy are separate but interrelated concepts. In particular output legitimacy requires a certain capacity to enact effective change.

frequently do not have the knowledge and regulatory expertise required to make sensible rules for technology or to enforce them, thus further reducing the value of their organisation and nodality (Lessig, 2006; Pasquale, 2018). While some states— such as the US and China—have surveillance access to user data collected by technology firms, most states are disrupted by globally operating tech firms, which increasingly hold structural power over information, beliefs and technology (Snowden, 2019; Strange, 2015; Webb, 2019). While many of these trends are not only driven by technology but also by globalisation, they indicate that the authority, organisation, production, and nodality of many states are no longer superior to those of the most powerful firms. In the governance literature it is well-established that even if states are still indispensable governance actors (DeNardis, 2014; Leung, 2019), "the state is far from the only game in town, and may no longer be the most important game in town" (Abbott & Snidal, 2009, p. 87). As a result, states currently do not have the necessary capacity to govern AI by themselves.

Let us turn to the second dimension of legitimacy. While liberal democracies remain the only agents who can claim democratic input legitimacy and the procedural legitimacy of the rule of law, some argue that their lack of AI capacity challenges their output legitimacy, i.e. their ability to effectively improve the lives and protect the rights of citizens (Lee, 2018; Owen, 2016). Others argue that states could reassert their power by breaking up or socialising technology firms, which would re-establish them as the regulatory force that can protect citizens (Morozov, 2013; Pasquale, 2018; Zuboff, 2019). However, besides the feasibility complications of break-ups in markets with scale, network and feedback effects, there is a fear that if the state tries to reassert its power, its lack of expertise could lead to rigid overregulation, which would cripple the beneficial innovations that could arise from the development of AI (Floridi & Taddeo, 2016; Mayer-Schönberger & Ramge, 2018). Their tools may be ineffectual and sometimes even risk doing more harm than good. In short: Democratic states still have unique input and procedural legitimacy claims, yet struggle to effectively deliver outputs for citizens (Danaher, 2016).

3.2   Evaluating Corporate-Driven AI Governance

Since on some dimensions power has increasingly migrated from states to companies, which are better informed, more flexible, and control key resources, some argue that the governance of AI is best left to technology firms. So let us consider the capacity of firms to govern AI. Firms have always had enormous power over production and finance, yet through the new scale of the largest technology firms, this dominance has reached new heights (Birkinshaw, 2018). At the same time, their control over information—their nodality—has increased by orders of magnitude and frequently surpassed that of states (Webb, 2019; Zuboff, 2019).

Their business decisions are also political decisions, due to their enormous economic assets, information asymmetries, and control over technology (Broeders & Taylor, 2017). This has allowed them to bolster their de facto authority, since it is hard for others, even states, to challenge their decisions in some domains (Coase, 1937; Pasquale, 2018). Leading AI firms not only act as norm entrepreneurs for soft norms as well as for hard law mechanisms; they also set technical and behavioural standards and are developing private self-governance regimes that do not rely on states to operate (Fairbank, 2019; Gorwa & Peez, 2020; Hurel & Lobato, 2020; Wagner, 2018). While they do not have the legislative authority to make generally applicable rules, the decisions of dominant players in markets with scale, network and feedback effects can reach as many or even more people than a national government (Mayer-Schönberger & Ramge, 2018). Facebook has surpassed Christianity as the human organisation with the most active users and it knows more about them than any state does about its population (Fletcher, 2016). In short: technology firms increasingly have superior control over production, finance, information and in some decisions even authority compared to states.

Since the recognition of this power of the most valuable tech companies in the world has led to calls for more corporate responsibility, we next need to consider the legitimacy of tech firms. They are the actors with the best understanding of technology and in some cases the ability to enforce norms and policies swiftly in product development and management (Cohen, 2019; Zittrain, 2008). Yet, companies are interest-driven actors which lack the independence and representativeness of states, since by law and culture they are focused on profits. While company-led private governance regimes sometimes borrow input legitimacy by involving civil society organisations and democratic states, they ultimately only prefer the regulation that serves their profit goal (Abbott & Snidal, 2009; Fairbank, 2019; Mena & Palazzo, 2012; Russell, 2019). For regulatory proposals with other goals they prefer self-regulation with minimal compliance costs.

Some hope to push, shame, and entrap tech firms in ethical commitments to improve their behaviour (Raji & Buolamwini, 2019). However, sceptics fear that private governance regimes merely allow companies to engage in 'ethics washing' (Barnett, 2018; Wagner, 2018). Others worry that while firms may now have become endowed with largely unchecked levers of power, they lack the procedural accountability mechanisms to legitimately wield this power (Schaake, 2020a; Cohen, 2019). Thus, even if firms have the capacity to deliver better outcomes than states, their motivations are not aligned with the goals of better governance outcomes and they lack the input and procedural legitimacy that is crucial for legitimate governance (Abbott & Snidal, 2009; Mena & Palazzo, 2012).

A preliminary conclusion to the first question may hence sound somewhat disheartening: while states are legitimate but not capable to govern technology unilaterally, technology firms have expertise but lack legitimacy. Prima facie this may seem to be a dilemma since neither of the available actors is able to govern AI by combining both power and legitimacy. However, as the following sections attempt to show, together, firms and states could successfully govern AI.

3.3 The Case for a Combined Approach to AI Governance

While many AI governance discussions focus on whether states or firms should set the norms for AI, I argue that both public and private governance regimes are required. This is primarily the case because they can make different contributions to the governance of emerging technologies. Governments make binding laws in the domain of hard ethics, which is about the boundaries of what is morally wrong or required. Companies need to comply with hard ethics and decide what to do within the scope of what is legal. For this they can draw on soft ethics, which provides guidance for what is morally preferable (Floridi, 2018). Neither approach can work unilaterally: hard ethics is necessary but insufficient and soft ethics only works if it operates within hard ethics.

Floridi compares this to a game: hard ethics determines the rules of the game, i.e. the legal and illegal moves, while soft ethics is about the winning strategies, i.e. the best moves (Floridi, 2018). Additionally, both mechanisms are necessary since they complement each other's regulatory benefits and drawbacks. This argument builds on the insights from the study of transnational governance that single actor schemes are implausible and that transnational regulatory schemes can assemble needed capacities from diverse partners (Abbott & Snidal, 2009). Law is enforceable in court but often moves slowly and nationally, while private governance regimes can be more global, flexible and agile but are often unenforceable (Susskind, 2018).

Arguably, the governance of emerging technologies like AI, in which the sizable benefits and risks are subject to significant uncertainty, any promising governance approach needs to combine (a) enforceability to avoid the gravest risks and (b) flexibility to avoid choking valuable innovations (Floridi & Taddeo, 2016). Private and public governance regimes can govern where their relative advantages prevail. Both need to take hard and soft ethics into account, yet without states hard ethics rules cannot be set and without firms soft ethics goals will not be reached. A successful approach to AI governance needs to combine the power and legitimacy of both states and tech firms. Both private and public governance regimes are necessary but individually insufficient for AI governance. Let us next begin to explore the implications of this finding.

Thus far this may sound like an unexciting position: everybody should work together and collaborate. Yet, this raises a number of further questions, the answers to which require more radical approaches. Let us next turn to the elephant in the room: how can the power and legitimacy of states and firms be combined? While it may in principle be desirable to combine the capacity and legitimacy of states and firms, it may turn out to be unfeasible. Some fear that such attempts of a cooperative relationship between states and firms may end up in a version of a state-controlled economy, while others fear a crony capitalist state capture (Cann & Balanyá, 2019; Mayer-Schönberger & Ramge, 2018). I argue that companies and states can work together, while at the same time remaining accountable to their respective stakeholders. However, avoiding the corruption of such a combined approach requires fundamental good governance reforms.

What matters is not so much the speed of technological development, but the extent to which the interests of those driving technological developments and governance are aligned with ethics (Floridi, 2018). For AI governance to succeed and technology firms to be reliably responsible governance actors, they need to become intrinsically and reliably motivated to act with benevolence. *Rather than only regulating technology itself, it is necessary to systematically change the governance of those firms that perform a core social function in the digital transformation and drive the development of technology.*

4 How to Achieve Responsible AI Governance

The chapter now turns to the second question of how an AI governance approach that relies on states and firms can succeed. Admittedly, if states and firms were to join forces in the way they exist today, they would be unlikely to succeed. There are still many possible pitfalls in the relationship between states and firms, even if they acknowledge their interdependence (Owen, 2016). Only under certain conditions will the state-firm relationship bring about ethical AI governance. While many argue that it is primarily states that need to be reformed and learn from tech firms (Margetts & Dorobantu, 2019), I argue the opposite. The state does need to update its capacities, yet the more fundamental reform is required for tech firms, since they need to reboot their purpose. Let's begin by briefly considering the necessary reform of the state.

4.1 Updating the State for the Digital Age

The task of the state in the transition to advanced AI systems is to maintain its democratic input legitimacy, while at the same time reforming its capacity to be able to effectively respond to the challenges of AI governance, particularly those which require hard ethics rules. Admittedly, this is easier theoretically said than practically done. Let us thus consider a number of concrete measures states can take to improve their capacity. First and foremost, states need to improve their understanding of the emerging technologies. By recruiting new talent, improving the technical proficiency of their own administration and building more aligned partnerships for knowledge transfer with academia and civil society, states can improve their capacity and ability to serve citizens (Cath, 2018; Margetts & Dorobantu, 2019). Additionally, states can reduce the knowledge asymmetry with firms and improve accountability by obliging firms to become more transparent and share their data (Mayer-Schönberger & Ramge, 2018; Schaake, 2020b).

Even if states cannot govern unilaterally, they still can lead as norm entrepreneurs and governance orchestrators (Abbott & Snidal, 2009; Bradford, 2020). As regulators, states need to recognise that their role is not to fix value-neutral market failures but to set missions to create value-laden markets (Mazzucato, 2014, 2021). Moreover, through innovative regulatory models, such as sandboxes, policy proto-typing, and new forms of citizen participation they can improve their ability to enforce hard ethics without preventing valuable innovations (Kimbell & Bailey, 2017; Mazzucato, 2014). Lastly, while firms can outmanoeuvre states individually in the currently fragmented global AI governance regimes, the most feasible mecha-nism for states to regain leverage over technology firms is to coordinate among themselves and pool their governance capacity in communities of shared interests and values (Bendiek, 2018; Bradford, 2020; Maas, 2021).

Yet, the state not only needs to improve its capacity but also take measures to maintain its legitimacy. The central democratic ability of citizens to 'throw out the bums' in free and fair elections needs to be safeguarded by effectively limiting mis-information (Bradshaw & Howard, 2019; Mounk, 2018). At the same time, the pub-lic should truthfully be informed about what data the state is collecting, in order to be able to evaluate which forms of data-collection it wants to allow (Snowden, 2019). The protection of fundamental rights and the rule of law need to be safe-guarded but also updated and expanded to new circumstances (Muller, 2020; Cohen, 2019; Lessig, 2006). If the state is to work closer with private corporations without being corrupted, the state needs to impose (a) a strict transparency regime on its activities, for instance about contacts with lobbyists, (b) cool-down periods to pre-vent revolving doors and conflicts of interests between regulators and their counter-parts in the industry, and (c) strict limitations of campaign finance to prevent systemic corruption of law making (Cann & Balanyá, 2019; Crouch, 2005; Lessig, 2015). These measures are necessary to ensure democratic institutions not only exist formally but that citizens enjoy meaningful self-governance. States which implement these measures can establish the power and legitimacy, to not chase but lead in the governance of AI.

4.2   Rebooting Tech Firms with Purpose

The current generation of tech giants was established at a time when the Friedman doctrine was dominant (The British Academy, 2018; Zuboff, 2019). It posits that the sole goal of firms is to maximise profits for shareholders as long as they respect the law. This doctrine argues that firms should not take the public interest into account, since their executives are not accountable to the public and the public interest is exclusively the function of the state (Friedman & Friedman, 2002). Crucially, the Friedman doctrine acknowledges hard ethics but purports value-free behaviour within its boundaries and only regards the needs of one group (to the detriment of all others). However, firms are also moral agents that are not only subject to hard ethics rules, but also soft ethics (Floridi, 2018; Floridi & Sanders, 2004). Currently hard and soft ethics already have some influence on tech companies through stakeholder pressure.[2] Yet, soft ethics is still fighting an uphill battle in the tech industry, since despite any idealistic beginnings of their founders, publicly tech listed firms by law and culture are held accountable to profit maximisation (Abbott & Snidal, 2009; Foroohar, 2019; Zuboff, 2019). In short, currently the corporate objective function and governance is not aligned with soft ethics but shareholder value.

This is a problem for three reasons. First, it naively assumes harmony between shareholders and everyone else (Jensen, 2001; Ulrich, 2008). This has led firms to prioritise financial performance and shareholders over social and environmental responsibilities, and led to conflicts with stakeholders as well as an erosion of integrity (The British Academy, 2018; Zuboff, 2019; Freeman et al., 2010).[3] Secondly, while public pressure on profit-driven firms can help to hold firms accountable, it has limited reliability as a governance mechanism, since it depends on scandalisation and simplification in the limited attention span of the news cycle (Floridi, 2018; Raji & Buolamwini, 2019). If tech firms establish private ethics regimes but ultimately still prioritise profit, they merely engage in 'ethics washing' to improve their image and dissuade stricter regulation (Wagner, 2018). In other words, under the currently dominant shareholder model they dodge their soft ethics responsibilities while at the same time working against hard ethics rules. Thirdly, it is not enough for the state to try to correct grievances through ex post regulation after they arise, but firms are uniquely positioned to try to ethically steer the development of technology ex ante (Floridi, 2018). Especially powerful firms that perform important social functions and to which few alternatives exist, need a corporate governance that reflects an alignment with the public interest (The British Academy, 2018). Thus, the decision-making structure of tech firms needs to be ethically realigned.

This shift has crucial implications. *While currently the dominant paradigm for tech firms is to aim to maximise profits (optimisation goal) while being socially acceptable (bounding condition), they need to shift to do what is socially preferable*

---

[2] To illustrate how hard and soft ethics influence corporations, let us consider the example of Google. On the one hand, Google was fined € 7 billion by the European Commissions for anti-competitive behaviour in the sectors of online search, advertising and mobile operating systems (Warren, 2018). This is an example of regulation based on hard ethics in action, effectively ensuring that a dominant position is not abused. On the other hand, due to pressure by its employees and the public, Google backed out of the controversial Chinese censorship and US military projects Dragonfly and Maven (Wolverton, 2018) and introduced consumer protection features for 'digital well-being' (Pardes, 2018). These are mechanisms of soft ethics, since the point of contention was not what is legal but what is socially preferable. Yet, these examples also illustrate that corporate governance is not intrinsically aligned with soft ethics and that soft ethics is mostly pushed onto the agenda by stakeholders. A solution requires building a more constructive normative cascade than the status quo, where users have to lobby ex post to change corporate behaviour (Floridi, 2018).

[3] While much of this holds for firms in general, it is particularly pressing for tech firms as they have become the most powerful firms, especially when it comes to the governance of emerging technologies (Birkinshaw, 2018).

(optimisation goal) while being commercially viable (bounding condition).[4] As pioneering companies that already practice this institutional logic of responsible stewardship and accountability demonstrate, aligning the corporate objective with soft ethics is not mutually exclusive with profits (Raisher, 2019; Rieback, 2019; The Economist, 2019; Yunus & Weber, 2010). However, it does require giving up the privileged status of shareholders, in favour of a stakeholder model, which equally takes the needs of all those affected by a company's actions into account. The stakeholder approach sees profit more analogous to blood in a body: blood needs to flow to ensure survival but does not need to be maximised and is not the existential purpose (Freeman et al., 2010). If tech companies are serious about playing the game of soft ethics, they need to cease viewing ethics as a constraint on their commercial activity, but as a goal to be enabled by it. If they do not, they may be powerful but not legitimate.

4.3   Avenues Toward Systemic Change

Even if a reader agrees that ideally AI governance requires both public and private action and that for this to be successful tech firms need to be fundamentally transformed, they may still doubt if this is feasible given that vested interests control most sources of structural power (Strange, 2015). This is admittedly the major challenge to corporate reform. In this last section, I want to provide some suggestions for how determined system entrepreneurs can activate levers to bring about this change. Five of those levers to reform corporations around public purposes are identified in a report of the British Academy (2018): (1) making ownership more diverse and redefining it to no longer be synonymous with shareholders, (2) reforming corporate governance to make firms accountable to their stakeholders, for instance by including their representatives on the board, (3) use regulation to align corporate activities with the common good, (4) coordinating taxation globally and reward responsible behaviour, (5) redirecting investment, for instance through restructuring procurement and subsidies. Vested interests will not give up without a fight, but systemic change also needs to be made attractive as it is also a paradigmatic fight for hearts and minds (Yunus & Weber, 2010). If regulatory hard and soft power are combined in a smart way, systemic change in corporate governance is indeed possible (Mayer, 2018; Nye, 2011).

---

[4] What remains the same is the following: (a) when there is an option that maximises profits and is socially preferable companies will do it, (b) when what is socially desirable is incompatible with the survival of the company, companies will not do it (Porter & Kramer, 2011). Here is what changes with the new paradigm: previously when companies had to make a choice between doing (a) what was socially less desirable but would maximise profits and (b) what was socially desirable and commercially viable, they previously would have chosen (a) while when considering soft ethics they would choose (b).

However, systemic change does not only rely on whether existing institutions can be changed, but also on whether new ones can be created and established (Meadows, 2015). Despite the efforts of tech giants to limit and suppress competition, they are also likely to be subject to creative destruction by a new corporate generation (Mazzucato, 2014; Wu, 2011). Since new institutions are more malleable, it is crucial that the next generation of tech start-ups takes these corporate governance lessons to heart. While the dominant start-up ideal, so-called unicorns, still idolises high valuations, new start-up models are gaining momentum. An increasing number of founders wants to create zebras, firms that are collaborative, refuse venture capital and the shareholder pressure that comes with it, and crucially aim to not just do what is socially acceptable but socially preferable (Griffith, 2019; Rieback, 2019; Waters & Kruppa, 2021). They also face tailwinds from a new generation of institutional investors and family offices that embrace a three-pronged investment strategy paradigm of risk-return-impact (Cohen, 2020). All these levers of change can be combined to align the objective function of tech firms with soft ethics.

5 Conclusion

This chapter has argued that from the perspective of power and legitimacy, the state-driven and corporate-driven approach to AI governance are both unlikely to succeed and that the supposed trade-off between private and public governance is a false dichotomy. Combining public governance for hard ethics and private governance for soft ethics are necessary conditions for legitimate AI governance. For this to succeed, both firms and states need to be responsible stewards and be held accountable to their citizens and stakeholders. It is the contention of this chapter that AI governance requires an update in the governance capacity of states and an ethical realignment of the purpose of most tech firms. In the last section, I proposed how such a reform is possible if the available levers for change are engaged wisely. These measures may not be sufficient to successfully govern emerging technology, yet a sensible combination of these steps is necessary for capable and legitimate digital governance regimes.

While this chapter remained within the frame of liberal democratic states, future scholarship could complicate the picture by including interactions with authoritarian regimes into the analysis or by investigating the role of power inequalities among different states. While both the state and the firm are required for AI governance, important differences remain in their hard and soft ethics responsibilities. Crucially, firms need to step up their game of soft ethics so both can effectively reinforce each other. AI can be governed if states and companies combine their relative advantages of flexibility, expertise, enforceability and effectiveness. Nota bene, this requires a mindset shift and can only happen if companies see soft ethics not as a constraint but as a primary objective of their activities. Holding companies accountable through hard ethics is necessary but not sufficient. The increasingly shared roles of states and corporations also require corporate governance innovation to meet the requirements of responsible, legitimate and accountable use of power. In a combined approach, states and firms can complement each other and work together to make their respective contributions to the governance of AI.

References

Abbott, K. W., & Snidal, D. (2009). The governance triangle: Regulatory standards institutions and the shadow of the state. In *The politics of global regulation* (pp. 44–88). Princeton University Press.

Barnett, M. L. (2018). *Limits to stakeholder influence: Why the business case won't save the world.* Edward Elgar Publishing.

Bendiek, A. (2018). *The EU as a force for peace in international cyber diplomacy* (SWP comment). German Institute for International and Security Affairs. https://www.swp-berlin.org/en/publication/the-eu-as-a-force-for-peace-in-international-cyber-diplomacy/

Birkinshaw, J. (2018). How is technological change affecting the nature of the corporation? *Journal of the British Academy, 6*(s1), 185–214. https://doi.org/10.5871/jba/006s1.185

Bradford, A. (2020). *The Brussels effect: How the European Union rules the world.* Oxford University Press.

Bradshaw, S., & Howard, P. N. (2019). Social media and democracy in crisis. In M. Graham & W. Dutton (Eds.), *Society and the internet* (pp. 212–227). Oxford University Press. https://doi.org/10.1093/oso/9780198843498.003.0013

British Academy. (2018). *Reforming business for the 21st century: A framework for the future of the corporation.* https://www.thebritishacademy.ac.uk/publications/reforming-business-21st-century-framework-future-corporation

Broeders, D., & Taylor, L. (2017). Does great power come with great responsibility? The need to talk about corporate political responsibility. In M. Taddeo & L. Floridi (Eds.), *The responsibilities of online service providers* (Vol. 31, pp. 315–323). Springer International Publishing. https://doi.org/10.1007/978-3-319-47852-4_17

Cann, V., & Balanyá, B. (2019). *Captured states: When EU governments are channels for corporate interests.* Corporate Europe Observatory. https://corporateeurope.org/sites/default/files/ceo-captured-states-final_0.pdf

Cath, C. (2018). Governing artificial intelligence: Ethical, legal and technical opportunities and challenges. *Philosophical Transactions of the Royal Society A: Mathematical, Physical and Engineering Sciences, 376*(2133), 20180080. https://doi.org/10.1098/rsta.2018.0080

Coase, R. H. (1937). The nature of the firm. *Economica, 4*(16), 386–405. https://doi.org/10.1111/j.1468-0335.1937.tb00002.x

Cohen, J. E. (2019). *Between truth and power: The legal constructions of informational capitalism.* Oxford University Press.

Cohen, R. (2020). *A guide to the impact revolution.* On Impact. https://www.onimpactnow.org/the-complete-guide-index

Crouch, C. (2005). *Post-democracy.* Polity.

Dafoe, A. (2018). *AI governance: A research agenda.* University of Oxford. https://www.fhi.ox.ac.uk/wp-content/uploads/GovAI-Agenda.pdf

Danaher, J. (2016). The threat of algocracy: Reality, resistance and accommodation. *Philosophy & Technology, 29*(3), 245–268. https://doi.org/10.1007/s13347-015-0211-1

DeNardis, L. (2014). *The global war for internet governance.* Yale University Press.

Ebert, I. (2019). The tech company dilemma. Ethical managerial practice in dealing with government data requests. *Zeitschrift Für Wirtschafts-Und Unternehmensethik, 20*(2), 264–275. https://doi.org/10.5771/1439-880X-2019-2-264

Estlund, D. (2009). *Democratic authority: A philosophical framework.* Princeton University Press.

Fairbank, N. A. (2019). The state of Microsoft?: The role of corporations in international norm creation. *Journal of Cyber Policy, 4*(3), 380–403. https://doi.org/10.1080/23738871.2019.1696852

Fletcher, T. (2016). *Naked diplomacy: Power and statecraft in the digital age.* William Collins.

Floridi, L. (2017). Digital's cleaving power and its consequences. *Philosophy & Technology, 30*(2), 123–129. https://doi.org/10.1007/s13347-017-0259-1

Floridi, L. (2018). Soft ethics and the governance of the digital. *Philosophy & Technology, 31*(1), 1–8. https://doi.org/10.1007/s13347-018-0303-9

Floridi, L., & Sanders, J. W. (2004). On the morality of artificial agents. *Minds and Machines, 14*(3), 349–379. https://doi.org/10.1023/B:MIND.0000035461.63578.9d

Floridi, L., & Taddeo, M. (2016). What is data ethics? *Philosophical Transactions of the Royal Society A: Mathematical, Physical and Engineering Sciences, 374*(2083), 20160360. https://doi.org/10.1098/rsta.2016.0360

Foroohar, R. (2019). *Don't be evil: How big tech betrayed its founding principles-and all of us* (1st ed.). Currency.

Freeman, R. E., Harrison, J. S., Wicks, A. C., Parmar, B. L., & de Colle, S. (2010). *Stakeholder theory: The state of the art.* Cambridge University Press.

Friedman, M., & Friedman, R. D. (2002). *Capitalism and freedom* (40th anniversary ed). University of Chicago Press.

Gorwa, R., & Peez, A. (2020). Big tech hits the diplomatic circuit: Norm entrepreneurship, policy advocacy, and Microsoft's cybersecurity tech accord. In D. Broeders & B. van den Berg (Eds.), *Governing cyberspace: Behavior, power, and diplomacy* (pp. 263–284). Rowman & Littlefield.

Griffith, E. (2019, January 11). More start-ups have an unfamiliar message for venture capitalists: Get lost. *The New York Times.* https://www.nytimes.com/2019/01/11/technology/start-ups-rejecting-venture-capital.html

Hare, S. (2016). For your eyes only: U.S. technology companies, sovereign states, and the battle over data protection. *Business Horizons, 59*(5), 549–561. https://doi.org/10.1016/j.bushor.2016.04.002

Haugan, F. (2021). *Statement of Frances Haugen.* Whistleblower Aid. United States Senate Sub-Committee on Consumer Protection, Product Safety, and Data Security. https://www.commerce.senate.gov/services/files/FC8A558E-824E-4914-BEDB-3A7B1190BD49

Hood, C., & Margetts, H. (2007). *The tools of government in the digital age* (New ed.). Palgrave Macmillan.

Hurel, L. M., & Lobato, L. C. (2020). Cyber-norms entrepreneurship? Understanding Microsoft's advocacy on cybersecurity. In D. Broeders & B. van den Berg (Eds.), *Governing cyberspace: Behavior, power, and diplomacy* (pp. 285–314). Rowman & Littlefield.

Jensen, M. C. (2001). Value maximization, stakeholder theory, and the corporate objective function. *Journal of Applied Corporate Finance, 14*(3), 8–21. https://doi.org/10.1111/j.1745-6622.2001.tb00434.x

Kimbell, L., & Bailey, J. (2017). Prototyping and the new spirit of policy-making. *CoDesign, 13*(3), 214–226. https://doi.org/10.1080/15710882.2017.1355003

LaJeunesse, R. (2020, January 2). *I was Google's head of international relations. Here's why I left.* Medium. https://medium.com/@rossformaine/i-was-googles-head-of-international-relationshere-s-why-i-left-49313d23065

Lee, K.-F. (2018). *AI superpowers: China, Silicon Valley, and the new world order.* Houghton Mifflin Harcourt.

Lessig, L. (2006). *Code: And other laws of cyberspace* (version 2.0). Basic Books.

Lessig, L. (2015). *Republic lost: The corruption of equality and the steps to end it* (revised edition). Twelve.

Leung, J. (2019). *Who will govern artificial intelligence? Learning from the history of strategic politics in emerging technologies* [DPhil Thesis, University of Oxford]. https://ora.ox.ac.uk/objects/uuid:ea3c7cb8-2464-45f1-a47c-c7b568f27665/download_file?file_format=pdf&safe_filename=JADE%2BLEUNG%2B-%2BDPHIL%2BTHESIS%2B-%2BSep19.pdf&type_of_work=Thesis

Maas, M. (2021). *Artificial Intelligence Governance under Change: Foundations, Facets, Frameworks* [PhD Thesis, University of Copenhagen]. https://drive.google.com/file/d/1vIJUAp_i41A5gc9Tb9EvO9aSuLn15ixq/view

Margetts, H., & Dorobantu, C. (2019). Rethink government with AI. *Nature, 568*(7751), 163–165. https://doi.org/10.1038/d41586-019-01099-5

Mayer, C. (2018). *Prosperity: Better business makes the greater good.* Oxford University Press.

Mayer-Schönberger, V., & Ramge, T. (2018). *Reinventing capitalism in the age of big data.* John Murray.

Mazzucato, M. (2014). *The entrepreneurial state: Debunking public vs. private sector myths* (revised edition). Anthem Press.

Mazzucato, M. (2021). *Mission economy: A moonshot guide to changing capitalism.* Allen Lane.

McNamee, R. (2018, January 29). Why not regulate social media like tobacco or alcohol? *The Guardian.* https://www.theguardian.com/media/2018/jan/29/social-media-tobacco-facebook-google

Meadows, D. (2015). *Thinking in systems: A primer.* Chelsea Green Publishing.

Mena, S., & Palazzo, G. (2012). Input and output legitimacy of multi-stakeholder initiatives. *Business Ethics Quarterly, 22*(3), 527–556. https://doi.org/10.5840/beq201222333

Morozov, E. (2013). *To save everything, click here: The folly of technological solutionism* (1st ed.). PublicAffairs.

Mounk, Y. (2018). *The people vs. democracy: Why our freedom is in danger and how to save it.* Harvard University Press.

Muller, C. (2020). *The Impact of Artificial Intelligence on Human Rights, Democracy and the Rule of Law.* Council of Europe. https://rm.coe.int/cahai-2020-06-fin-c-muller-the-impact-of-ai-on-human-rights-democracy-/16809ed6da

Nemitz, P. (2018). Constitutional democracy and technology in the age of artificial intelligence. *Philosophical Transactions of the Royal Society A: Mathematical, Physical and Engineering Sciences, 376*(2133), 20180089. https://doi.org/10.1098/rsta.2018.0089

Nye, J. S. (2009). *Soft power: The means to success in world politics.* PublicAffairs.

Nye, J. S. (2011). *The future of power.* PublicAffairs.

Owen, T. (2016). *Disruptive power: The crisis of the state in the digital age.* Oxford University Press.

Pardes, A. (2018, September 5). Google and the Rise of "Digital Well-Being." *Wired.* https://www.wired.com/story/google-and-the-rise-of-digital-wellbeing/

Pasquale, F. A. (2018). Tech platforms and the knowledge problem. *American Affairs, 2*(2) https://americanaffairsjournal.org/2018/05/tech-platforms-and-the-knowledge-problem/

Porter, M. E., & Kramer, M. R. (2011, January 1). Creating shared value. *Harvard business review*, January–February 2011. https://hbr.org/2011/01/the-big-idea-creating-shared-value

Raisher, J. (Ed.). (2019). *Steward-ownership: Rethinking ownership in the 21st century.* Purpose Foundation. https://purpose-economy.org/content/uploads/purposebooklet_en.pdf

Raji, I. D., & Buolamwini, J. (2019). Actionable auditing: Investigating the impact of publicly naming biased performance results of commercial AI products. *Proceedings of the 2019 AAAI/ACM Conference on AI, Ethics, and Society*, 429–435. https://doi.org/10.1145/3306618.3314244.

Rieback, M. (2019). *Post-growth entrepreneurship.* Codemotion Rome. https://www.codemotion.com/magazine/dev-hub/cto/post-growth-entrepreneurship/

Russell, S. J. (2019). *Human compatible: Artificial intelligence and the problem of control.* Viking.

Schaake, M. (2020a, February 20). Big Tech companies want to act like governments. *Financial Times.* https://www.ft.com/content/36f838c0-53c5-11ea-a1ef-da1721a0541e

Schaake, M. (2020b, August 7). AI's invisible hand: Why democratic institutions need more access to information for accountability. *The Rockefeller Foundation.* https://www.rockefellerfoundation.org/blog/ais-invisible-hand-why-democratic-institutions-need-more-access-to-information-for-accountability/

Snowden, E. (2019). *Permanent record.* Henry Holt and Company.

Strange, S. (2015). *States and markets.* Bloomsbury Publishing.

Susskind, J. (2018). *Future politics: Living together in a world transformed by tech.* Oxford University Press.

The Economist. (2019). What open-source culture can teach tech titans and their critics. *The Economist.* https://www.economist.com/business/2019/07/20/what-open-source-culture-can-teach-techtitans-and-their-critics

Ulrich, P. (2008). *Integrative economic ethics: Foundations of a civilized market economy.* Cambridge University Press.

Wagner, B. (2018). Ethics as an escape from regulation: From ethics-washing to ethics-shopping. In *Being profiled: Cogitas ergo sum* (pp. 84–90). Amsterdam University Press.

Warren, T. (2018, July 18). Google fined a record $5 billion by the EU for android antitrust violations. *The Verge.* https://www.theverge.com/2018/7/18/17580694/google-android-eu-fine-antitrust

Waters, R., & Kruppa, M. (2021, May 29). Rebel AI group raises record cash after machine learning schism. *Financial Times.* https://www.ft.com/content/8de92f3a-228e-4bb8-961f-96f2dce70ebb?sharetype=blocked

Webb, A. (2019). *The Big Nine: How the tech titans and their thinking machines could warp humanity.* PublicAffairs.

Wolverton, T. (2018). Google's recent behavior shows the troubling reality of an internet superpower that abandoned its vow to not "be evil." *Business Insider.* https://www.businessinsider.com/googlessecurity-bug-dragonfly-and-maven-show-its-not-trustworthy-2018-10

Wu, T. (2011). *The master switch: The rise and fall of information empires* (1st ed.). Vintage Books.

Yunus, M., & Weber, K. (2010). *Building social business: The new kind of capitalism that serves humanity's most pressing needs.* Public Affairs.

Zittrain, J. (2008). *The future of the Internet and how to stop it.* Yale University Press.

Zuboff, S. (2019). *The age of surveillance capitalism: The fight for the future at the new frontier of power.* Profile Books.

The Impact of Australia's News Media Bargaining Code on Journalism, Democracy, and the Battle to Regulate Big Tech

Emmie Hine

Abstract The Australian News Media and Digital Platforms Mandatory Bargaining Code, legislation requiring large digital platforms to pay news publishers for content that the platforms display, represents the next front in the war to regulate Big Tech. Its ostensible purpose is to provide financial support to publishers and journalists outcompeted by the dominance of digital platforms—namely Google and Facebook—and, by extension, protect democratic institutions. In this chapter, I outline the different approaches Google and Facebook used in their counterattacks to the Code and how their different business models resulted in a more advantageous outcome for Facebook. There are many concerns regarding the Code's impact on small publishers and net neutrality, and while it is a promising first step to support publishers, I argue that the Code may not achieve its stated goals. Furthermore, Facebook's aggressive actions may have harmed its overall position when considering possible future regulation.

Keywords News media and digital platforms mandatory bargaining code · Australia · Big tech · Journalism · Democracy · Platform regulation

1 Introduction

In February of 2021, the Australian Parliament passed the News Media and Digital Platforms Mandatory Bargaining Code. The code was promoted by the Australian Competition and Consumer Commission (ACCC) and is intended to "[address] bargaining power imbalances between digital platforms and Australian news businesses" (Frydenberg & Fletcher, 2020). It requires large digital platforms to

---

This chapter was adapted and expanded from (Hine, 2021), the author's newsletter on technology ethics.

E. Hine (✉)
University of Oxford, Oxford, UK

© The Author(s), under exclusive license to Springer Nature Switzerland AG 2022
J. Mökander, M. Ziosi (eds.), *The 2021 Yearbook of the Digital Ethics Lab*, Digital Ethics Lab Yearbook, https://doi.org/10.1007/978-3-031-09846-8_5

negotiate compensation with local news providers for displaying their content (Boom, 2021). The government's argument is that, by providing access to headlines and snippets of news, online platforms are unbundling the news platforms provide from the advertisements they depend on, impacting newsroom profits and journalists' livelihood (Stilgherrian, 2021). It allows registered news businesses to demand deals with "designated digital platform [services]" that "reproduce," link to, or provide an extract of their content. It also requires designated platforms to give news businesses 14 days' notice regarding algorithmic changes that impact the ranking of news. If the news business and platform cannot come to an agreement, the ACCC oversees a final-offer arbitration process. Treasury Laws Amendment (News Media and Digital Platforms Mandatory Bargaining Code) Bill 2021, 2021. "Designated digital platforms" are determined by the Minister for Communications, providing a bargaining chip to hold over the heads of platforms (Stilgherrian, 2021).

This conflict represents a new front in the global battle to reclaim power from Big Tech. Big Tech platforms "thrive on network effects" and are designed to extract market share and revenue as fast as possible, creating "self-reinforcing market-conquering logics" that change entire industries (Fernandez et al., 2021). This is building their power to rival that of some governments. When it comes to journalism, this power, which threatens to quash the industry, is especially concerning. Journalism, to put it simply, makes news (Schudson, 2020). We in democratic societies trust this news to reflect reality because of journalists' adherence to journalistic ethics, so we rely on it to inform how we participate in democracy (Schudson, 2020). Threats to journalism are thus threats to democracy itself, so embedded in my argument is the assertion that because a healthy press is crucial to a healthy democracy, protecting the press is vital.

In an attempt to preserve their power, Google and Facebook actively lobbied against the bill, winning exemptions for YouTube and Instagram (Barbaschow, 2020). Google and Facebook both threatened to pull services from the Australian market. Google, however, ultimately voluntarily signed agreements with Australian publishers. Facebook followed through on its threats, removing all news from Australian Facebook and restricting international users from seeing Australian news on the platform, before also eventually signing deals with news outlets (Choudhury, 2021; Easton, 2021). In this piece, I will argue that the differences in business model between the platforms precipitated these different approaches, but that Facebook may have ultimately harmed its position in the battle to regulate Big Tech through its strong-arm tactics. Additionally, I will posit that the Bargaining Code may not achieve its ostensible goal of fostering a healthier news media ecosystem with less influence from Big Tech.

2   Outcompeted

One reason the Australian government has put forward for regulating digital platforms, specifically Facebook and Google, is that big digital platforms have created a power imbalance with news media providers through their dominance of the advertising ecosystem. Facebook and Google effectively have a duopoly on advertising dollars on the Internet, which is growing as online ads prove more effective than traditional print ads. Together, they take 81 cents of every dollar spent on online advertising in Australia (Associated Press, 2021). Facebook claims that it does a net good for publishers, referring what it estimates to be A$407 million ($315 million) worth of traffic to Australian publishers in 2020 (Easton, 2021). And yet, news platforms are still struggling. The pool of ad dollars available for traditional media, even publications with an online presence, is shrinking as print ad spending is shifted to online ad spending, which benefits Facebook and Google far more than the media platforms; even when ads are displayed on news sites, the ad platform gets most of the revenue (Letts, 2016). This is an existential threat to journalism.

Given how much damage online platforms have done to newsrooms, the claim of benefits from clickthroughs seems spurious. Besides siphoning away ad revenues, Facebook drove the disastrous "pivot to video" by inflating the view counts of videos, which may have directly cost hundreds of journalists their jobs (Meyer, 2018). Permitting the unchecked spread of fake news has damaged trust in news organisations and threatened democracy (Binkowski, 2019), as the storming of the US Capitol—egged on by lies spread by tech platforms—demonstrated all-too-viscerally. While it could be argued that platforms should have seen the writing on the wall and found a new model to ensure their future, the fact remains that they are being outcompeted by online platforms that fundamentally changed the game and now are taking measures to maintain their dominance.

The Australian government is specifically targeting Facebook and Google because of their power, acknowledging that "Digital platforms have fundamentally changed the way that media content is produced, distributed and consumed" (Frydenberg & Fletcher, 2020). The press release on the code's introduction into Parliament states that its goal is to "ensure that news media businesses are fairly remunerated for the content they generate" and to "support a diverse and sustainable Australian news media sector" (Frydenberg & Fletcher, 2020). On its face, the Code appears to be designed to give Australian media companies more money (siphoned from Facebook and Google) and negotiating power. In the next section, I will discuss Google and Facebook's unique situations and how the Code's logic may be flawed.

3 Google: Pre-emptive Dodging

In January of 2021, Google published a blog post claiming that the Media Bargaining Code would "break Google Search as you know it" because it would make Google pay to link to news content (Silva, 2021b). Google claims that it is merely carrying out its role in serving information as defined by the Internet economy, arguing that "Digital platforms do not owe publishers compensation for the emergence of an internet-based economy" and that the arbitration process unfairly favours publishers (Cerf, 2020), which is by design. The code is meant to give a leg up to legacy media, which has struggled in the internet-based economy that Google and Facebook dominate. While its critics argue that the Media Code is necessary to protect journalism and thus democracy, Google (and, as we will see, Facebook) yoked its counter-arguments to the need for a free and open internet founded on the principles of net neutrality.[1]

Google argues that forcing it to treat news links differently (though it should be noted that it does not have to pay per link it displays, just define a dollar value for the news content it makes available) violates net neutrality, which threatens Google's business model in particular. Despite its power, Google's business model left it especially vulnerable to pressure from the News Bargaining Code. Google's only leverage was to withdraw from the Australian market altogether. Google claims that news represents 1% of searches in Australia, but blocking those searches or preventing a website from appearing in results would violate principles of net neutrality, which Google touts its commitment to (Silva, 2021a). However, ceasing to providing service in Australia altogether would open the door to rivals, such as Microsoft's Bing (Newton, 2021b). Bing only controls 3.62% of the Australian search market to Google's 94.45% (Dunne, 2021), but signalled its aim to become Australian's news-friendly search engine if Google pulled out in a blog post by president Brad Smith. Titled "Why an Australian proposal offers part of what's needed for technology, journalism and American democracy itself," Smith writes that the Australian model should be expanded to other countries, including the United States (Smith, 2021). This is grounded in his argument that the disinformation campaign waged by Trump demonstrates the dangers of the erosion of independent journalism and that "the internet and social media have not been kind to the free press" (Smith, 2021).

Microsoft took the moral stance that democracy should be prioritised above all else, countering Google's argument that net neutrality is more important. Ultimately, Google found that it did not have a moral or economic leg to stand on. Because pulling out of markets is not a sustainable business strategy, Google instead found a strategy that allowed it to pre-empt the bargaining requirements. Instead of waiting to be forced into arbitration, Google began signing publishers to its Google News Showcase, a tab in its Google News product that displays content from official partners (Newton, 2021b). Google boasted that this "financial [investment] in the future of journalism" preserves "a free and open web which works the same way for everyone" and is a triumph for Australian journalism (Silva, 2021b). The Verge, however, characterised these deals as a face-saving move performed "under duress" (Newton, 2021b). Regardless of spin, these deals were successful in preventing Google from becoming a "designated digital platform" bound by the arbitration agreements (Kohler, 2021), and the Australian government succeeded in directing money to news publishers.

---

[1] Net neutrality is the principle that all links on the internet should be treated equally and free to access. The business model of search engines, which provide links to users, is possible because search engines do not have to pay to display certain links.

4   Facebook: Strongarm Tactics

Facebook's business model is significantly different from Google's, which gave it more leverage over the Australian government. Unlike Google, Facebook's business model is not to provide lists of links to users. While organizations can de-list their pages from Google, doing so is business suicide,[2] so allowing Google's web crawlers to index one's site is effectively mandatory. Posting on Facebook, though, is voluntary. Facebook's argument is that Facebook should not have to pay for "content it didn't take or ask for," given that news providers use Facebook's free platform to disseminate content. By Facebook's reckoning, the traffic they send to news publications is worth far more than the "business gain" to Facebook from news, estimated at 4% of users' News Feeds (Easton, 2021). Facebook says it is doing publishers a service by allowing them to use Facebook to promote their content, and that if they have to pay publishers, providing this "service" will not be worth it for Facebook (Easton, 2021). Facebook has never promised neutrality in the service it provides and, given the small proportion of content that news represents on Facebook, was able to carry through on its threat to block Australian users from viewing or sharing all news (domestic and international) and prevent users worldwide from seeing and sharing Australian news, without impact to its business (Easton, 2021). In fact, Facebook disabled the pages of Australian publishers altogether. This led to an overall 13% drop in traffic to Australian news sites from within the country (Purtill, 2021c), and thus a corresponding drop in revenue. Facebook only restored news access after the code passed with amendments negotiated between Treasurer Josh Frydenberg and Facebook CEO Mark Zuckerberg (Purtill, 2021b).

Facebook's strong-arming won them several key concessions from the government in the final bill. Under the version that passed the Senate, the Treasurer must consider whether a platform has already reached commercial agreements with news businesses before declaring them a "designated digital platform" and forcing them to negotiate deals. In addition, they must be given a one-month notice that they will be subject to the code (Purtill, 2021b), providing an opportunity for further negotiations. Other concessions include the assurance that provisions in the code prohibiting platforms from differentiating between both registered and non-registered news organisations at various stages of the registration, negotiation, and remuneration processes will not apply if a commercial agreement has been reached between the platform and organisation (even if the remuneration provided by the agreement differs from "usual business practices"); and a statement that final arbitration is only a last resort for when commercial deals cannot be reached (Treasury Laws Amendment, News Media and Digital Platforms Mandatory Bargaining Code, Bill 2021, 2021; Barbaschow, 2021). Furthermore, Facebook retains the right to shut off news in

---

[2] An experiment by deals site Groupon showed that de-listing their site from Google showed that search-engine-attributable search dropped to near-zero, while "direct" visits also dropped by 60% (McKenna, 2014).

Australia again, keeping the nuclear option on the table (Brown, 2021). Safety net secured, Facebook proceeded to start signing news organisations to its Facebook News silo, which operates similarly to Google News Showcase. However, though the two outcomes appear the same on their faces, Facebook is still coming out ahead of Google. While Facebook is still paying Australian news organisations, its strong-arm tactics mean that it will likely pay far less than it would have under arbitration, and media law researcher James Meese estimates that it is also less than what Google, with its much smaller leverage, is paying. He also points out that Facebook had already been rolling out Facebook News in other countries, and so may ultimately just pay organisations what it would have anyway, while also winning what is essentially an exemption from the News Bargaining Code (Purtill, 2021b).

5   Concerns over the Media Code

The Media Bargaining Code has—at least in the case of Google News Showcase—succeeded in securing additional revenue for Australian media companies (estimated at A$200 million (Smyth, 2021)), but at a potentially high cost to individual journalists, small publications, and the structure of the Internet as a whole.

From a financial side, one major concern is that the payments from Facebook and Google might benefit news companies, but not journalists. Small publishers with under A$150,000 in annual revenue aren't eligible for payments, and billionaire Rupert Murdoch owns News Corp., Australia's largest publisher (which has deals with Facebook and Google), so where the monetary benefits are going is unclear (Newton, 2021a). There is no requirement that the payments be used to directly help journalists, who have borne the brunt of the paradigmatic shifts in the media ecosystem, which is a major oversight. After all, journalism cannot be produced by faceless corporations; it is the journalists on the ground who deliver the crucial information. Rod Sims, chair of the ACCC, stated in June that "We are on track for deals all around… the media companies are happy—and that's the key point," but made no mention of whether journalists are also satisfied (Smyth, 2021).

Economically, small publications ineligible for payment are doubly impacted because they are more reliant on Facebook and Google for traffic (Samios, 2021). News may only be 4% of content on Facebook (Easton, 2021), but that is still a huge number of links, and for smaller publishers these platforms are their lifeline. One Australian youth publication said that Facebook and Google account for 75% of its traffic (Rigby, 2021). Smaller publications may also be less likely to get deals with Google News Showcase and Facebook News, leaving them in the same situation they were in before. Small publications are often targeted towards groups that can be overlooked by mainstream media. They are thus just as important to sustaining a healthy media ecosystem and should be offered support.

The other major concern—and the one that Facebook and Google piggybacked on in their responses—is that the Media Bargaining Code, by treating news links differently than other links, violates the principle of net neutrality. Furthermore, the final Code includes a provision that appears to allow media platforms to treat content from platforms that have signed agreements "preferentially" (Treasury Laws Amendment (News Media and Digital Platforms Mandatory Bargaining Code) Bill 2021, 2021). Sir Tim Berners-Lee (the inventor of the Web) thinks that these challenges to net neutrality could open the door to the death of the open Internet as we know it (Cellan-Jones, 2021); Facebook and Google piggybacked on that argument. However, Facebook ignored that in its furious counterattack, it violated the very principle it claims to hold dear. For many, Facebook is the Internet. This is true both figuratively in terms of the attention share it has and literally when it provides Internet access to developing areas through its Free Basics program,[3] and so banning categories of content wholesale is itself threatening net neutrality (Binkowski, 2019). The concern over net neutrality is valid; having free and equal access to information online is important to ensure that the democratic public stays informed. However, because Facebook and Google have been striking independent deals and the Media Bargaining Code is not actually being enforced, these concerns may be obviated for now. Future work should examine the knock-on effects of the passage of this bill and potential consequences should the Australian government elect to enforce it in the future.

6   Predatory Aggression

Facebook, with far more leverage than Google, emerged as the main combatant in the battle over the News Media Bargaining Code. In blocking news content, Facebook demonstrated the enormous power it has over the survival of media platforms, illustrating the danger of having an entire media ecosystem dependent on a dominant power. Facebook's news ban shows that, with a snap of its fingers, it can single-handedly devastate small news providers, increase the prevalence of fake news, and potentially even cause physical harm, displaying its danger for the news industry and democracy itself.

With a single action, Facebook drastically reduced traffic to Australian news sites. There was an overall drop in domestic traffic to Australian news sites of 13%, which is likely to be even higher for small organisations that depend on Facebook for the majority of their traffic (Purtill, 2021c). Smaller publications are more likely to rely on Facebook for a larger proportion of their traffic—and thus revenue—and so were more likely to be hit hard by the news ban (Rigby, 2021). This directly

---

[3] Free Basics, a Facebook-developed app that gives users free access to a Facebook-selected list of websites, has itself been criticized for infringing on net neutrality, as well as encouraging clickbait and "digital colonialism." Mark Zuckerberg's position that "Arguments about net neutrality shouldn't be used to prevent the most disadvantaged people in society from gaining access or to deprive people of opportunity" shows that Facebook sees net neutrality as one tool in its efforts for platform supremacy that can be deployed and discarded at will (Solon, 2017).

displays how Big Tech platforms have altered the news industry and made it dependent on its suffocating presence.

In addition to impacts on publishers, Facebook's ban on news negatively impacted the larger online news ecosystem. Fake news quickly rushed in to fill the void created by the news purge. In a small experiment, the BBC found that after the ban, searches for posts related to COVID-19 and vaccines revealed more posts with "misleading content" than before (Reality Check & BBC Monitoring, 2021). The Verge dismissed concerns about the rise in fake news that will result by removing high-quality news sources, asking, "But what if, in the meantime, Australians simply… visit websites? Subscribe to newsletters? Read… books?" (Newton, 2021b). The statement's acknowledged naivety is an understatement, as Facebook is a major source of news for a large number of people. 39% of Australians use Facebook for general news, and 49% use it for news about COVID-19; for many, social media is the only way they access the news (Purtill, 2021a). In a poll by The Guardian, only 30% of respondents agreed that "If Facebook stopped offering news on its platform, I would use it less often." Three-quarters claimed that they would "Go directly to news sites to read content," but Facebook is designed to trap users' attention, decreasing the likelihood that users will venture off the platform even if the balance of content shifts. A whopping 69% of respondents agreed that in the event of a Facebook-Google news ban, they would "Continue to use Google and Facebook and read less news" (Lewis, 2020). Facebook and Google are major drivers of traffic for a reason: they give users news they are interested in, obviating the need to go directly to the site and sift through the headlines. Internet denizens have gotten used to having the news delivered to them, rather than having to seek it out. This may be doubly true for international users, who may be far less likely to directly navigate to an Australian publication unless it is presented directly in their News Feed or Google search. Seeing less genuine news and more fake news injures the informational health of the public (Floridi, 2016), and when the public's information has been poisoned, democracy suffers.

Facebook also threatened direct physical harm in its purge. Claiming to be complying with the letter of the law, Facebook disabled the pages of local health departments, hospitals, emergency services, and even suicide and domestic violence services (Dye & McGuire, 2021). When protest arose, Facebook stated that Australia had gotten what it was asking for: "As the law does not provide clear guidance on the definition of news content, we have taken a broad definition in order to respect the law as drafted" (Taylor, 2021). However, perhaps realising the danger of disabling state health department pages in the midst of a pandemic, Facebook took action to restore those pages, but faced immense criticism. Facebook actually impeded the function of the Australian state, displaying its power over an entire democratic nation.

7 Conclusion

Ultimately, this fight is not about who is posting what links on the Internet. It is about the fact that Facebook and Google dominate the Internet and have fundamentally changed the media and communications ecosystem such that news organisations are outcompeted before they can adapt, which threatens democracy. These platforms are a rogue wave that have capsized the ship; there is nothing the publications can do. Publications, stuck in a toxic lockstep with data platforms, are more and more reliant on the platforms that do not need them. Publications will always lose in this vampiric codependency. To maintain their power in this relationship, Google and especially Facebook worked to assert authority over a world government. Clearly, some action is necessary. Though the Media Bargaining Code is promising, compelling payments only to large publications is not the way forward. It does not provide enough support for the journalists impacted by the changing tides of the media ecosystem and fundamental to the health of the journalism ecosystem or the small publications most affected, and it threatens the principles of net neutrality. A cynic might say that the Australian government devised the Media Bargaining Code to provide pay-outs to Australia's large media corporations, and if that was the goal, then it has succeeded. If Australia and other governments truly want to balance power between platforms and media outlets and foster a healthier news ecosystem, protecting democracy in the process, they could tax data platforms and use the revenue to directly support journalism and address the externalities these platforms have caused. Alternatively, social media companies could set up an independent foundation to support journalism, as suggested by Brooke Binkowski, former managing editor of Snopes (Binkowski, 2019). Ultimately, the Code does not address that Big Tech has changed how people access and consume news, nor does it curtail its power to mediate what news we see. This would require states to take action to regulate platforms' recommendation algorithms, the core of their business, and something they would not take lightly.

This battle is only just beginning. Canada is pushing to create a similar payment structure (Karadeglija, 2021), and Microsoft, after breaking with Big Tech to side with the Australian government, is now teaming up with EU publishers to lobby for journalism payments, which could impact initiatives in the EU and US (Chan, 2021). As shown by its forced acquiescence, Google may be a sitting duck for these initiatives. Facebook, on the other hand, has a larger playbook. However, its aggressive actions in Australia may have increased regulatory scrutiny on it, with politicians from the US and UK criticising the news ban (Cellan-Jones, 2021). Its borderline authoritarian tactics, rather than forcing governments to back down, may instead encourage regulators to adopt even more aggressive tactics in the war to regulate Big Tech. Whether this will see shots fired at the inner keep—content recommendation algorithms—remains to be seen.

References

A Bill for an Act to amend the Competition and Consumer Act 2010 in relation to digital platforms, and for related purposes, no. 177/20, Parliament of Australia. (2021). https://parlinfo.aph.gov.au/parlInfo/search/display/display.w3p;query=Id:%22legislation/bills/r6652_aspassed/0000%22

Associated Press. (2021, February 17). *Google strikes deals with Australian news publishers, while Facebook cuts off sharing*. MarketWatch. https://www.marketwatch.com/story/google-strikes-deals-with-australian-news-publishers-while-facebook-cuts-off-sharing-01613600522

Barbaschow, A. (2020, December 10). *Media Bargaining Code enters Parliament despite Google and Facebook's best efforts*. ZDNet. https://www.zdnet.com/article/media-bargaining-code-enters-parliament-despite-google-and-facebooks-best-efforts/

Barbaschow, A. (2021, February 23). *News to remerge on Facebook in Australia after deal struck with government*. ZDNet. https://www.zdnet.com/article/news-to-remerge-on-facebook-in-australia-after-deal-struck-with-government/

Binkowski, B. (2019, February 8). *Opinion: Fact-checking Facebook was like playing a doomed game of whack-A-mole*. BuzzFeed News. https://www.buzzfeednews.com/article/brookebinkowski/fact-checking-facebook-doomed

Boom, D. V. (2021, February 14). *Google's fight in Australia could change the future of media*. CNET. https://www.cnet.com/news/googles-fight-in-australia-could-change-the-future-of-media/

Brown, C. (2021, February 23). *The Value of News on Facebook*. https://www.facebook.com/journalismproject/news-australia-decision

Cellan-Jones, R. (2021, February 19). Tech Tent: Facebook v Australia - two sides to the story. *BBC News*. https://www.bbc.com/news/technology-56120281

Cerf, V. (2020, November 23). *A fair code for an open internet*. Google: The Keyword. https://blog.google/around-the-globe/google-asia/australia/fair-code-open-internet/

Chan, K. (2021, April 23). *Microsoft, EU publishers seek Australia-style news payments*. AP News. https://apnews.com/article/europe-media-news-industry-europe-f33528aa575b2acf75c98ad99ecb0975

Choudhury, S. R. (2021, February 25). *Australia passes new media law that will require Google, Facebook to pay for news*. CNBC. https://www.cnbc.com/2021/02/25/australia-passes-its-news-media-bargaining-code.html

Dunne, R. (2021, February 3). Bing steps up to replace Google in Australian search showdown. *Search Engine Journal*. https://www.searchenginejournal.com/bing-replace-google-australia-search/394544/

Dye, J., & McGuire, A. (2021, February 18). *Facebook news ban: BOM, Queensland Health, Cricket Australia, others removed*. Sydney Morning Herald. https://www.smh.com.au/national/facebook-news-ban-hits-emergency-services-and-government-health-departments-20210218-p573ks.html

Easton, W. (2021, February 17). *Changes to sharing and viewing news on facebook in Australia*. About Facebook. https://about.fb.com/news/2021/02/changes-to-sharing-and-viewing-news-on-facebook-in-australia/

Fernandez, R., Adriaans, I., Klinge, T. J., & Hendrikse, R. (2021, February 5). *How Big Tech is becoming the Government*. SOMO. https://www.somo.nl/how-big-tech-is-becoming-the-government/

Floridi, L. (2016, November 29). Fake news and a 400-year-old problem: We need to resolve the 'post-truth' crisis. *The Guardian*. http://www.theguardian.com/technology/2016/nov/29/fake-news-echo-chamber-ethics-infosphere-internet-digital

Frydenberg, J., & Fletcher, P. (2020, December 8). *News media and digital platforms mandatory bargaining code | Treasury ministers*. Australian Government: The Treasury. https://ministers.treasury.gov.au/ministers/josh-frydenberg-2018/media-releases/news-media-and-digital-platforms-mandatory-bargaining

Hine, E. (2021, February 23). *ER4: An Australian love triangle/shipwreck*. The Ethical Reckoner. https://ethicalreckoner.substack.com/p/er4-an-australian-love-triangleshipwreck

Karadeglija, A. (2021, February 11). *Heritage Minister Steven Guilbeault pledges to press ahead with forcing tech giants to pay for news*. National Post. https://nationalpost.com/news/heritage-minister-steven-guilbeault-pledges-to-press-ahead-with-forcing-tech-giants-to-pay-for-news

Kohler, A. (2021, March 16). *The News Bargaining Code is officially dead*. The New Daily. https://thenewdaily.com.au/news/2021/03/17/alan-kohler-news-bargaining-code-dead/

Letts, S. (2016, January 29). Global internet giants crushing Australian media. *ABC News*. https://www.abc.net.au/news/2016-01-29/global-internet-giants-crushing-australian-media/7125458

Lewis, P. (2020, September 8). The stakes are high for Facebook and Google if Australians decide to get their news elsewhere. *The Guardian*. https://www.theguardian.com/australia-news/commentisfree/2020/sep/08/the-stakes-are-high-for-facebook-and-google-if-australians-decide-to-get-their-news-elsewhere

McKenna, G. (2014, July 8). *Experiment shows up to 60% of 'direct' traffic is actually organic search*. Search Engine Land. https://searchengineland.com/60-direct-traffic-actually-seo-195415

Meyer, A. C. M., Robinson. (2018, October 18). *How Facebook's chaotic push into video cost hundreds of journalists their jobs*. The Atlantic. https://www.theatlantic.com/technology/archive/2018/10/facebook-driven-video-push-may-have-cost-483-journalists-their-jobs/573403/

Newton, C. (2021a, February 17). *Australia's bad bargain with platforms*. Platformer. https://www.platformer.news/p/australias-bad-bargain-with-platforms

Newton, C. (2021b, February 18). *Why Google caved to Australia, and Facebook didn't*. The Verge. https://www.theverge.com/2021/2/18/22288510/google-facebook-australia-news-media-bargaining-code

Purtill, J. (2021a, February 18). 'The only news left': Anti-vaccine pages unscathed by Facebook news ban. *ABC News*. https://www.abc.net.au/news/science/2021-02-18/facebook-news-ban-misinformation-spread-covid-vaccine-rollout/13167318

Purtill, J. (2021b, February 25). There was 'definitely one loser' in Facebook's battle with the government. *ABC News*. https://www.abc.net.au/news/science/2021-02-26/facebook-google-who-won-battle-news-media-bargaining-code/13193106

Purtill, J. (2021c, March 2). These graphs tell the story of Facebook's news ban—And what happened after. *ABC News*. https://www.abc.net.au/news/science/2021-03-03/facebook-news-ban-australian-publisher-page-views-rebound/13206616

Reality Check & BBC Monitoring. (2021, February 20). *Facebook in Australia: What happened after news was blocked? - BBC News*. BBC News. https://www.bbc.co.uk/news/56127158

Rigby, B. (2021, February 1). *'What are we doing here? Trying to save the titanic that's sinking?': Small publishers testify on 'potentially fatal' impacts if platforms 'bugger off'*. Mumbrella. https://mumbrella.com.au/trying-to-save-the-titanic-thats-sinking-small-publishers-front-senate-in-media-code-battle-666638

Samios, Z. (2021, February 18). *'Squashed': Smaller publishers fear fatal consequences from Facebook's news ban*. The Sydney Morning Herald. https://www.smh.com.au/business/companies/squashed-smaller-publishers-fear-fatal-consequences-from-facebook-s-news-ban-20210218-p573sw.html

Schudson, M. (2020, October 5). *The vital role of journalism in a liberal democracy*. The MIT Press Reader. https://thereader.mitpress.mit.edu/journalism-in-a-liberal-democracy/

Silva, M. (2021a, January 6). *Open letter—Update on the news media bargaining code in Australia—Google*. https://about.google/google-in-australia/jan-6-letter/

Silva, M. (2021b, February 26). An update on the news media bargaining code—Google. About Google. https://about.google/google-in-australia/an-open-letter/

Smith, B. (2021, February 11). *Microsoft's endorsement of Australia's proposal on technology and the news*. Microsoft On the Issues. https://blogs.microsoft.com/on-the-issues/2021/02/11/endorsement-australias-proposal-technology-news/

Smyth, J. (2021, June 1). Australian regulator claims victory in scrap with Big Tech over news. *Financial Times.* https://www.ft.com/content/ad706bd3-2aed-49da-b4f4-862f15a2e601

Solon, O. (2017, July 27). 'It's digital colonialism': How Facebook's free internet service has failed its users. *The Guardian.* http://www.theguardian.com/technology/2017/jul/27/facebook-free-basics-developing-markets

Stilgherrian. (2021, March 5). *Australia's news media bargaining code is a form of ransomware, and someone paid up.* ZDNet. https://www.zdnet.com/article/australias-news-media-bargaining-code-is-a-form-of-ransomware-and-someone-paid-up/

Taylor, J. (2021, February 18). Facebook's botched Australia news ban hits health departments, charities and its own pages. *The Guardian.* http://www.theguardian.com/technology/2021/feb/18/facebook-blocks-health-departments-charities-and-its-own-pages-in-botched-australia-news-ban

App Store Governance: The Implications and Limitations of Duopolistic Dominance

Josh Cowls and Jessica Morley

Abstract Much focus of the emerging research front known as platform governance is concerned with content moderation and other ways in which platform companies govern users and activity on their own platforms, as well as with how platforms are themselves governed by states (Gorwa, Inf Commun Soc 22(6): 854–871, 2019). Meanwhile, a related but distinct internet governance literature is concerned with what can be called "stack governance": important but often unseen governance activities occurring elsewhere on the "tech stack", beyond or behind platforms, apps and the open web (Donovan, Navigating the tech stack: When, where and how should we moderate content?. Centre for International Governance Innovation, 2019). A notable point of intersection between platform governance and stack governance is represented by app stores. Apple and Google hold an effective duopoly over smartphone software, with 99.2% of devices worldwide running on either Apple's iOS or Google's Android software. Both provide a pre-installed app store through which most apps are discovered and downloaded. In addition to the oft-discussed competition concerns that this duopoly poses, Apple and Google's dominance over app store governance poses broader societal dangers. This includes the risk of harms arising from the spread of dangerous content via apps, such as those claiming to deliver impossible health outcomes, or permitting hate speech, as well as concerns for democracy and accountability that arise from this duopolistic dominance. In order to illuminate these risks, we explore four emblematic episodes of app store governance: the prevalence of inefficacious health apps on Apple's App Store; the decision to remove far-right platform Parler from app stores following the Capitol Riots in January 2021; Apple and Google's initial rejection of a UK Covid-19 contact tracing app in 2020; and both companies' removal of the Russian opposition "Smart Voting" app in 2021. Taken together, the cases we cover surface several considerations that we argue ought to shape legal, ethical, and technical approaches to app store governance—encompassing both how Apple and Google

J. Cowls (✉) · J. Morley
Oxford Internet Institute, University of Oxford, Oxford, UK
e-mail: josh.cowls@oii.ox.ac.uk; Jessica.morley@oii.ox.ac.uk

© The Author(s), under exclusive license to Springer Nature
Switzerland AG 2022
J. Mökander, M. Ziosi (eds.), *The 2021 Yearbook of the Digital Ethics Lab*,
Digital Ethics Lab Yearbook, https://doi.org/10.1007/978-3-031-09846-8_6

govern their app stores, and how the two companies themselves are and ought to be governed by states.

Keywords Apple · App stores · Covid-19 · Google · Parler · Platform governance · Stack governance

1 Introduction

The governance of the internet and related digital technologies involves an almost bewildering range of laws, rules, policies, standards and mechanisms, and implicates a heterogenous array of actors, making it "a difficult horse to catch" (Ziewitz & Pentzold, 2014). In addition to high-level multilateral standards-setting bodies such as the Internet Engineering Task Force (Cath, 2021), control of different aspects and functions of online life is exercised by entities including governments and, increasingly, private companies (DeNardis, 2014; Sharon, 2020; Taylor, 2021).

One aspect of internet governance which commands considerable and growing attention is platform governance (Gillespie, 2018; Gorwa, 2019; van Dijck et al., 2018). The millions of decisions about individual pieces of content on social media taken every day—by some combination of automated systems (Gorwa et al., 2020; Katzenbach & Ulbricht, 2019) and contracted and volunteer human moderators (Jhaver et al., 2019; Matias, 2019; Roberts, 2019)—constitutes a vast governance "engine", enforcing rules and policies variously set by states (Schulz, 2018), platform operators (Klonick, 2017), and quasi-autonomous oversight bodies (Douek, 2019), operating across national borders (Bloch-Wehba, 2019). At its core, platform governance can be taken to encompass two key facets of internet governance: the governance of platform operators by states, and the governance of platform users by platform operators (Gorwa, 2019).

Given the scale and centrality of social media platforms in contemporary societies, the scholarly and policy focus on platform governance is neither surprising nor unwelcome. Nonetheless, the governance decisions taken above and beyond (and sometimes also *about*) social media platforms by other entities—including but not limited to states—have received somewhat less attention, yet are arguably of just as much societal significance. Organisations which provide infrastructural services such as cloud computing, online payment processing, domain registration, and other functions elsewhere on the "tech stack" (Donovan, 2019), have been drawn into de facto governance roles when deciding whether to permit or remove variously offensive sites and platforms on their services. This "infrastructural turn" in internet studies is not new (DeNardis, 2012; Musiani et al., 2016), but has taken on greater prominence of late, as what DeNardis in 2012 dubbed the "hidden levels of internet control" have come increasingly into view amidst a series of controversial events. These flashpoints have included, for example, Cloudflare's decision to terminate its infrastructure services to fringe alt-right message board site 8chan after the manifesto and video footage of the 2019 Christchurch mosque massacre were spread on the site (Kelly, 2019); social network OnlyFans' initial decision—later reversed—to ban adult content after pressure from banking services (Bronstein, 2021); and Google and GoDaddy's decision to terminate domain registration services for neo-Nazi site The Daily Stormer after the Charlottesville riot (Romano, 2017). These and other incidents have helped to highlight the importance of what we could call, for shorthand, "stack governance".

Thus, while platform governance (in the "platforms govern" sense of the term: Gorwa, 2019) centres on the relationship between platforms and their own users, stack governance involves cases where the policies and practices of platforms themselves are governed by other (non-state) entities. But the two phenomena are not easily disentangled in practice. Take the case of app stores, specifically those operated by Apple and Google for their iOS and Android smartphone platforms. App stores themselves are platforms (Gillespie, 2018, p. 18) whose operators such as Apple or Google in turn purvey—and, via their submission and review processes, govern—other platforms, such as Snapchat or Instagram. Since app stores have the capacity—and, as we will see, sometimes the willingness—to reject or remove platforms from their app stores due to those platforms' inadequate content moderation practices, app store governance is an instance of *platform governance of platforms..* This involves an indirect form of content moderation, or "meta-moderation", wherein the subject platform's own content moderation and governance practices are assessed, and potentially sanctioned, by app store operators.

Situated at the intersection of platform governance and stack governance, the complex nature and scope of app store governance (hereafter "ASG"), and its high social stakes (in terms of its impact on freedom of expression, for example), necessitate careful consideration. That is the task of this chapter, in which we use a series of notable episodes in ASG to shed light on the challenges and tensions inherent to ASG. In Sect. 2, we provide further background to app stores and how they are governed. We also assess the implications of the present duopoly that Apple and Google hold over app store operation in most territories globally. In Sect. 3, we present three short case studies of ASG, namely the removal of fringe social media platform Parler from app stores, the development and rejection of Covid-19 contact tracing apps, and the removal of the opposition "Smart Voting" app from app stores in Russia. In Sect. 4, we conclude by identifying the common themes and contradictory insights to emerge from the case studies and consider their implications for app store governance going forward.

2 App Store Governance: Duopolistic Dominance

One of the most important affordances of modern personal computing devices, such as desktop and laptop computers and more recently smartphones and tablets, is their ability to run programmes, apps, and other pieces of software developed by parties other than the manufacturers of the hardware and developers of the core operating system (OS) software. Yet historically this affordance has not been without controversy. For instance, the most significant legal case of the early phase of personal computing, *United States v. Microsoft* (2001), turned on whether Microsoft was justified in bundling its Internet Explorer (IE) web browser with its dominant Windows operating system, given the advantage that this gave IE over rivals such as Netscape in the so-called "browser wars" of the late 1990s. The more recent rise of smartphones capable of downloading and executing third-party software packages, or apps, has likewise prompted debate over the appropriate level of control and range of responsibilities that smartphone manufacturers and OS developers ought to hold with respect to what users can do with their devices. In contrast to the monopolistic desktop era, however, the smartphone era features a duopoly, held by two "big tech" companies, Apple and Google, whose software together powers 99.2% of smartphones worldwide. Since the two companies have taken notably distinct approaches to handling third-party apps, we assess them in turn.

The first of Apple's iPhone smartphone devices, which was announced in January 2007 and released later that year, launched without an app store: the company's initial focus was on its own pre-installed apps, with third-party developers limited to developing web applications to be accessed through the in-built web browser (Apple, 2008). The ability to install and run third-party apps arrived a year later with Apple's launch of its App Store for the iPhone's iOS software. The App Store allows developers to submit apps for iPhone users to download, either for free or for a fee, and takes a 30% cut from App Store revenue raised. This model has received considerable criticism for Apple's excessive revenue cut, as well as over claims that Apple systematically disadvantages rival software developers through onerous restrictions, such as preventing developers of apps from even informing users that they can secure a cheaper subscription rate (i.e., minus Apple's fees) if they sign up for their service elsewhere. More broadly, Apple's iOS is designed to prevent users from "sideloading" apps—that is, installing apps to their devices in any way other than through the official App Store (e.g. direct downloads from the web, from another device, or through a third-party app store), marking a restrictive approach to device management. Facing renewed pressure to open up its app ecosystem (see Sect. 4), Apple recently published a "threat analysis of sideloading", in which the company claims that enabling the practice "would cripple the [iPhone's] privacy and security protections", suggesting that users would be left more exposed to harmful apps (Apple, 2021, p. 2). Apple's rejection of sideloading—which must of course be viewed sceptically, in the context of the enormous revenues it makes in its cut of developer income—rests on claims it makes about the alleged safety and tranquillity of its own App Store. Apple boasts that it "review[s] every app before it becomes available on the App Store to ensure it is free of malware and accurately represented to users, and [that it] swiftly remove[s] apps from the App Store if they are found to be harmful" (Apple, 2021, p. 4). It has also taken a strong stance against apps containing content such as pornography and drug use; former CEO Steve Jobs claimed that Apple had a "moral responsibility to keep porn off the iPhone" (Siegler, 2010). Outside analysts suggest, however, that despite these supposedly careful controls, the App Store is still littered with malicious apps such as those that scam unwitting users out of hundreds of dollars (Hollister, 2021; Lin, 2020). Despite modest concessions of late, such as reducing its revenue cut, generally the company has held firm to its tight control of how users can install third-party software on their devices. Because the official App Store remains the only way that users can install apps, this raises the stakes for Apple's responsible management of the App Store.

The Android mobile operating system, which has a roughly 70% global smartphone market share, is borne of a superficially different philosophy compared to Apple's iOS. Android's core code is based on open-source software, and it is developed and maintained by the Open Handset Alliance (OHA), a consortium of over 84 software and hardware manufacturers, which announced Android in 2007 as "first truly open and comprehensive platform for mobile devices" (Open Handset Alliance, 2007). From the start, however, Android has been dominated by Google. The Android name is trademarked by Google, and most Android smartphones come pre-installed with Google's proprietary apps such as its cross-platform browser Google Chrome and most significantly Google's app store, known as the Play Store. Though Android users are, unlike iPhone users, permitted to sideload apps from outside the Play Store, the process is less convenient and the ability to do so is disabled by default. Moreover, members of the OHA are forbidden from forking Android to create alternative versions, as Amazon (a non-OHA member) has done with its FireOS. This approach resulted in the $5.05bn fine handed down by the European Commission to Google in 2018 (Brodkin, 2018), on the basis of Google's requirements that manufacturers pre-install Chrome as a condition for licensing the Play Store (and the millions of apps it offers), as well as the inducements it offered manufacturers to pre-install Google apps and its prevention of other manufacturers from pre-installing apps on devices running other Android forks (European Commission, 2018).

Therefore, despite the underlying differences between the world's two dominant smartphone operating systems, there are several important commonalities between them. Apple and Google exert considerable control over the respective systems, albeit in ways that are commercially and technically distinct, which has landed them considerable criticism and sizeable financial penalties. Moreover, the feasibility and sustainability of both smartphone ecosystems rely heavily on well-managed app stores. Apple's App Store is the only safe and remotely straightforward way to download apps to iPhones, while Google's Play Store is not the only way, but certainly the simplest way, to download apps to Android devices. And Google's costly positioning of the Play Store as a necessary pre-installation further demonstrates the importance it places on its app store as central to the Android ecosystem.

The central significance to the smartphone experience of well-managed app stores raises questions about how Apple and Google govern their app stores—including the overarching policies and guidelines that determine what sorts of apps are allowed, and the processes that are put in place to ensure that these standards are enforced. The four case studies that follow highlight different aspects of app store governance and its shortcomings.

3   Instructive Episodes in App Store Governance

3.1   *Babylon*

Babylon, founded in 2013, is a UK-based, online-first health service provider offering potential patients 24/7 access to GPs, physiotherapists, nurses, and pharmacists, and an algorithmically driven symptom checker (or triaging tool) via the app. Patients can either register with the app as a private customer and pay for their virtual appointment, or de-register from their current National Health Service (NHS) GP, re-register with Babylon's 'GP at hand' practice, and use the service for free. As the NHS is a state-funded public health service, offered free at the point of care, the availability of this latter option places Babylon in the category of a quasi-public service provider.

At first glance, this may appear to present no particular ethical or governance-related concerns. After all, GP practices have always been private businesses that contract with the NHS to provide care to NHS patients. From this perspective, Babylon is just another GP practice simply offering appointments, albeit via video-consultation rather than in-person. However, if one dives deeper into the way in which NHS services are governed, it becomes clear that in offering NHS services to anyone with an Android or iOS smartphone, Babylon is an app that is not only technically disruptive, but morally disruptive too.

To understand this, we must first define the meaning of moral disruption. In his book *Before Bioethics*, Baker (2013) defines morally disruptive technologies as those which 'undermine established moral norms or ethical codes' and thus cause moral uncertainty. Nickel (2020) argues that this moral uncertainty can be harmful because it hinders the ability of individuals (in this instance clinicians, patients, and governing bodies) to identify their moral obligations, i.e., their rights and responsibilities. This prevents individuals from being able to exercise their moral agency and from being able to hold others morally accountable for their actions. This applies to Babylon's GP at Hand offering because, in launching in a direct-to-consumer fashion via app stores, it completely undermined the systems put in place to monitor the quality, safety and effectiveness of NHS services, largely operating instead as a 'black box'—and so making it infinitely harder for appropriate governing bodies to identify situations in which patients were being exposed to risk, and harder to hold those responsible accountable. Babylon did this in three separate ways.

First, GP practices are primarily paid and monitored for variation in care according to the make-up of their 'list.' The list is of all the patients registered with that practice. In general, the larger the list size, the more the practice is paid by the NHS for the care it provides to the individuals. However, this model has some nuances. This is because the payment (known as the Global sum payment) is calculated using the Carr-Hill formula. The formula takes into consideration factors such as individual patients' age, gender and health conditions, and calculates a "weighted" count of patients according to need. This means that two practices with the same number of patients may have very different weighted patient numbers due to varying patient characteristics and, as a result, practices which may be similar in terms of list size, could receive very different levels of funding. In general, noting that this is a simplification, older and sicker patients have a higher weighting and so generate a greater payment, but they also cost the practice considerably more to care for. As such, much like gyms make profits off members who do not go the gym and therefore do not cause wear and tear to the equipment, GP practices typically make profit off younger and healthier patients who still generate a payment but rarely 'cost' the practice anything. Babylon exploited this model by appealing primarily to the young and healthy demographic (characteristics that also correlate with smartphone ownership) and encouraging them to de-register from their existing practice and re-register with the GP at Hand service. Not only did this undermine the payment model, but from a governance and accountability perspective, payments are also used to monitor GP performance. If a practice appears to be earning significantly more or significantly less, compared to established comparative baselines for a particular service (for example Warfarin monitoring clinics) this can be an indicator of unwarranted variation in care. By potentially deviating so far from an expected baseline, Babylon posed a threat to the ability of Governing bodies, for example NHS England, to effectively monitor its performance.

Second—continuing with the theme of undermining existing governance mechanisms—Babylon challenged the method used by the Care Quality Commission (CQC, the independent regulator of all health and social care services in England) to monitor the performance of NHS services. The CQC has traditionally worked by conducting in-person inspections of physical GP practices against a set of indicators designed to assess whether a practice is: safe; effective; caring; responsive to people's needs; and well-led, and ultimately whether it is meeting the legal requirements and regulations associated with the Health and Social Care Act 2008. Babylon's digital-first offering, and its public-private split, seriously challenged the CQC's ability to assess its performance against the traditionally used criteria.

The CQC did not even begin to develop a regulatory approach that would be suitable for online providers until 2015—two years after Babylon first appeared—and it did not conduct an inspection of any online providers until 2016 and still only did so after receiving intelligence about risk to patients. It did not fully refine its processes until 2017. The impact this had on CQC's ability to effectively evaluate the quality of care provided by Babylon is evident from the considerable variations in findings between inspections. In 2016, when Babylon was first 'inspected', CQC only inspected the GP video consultation service and relied almost exclusively on conversations with the Chief Executive of Babylon and on 13 emails from satisfied patients which Babylon had collected themselves. Unsurprisingly, this rather shallow 'inspection' found no issues with the service offered by Babylon, however, a year later—following criticism of the first inspection—a second inspection was carried out, this time entailing conversations with a much wider range of Babylon staff and external stakeholders, and a critical review of organisational documents and a review of medical records. This time the CQC found that Babylon was not providing effective care and, in some instances, was not providing safe care. Specifically,

CQC criticised Babylon for not always making appropriate prescribing decisions, for not ensuring medical records were updated (if, for example the patient was a non-NHS patient and Babylon did not write back to the NHS GP), and for not having a system in place to give assurance that patients' conditions were being appropriately monitored. By 2019, these issues were apparently found to be resolved, and the CQC rated Babylon as good in four out of the five areas and 'outstanding' when it came to being 'well-led.' Until there is a stable repeatable process for reviewing the care provided by Babylon, it is difficult to trust that this process is working as it should to protect patients.

Third, and finally, at no point has the NHS been able to exert any form of governing power over Babylon's symptom checker/triaging app. This is despite the fact that numerous academics, clinicians, and lay members of the public have raised concerns about the efficacy of an entirely 'black-box' diagnostic algorithm and its potentially consequential harms related to misdiagnosis, missed diagnosis, or misinformation (Morley et al., 2020a, b). The symptom checker *is* registered with the Medicines and Healthcare products Regulatory Agency (MHRA)—the body responsible for assessing the safety and efficacy of medical devices in England—as a regulated medical device. However, symptom checkers currently only need to be registered as a Class I medical device. Class I devices are assumed to pose limited threat to patient safety, and are therefore entirely self-certificated. In other words, whilst class I manufacturers are required to conduct an in-house clinical evaluation, and provide evidence that this has been done to the MHRA, there is very little opportunity for external bodies to hold Babylon (and others) to account for poor or even unsafe performance. This is concerning because, although Babylon has independently published evaluations comparing the performance of its app to the performance of GPs for specific clinical vignettes, these attempts—although to an extent laudable—have been found to be methodologically flawed (Fraser et al., 2018). Furthermore, even though a more recent evaluation found the performance of the symptom checker to be within one standard deviation of GP performance, this was only the case for a relatively small number of conditions. Indeed, out of eight available symptom checkers, Babylon was found to cover the smallest range of conditions (Gilbert et al., 2020).

Overall, it is clear that Babylon has presented a considerable challenge to the ability of the NHS—and by extension the UK Government—to govern the provision of its quasi-public service. This raises the question, of who can hold it accountable for its performance? And, most pertinently here, does this responsibility sit with the host app stores? The answer to the latter question is arguably, yes: app stores should at least absorb some responsibility for the performance of the services they are making available to the public. However, they take quite different stances in this regard. Whilst the Apple App Store Developer Guidelines do say, somewhat vaguely, that medical apps "may be reviewed with greater scrutiny" and that they must disclose data and methodology to support accuracy claims, Google's Play Store development guidelines place no such additional requirements on health app developers. This slightly varying yet in both cases apathetic approach from the app store providers may in part be explained by the ontology we discuss in the following case study: the app stores see themselves as responsible for the design of the apps themselves (for example, whether they take data protection seriously) but not necessarily for what is effectively 'user-generated' content and activity. It is, after all, arguably an individual's choice to decide to 'diagnose themselves' using a symptom checker app. However, this leaves the public exposed to harm and raises challenging questions about how Apple and Google, as effective duopolists, might try to compete (or collude) in this space. For example, does taking a firmer stance present a competitive advantage in that it might better protect Apple from a lawsuit, or does taking a more relaxed stance offer Google a competitive advantage in widening its potential 'customer' base? And beyond these distinctions, how—and where—can UK public authorities effectively intervene to ensure that patients are safe as smartphone-delivered healthcare grows?

3.2   Parler

Parler is a microblogging social network similar to Twitter, in which users are able to publish posts or "parleys" limited to 1000 characters, and to amplify or "echo" the posts of others. Parleys are presented in reverse-chronological order. Though ostensibly impartial, in practice Parler was established to provide an alternative "safe haven" for conservative Americans frustrated with the supposedly onerous and biased content moderation efforts of mainstream platforms, especially Twitter. Bursts of interest in Parler coincided with the increasing restrictions put on the accounts of leading conservative figures, including President Donald Trump, by Twitter and Facebook during 2020.

Parler was removed from the Apple's iOS App Store and Google's Play Store—and had its cloud hosting cut by Amazon Web Services—in the immediate aftermath of the riots at the United States Capitol building in January 2021. It was since restored by the iOS App Store in May 2021 after "negotiations" with Apple, yet at the time of writing remains suspended from the Play Store (though recall that Android users can more easily side-load apps to their devices).

The January 2021 suspension of Parler from app stores resembles earlier removals of similar apps, such as Gab, which was never allowed onto Apple's app store and was removed from the Google Play Store in 2017 for violating its policy against hate speech. Several features set the Parler case apart, however. First, compared with Gab and other extreme-right platforms and sites like 8Chan, Parler's executives made effort to portray the platform as at least superficially welcoming to all ideological groups. Parler co-founder and CEO John Matze made numerous appearances on US cable news and provided quotes to scores of journalists during 2020, in an attempt to construct a 'past imaginary' for Parler. These appearances portrayed Parler as a 'digital town square', designed to facilitate free-flowing debate and safeguard freedom of speech—often drawing an explicit contrast with larger, supposedly more censorious rivals. Second, the removal of Parler can be tied clearly to a specific 'real-world' event, namely the storming of the Capitol. Third, and related, the Parler removal reflected a seemingly coordinated approach, or at any rate over-lapping consensus, across both Apple and Google, which removed the app within hours of each other, and shortly after Amazon withdrew its cloud hosting support to the platform. Together these factors make the Parler removal a particularly significant instance of app store governance with several notable implications.

The key question raised by this episode concerns what might be thought of as 'platform ontology'. Parler can be seen as situated at a certain point on a spectrum ranging from mainstream platforms like Facebook and Twitter at one end, to singularly extreme far-right networks like Gab and 8Chan on the other. For present purposes, it matters less where exactly one Parler ought to be placed on this spectrum—the authors would personally place it closer to the latter end than the former—than the fact that it lies somewhere in between. As such, a key factor underlying Parler's removal in January 2021 (as well as its original acceptance onto both platforms, and its later reinstatement by Apple but not Google) concerns the threshold at which a platform ought to be defined in terms of its users, the content they share and behaviour they undertake. There is no doubt that Facebook, for example, houses millions of pieces of offensive content, such as material which would clearly constitute hate speech, nor that it facilitates illegal activities such as harassment, blackmail, and even sex trafficking. There are two principal reasons that it nonetheless remains on app stores: first, it attempts (with varying efficacy) to remove objectionable content and sanction offenders; and second, more fundamentally, because the proportion of the platform that contains objectionable material and illicit behaviour is sufficiently small. Were Facebook to consist mostly, or to a sufficiently large extent, of such material and behaviour, and/or were its efforts to prevent and remove these ills deemed insufficient, it would (one might hope) be removed from app stores.

In their respective guidelines to app developers, neither Apple nor Google note the existence of such a threshold, nor do they state the criteria they use to judge the sufficiency of efforts to moderate user-generated content. Apple's guidelines on these matters are vague almost to the point of satire. In the preamble to their App Store Review Guidelines, Apple notes as a "point to keep in mind" that:

> We strongly support all points of view being represented on the App Store, as long as the apps are respectful to users with differing opinions and the quality of the app experience is great. We will reject apps for any content or behavior that we believe is over the line. What line, you ask? Well, as a Supreme Court Justice once said, "I'll know it when I see it". And we think that you will also know it when you cross it.

The tongue-in-cheek reference to US Supreme Court Justice Potter Stewart's definition of pornography ("I know it when I see it") is not merely ironic given Apple's strict ban on pornographic material in App Store apps. For what is supposed to be a useful 'guideline' (and, ultimately, an enforceable standard), it is also the epitome of arbitrariness and opacity. The subsection of Apple's guidelines on "Objectionable content" (in the "Safety" section) are somewhat less vague. Under "objectionable content", Apple includes "content that is offensive, insensitive, upsetting, intended to disgust, in exceptionally poor taste, or just plain creepy". This may include "defamatory, discriminatory, or mean-spirited content, including references or commentary about religion, race, sexual orientation, gender, national/ethnic origin, or other targeted groups, particularly if the app is likely to humiliate, intimidate, or harm a targeted individual or group."

Meanwhile, in its Play Store Policy Centre, Google includes, as examples of "Inappropriate Content" that it prohibits, "Hate speech", which is taken to mean "apps that promote violence, or incite hatred against individuals or groups" based on a series of identifying characteristics; and "Violence", i.e. "apps that depict or facilitate gratuitous violence or other dangerous activities". Google also provides examples of common violations to illustrate these prohibited categories of inappropriate content.

Though Apple's guidelines are somewhat more concrete than its consciously vague preamble suggests, both its and Google's guidelines fall short with respect to platform apps such as Parler. Both make reference to "user-generated content"; Apple notes that "apps with user-generated content present particular challenges", and insists that such apps "include a method for filtering objectionable material from being posted to the app", as well as mechanisms for reporting content and blocking users. Likewise, Google asserts that apps that "contain or feature [user-generated content] must develop terms of use that "define objectionable content or behaviours in a way that complies with" the aforementioned guidelines.

Crucially, it is not clear from these guidelines where the line is drawn between apps that *themselves* transgress by e.g. "promot[ing] violence or incit[ing] hatred", and those that merely host significant amounts of such content. Returning to the spectrum described above, it seems that mainstream platforms like Facebook and Twitter fall into the latter camp, and are therefore subject to each app stores' standards for filtering, reporting and blocking (in short, content moderation), while apps such as Gab might well fall into the former camp given the sheer volume and proportion of objectionable content. This distinction is significant because it holds different implications for app store governance. Apps like Gab may be subject to what we could call "direct governance", whereby Apple and Google themselves take responsibility for judging the app's violations of the app stores' hate speech and violence policies. Apps like Facebook and Twitter by contrast enjoy "delegated governance" wherein the operators of those platforms are self-governing in the first instance.

Parler, as noted, sits somewhere between these two poles, making it difficult to know a priori which governance model it was subject to. Based on the statements released by Apple and Google explaining the reasons for Parler's removal, the evidence is mixed, but seems to suggest that Parler had initially been treated as a self-governing platform prior to the Capitol riots but was then placed under direct governance. In its statement, Google noted its "longstanding policies requir[ing] apps displaying user-generated content [to] have moderation policies and enforcement that removes egregious content like posts that incite violence". Apple, for its part, gave Parler 24 h to provide a "requested moderation improvement plan" and to "remove all objectionable content from your app", noting that "Parler is not effectively moderating and removing content that encourages illegal activity". Notably, Apple included in its letter a quote from John Matze that "I don't feel responsible ...

and neither should the platform, considering we're a neutral town square that just adheres to the law"; Apple contended instead that "Parler is in fact responsible for all the user-generated content present" on it.

It thus seems that Parler was assessed as a self-governing platform and that its moderation was deemed to have fallen short. Of course, when considering the justifications provided by Apple and Google, it is impossible to ignore the wider context, particularly the other factors noted above. These decisions were handed down immediately after the Capitol riots, and followed criticism of Apple and Google's inaction (in its letter to Parler Apple even cited specific tweets depicting objectionable material). And the two decisions were handed down within hours of each other, again suggesting in its letter to Parler a pressing need to be seen to act in the aftermath of political violence. Overall, this episode marks a significant chapter in app store governance, marking a moment when Apple and Google, facing public pressure, took measures into their own hands by ruling on the inadequacy of another platform's content moderation regime in an (inadvertently or otherwise) concerted fashion. Questions remain, however, as to whether action should have been taken sooner based on the available evidence prior to January 2021; whether Parler's reinstatement by Apple following ongoing dialogue between the two companies was warranted; and whether clearer guidance ought to be provided for what makes a platform a platform for the purpose of judging the suitability of an app entering an app store. We discuss the implications of these unanswered questions in the conclusion.

3.3   Contact Tracing for Covid-19

In spring 2020, when the novel coronavirus Covid-19 first began to spread rapidly across the world infecting millions and, sadly, killing hundreds of thousands, Governments and healthcare providers searched for public health measures they could rely on to slow the spread of the disease. Such measures included lockdowns, social distancing and, most relevant here, contact tracing i.e., identifying all those who have come into close contact with an infected individual and requiring them to self-isolate to avoid further spread of the virus. Traditionally, during public health emergencies, such as the Ebola epidemic, contact tracing has been conducted in a manual fashion; however, given the very high numbers of COVID-19 cases, it quickly became clear that a manual approach would be both implausible and ineffective. Instead, many governments—including the UK Government and the NHS—turned to digital solutions and 'contact tracing apps' became the preferred 'solution.'

Although variable in design, the basic premise behind all contact tracing apps is to record: (a) the infection status of an individual; and (b) their movements to identify when they had either encountered an infectious individual and so potentially become infected or when they may have infected another individual. When an individual is identified as having been a 'close contact,' the app in question should then notify them and inform them of their need to self-isolate. Given the sensitivity of the data involved, and the implications of being 'notified', the development of these apps raised significant ethical concerns related to privacy and autonomy (Morley et al., 2020a, b). In particular, a heated debate revolved around whether or not location data should be collected and, if so, whether it should be stored in a centralised or decentralised fashion.

In a bid to help resolve some of this tension, and to help governments in the midst of a crisis, in May 2020 Google and Apple collaborated to produce a privacy-preserving and interoperable API that contact tracing apps could use to trace potentially infected individuals, whilst ensuring the details of the individuals involved were only ever held locally on their own devices. Later this was developed into a Bluetooth-based contact tracing platform, which was made available for public use alongside full documentation detailing its Bluetooth and cryptography specifications. The NHS—specifically NHSX (the NHS body which, at the time, was responsible for digital, data and technology policy)—initially eschewed this option, in favour of developing their own 'in-house' NHS contact tracing app which would store pseudonymised data centrally. In part this was a political move; patients and the public have traditionally been very wary of private providers, including Apple and Google, having access to any data deemed to be medical and trust in the NHS has, in the past, been badly damaged by partnering with Google in the development of health apps (Morley et al., 2019), and so it seems likely that the NHS—and the Secretary of State for Health and Social Care—were keen to avoid this kind of association. It was also partly an issue of public health surveillance, with the NHS arguing that it needed access to a central repository of the data to effectively monitor the developing situation. Even if these arguments were reasonable, this plan was met with myriad technical difficulties and significant public pushback leading to multiple delays. Eventually, this pushed the NHS to launch the NHS Contact Tracing App using the tech giants' API—known as the Exposure Notification System—in Autumn 2020. This peaceful interlude was not, however, to last long.

In April 2021, the NHS planned to release an updated version of the app—to coincide with the easing of lockdown restrictions—which would collect and centralise data about users' exact locations via the 'check-in' feature of the app. This meant that the app would be collecting far more granular location data than before, and would not be relying on Bluetooth to do so. The intention was to identify COVID-19 hotspots or 'super-spreader' locations so that individuals who had simply been in a venue that later became identified as a COVID-19 hotspot could be notified and advised to isolate even *if* there was no Bluetooth data confirming that they were a close-contact. Despite the Government's conviction that this update would pose no problems, it was found to be in breach of the tech firms' Exposure Notification System guidelines which explicitly stated that apps 'must not share location data from the user's device with the public health authority, Apple, or Google.' For this reason, Google and Apple blocked the update and instead kept the old version live. In doing so, the duopolists demonstrated just how much of an influence they were (and remain) able to exert over elected Governments and—by extension—public life (Sharon, 2020).

3.4 "Smart Voting"

The *Умное голосование* (Smart Voting) app was created by a coalition opposed to the Putin regime in Russia, headed by Alexei Navalny. The app was designed to strengthen opposition to Putin's United Russia party at the ballot box, by coalescing support around specific alternative candidates. Google and Apple removed Smart Voting from their Russian app stores shortly before the 2021 elections to the lower house of the Russian national assembly, or Duma. The decisions followed what was reported as considerable pressure on both companies by the Russian state, which included the possibility of fines, a visit to Google's headquarters by Russian police, and even the threat of criminal prosecution of company executives. In related crackdowns, Russia had days earlier fined Facebook and Twitter for not deleting from their platforms content deemed illegal; Google also blocked access to a list of opposition voting endorsements Google Docs and complied with demands to remove several YouTube videos by Navalny. Meanwhile, popular messaging app Telegram was also forced to give ground by banning Smart Voting chatbots on its platform; notably its Russian-born CEO Pavel Durov cited Apple and Google's decision as justification for that of Telegram. Google and Apple's decisions were criticised by among others David Kaye, the former UN Special Rapporteur on Freedom of Expression, who described the move on Twitter as "clearly unlawful under international human rights law".

There are superficial similarities to the Parler case discussed in Sect. 3.2. There as here, Apple and Google acted in a concerted way to remove a guideline-violating app in the midst of an event in the outside world, on the basis of the impact of the app on that event. But whereas the pressure to suspend Parler from app stores arose from the media and public, demands to remove Smart Voting were driven by the Russian state with respect to breaches of Russian law. In explaining its decision, Apple notably pointed to a different section of its App Store guidelines than those used in decisions discussed above. Quoting the Legal section of its guidelines, it noted that "Apps must comply with all legal requirements in any location where you make them available" and cited the determination by the Russian Prosecutor's Office that the app constituted illegal election interference. The spuriousness or opportunism of the Russian state's claim notwithstanding, in a technical sense it provided the legal basis on which Apple and presumably Google justified their removal of the app.

This distinction between this and the Parler case is further evidence of the flexibility that Apple and Google have in interpreting app store guidelines to serve corporate purposes. In the case of Parler and of Smart Voting, the two companies would presumably have rather not found themselves in the position of being forced to act, and in the case of Smart Voting in particular, there is clear evidence that they did so under at least some degree of legal and even physical duress. But the two cases nonetheless highlight the utility of different aspects of app store guidelines when enforcing (politically uncomfortable) decisions, ranging from breaches of policies on Objectionable Content in one case to Legal violations in the other. This interpretive flexibility in turn invites further questions about the transparency, accountability and consistency of app store governance decisions. Whereas in the case of Parler, a reasonable question both companies may face was why the platform was permitted for so long prior to the riots at the Capitol, the Smart Voting case raises difficult questions about the complicity of app store governors in and/or their acquiescence to illiberal regimes. In both cases, Apple and Google might simply reply that they acted when further information became available: in the case of Parler, when its role in the Capitol riots became clear; and in the case of Smart Voting, when information about its illegality was presented to them. What each retort elides, though, is the arbitrariness of the process in place. In its explanation to Parler, Apple cited tweets highlighting objectionable material gathered by citizens (and publicised on a third platform, Twitter, no less), which would seem to undercut the care with which the company claims to review apps and updates thereof. In its explanation regarding Smart Voting, meanwhile, Apple parroted claims about the illegality of the app without acknowledging the arbitrariness of the Russian state's own interpretation and timing.

Similar challenges have arisen for various large platforms operating in hostile jurisdictions, such as TikTok's travails in India or Twitter's in Nigeria. But what makes the cases we discuss here distinct is that as app store governors, Apple and Google operate as gatekeepers between publics and the particular apps under scrutiny. The notion that app store guidelines are supposed to apply equally to all apps seems undercut by both outside-world events and the increased pressure that these events put on Apple and Google, rendering app store governance, in this case, entirely subservient to the whims of the Russian state.

4 Conclusions

The four episodes recounted above hold several implications for both the governance of app stores by Google and Apple, and the governance of Google and Apple by states. We close by identifying three of the most important.

First, each case involves different domains—public health and politics—as well as different jurisdictions. However, the points of convergence and of divergence between the episodes surface notable implications. The common factor across all four cases is, of course, the duopolistic dominance of Apple and Google. At various moments, the two tech giants have acted in concert, as with their collaboration on a privacy-protecting API for Covid-19 contact tracing, and in their concerted decision to remove the Smart Voting app as well as, initially, Parler. Though, at other times, they have diverged, as with the different approaches taken towards health apps like Babylon, and Apple's decision to reinstate Parler, these points of distinction are more fine-grained; the two companies diverge specifically on guidelines for health apps, but their guidelines in general are more similar, and Google's decision not to reinstate Parler may reflect the fact that Android users can still circumvent the Play Store and side-load Parler to their devices. On key decisions, when the social and political stakes are at their highest, Apple and Google's concerted decision-making underscores the power of their duopolistic domination over app store governance.

Second, while these companies seem inclined where possible to act in concert, this does not always amount to a decision to act with resolve. Whilst in the case of Covid-19 contact tracing, the two companies effectively forced a privacy-preserving API on national governments including the UK, in the case of the Smart Voting app, both companies relented under the face of Russian government pressure. These contrasting decisions were undoubtedly influenced by the different circumstances in each case; Apple and Google held sufficient control over their respective smartphone software to force governments into a particular model for contact tracing, while Russian government threats to fine the companies and even prosecute executives presumably influenced their acquiescence. This evident realpolitique holds lessons for policymakers in liberal democracies looking to regulate app store governors more forcefully, suggesting that fines or prosecution—as was floated in early discussions over the Online Harms Bill in the UK—may be an effective way to bend app stores to their will.

Finally, the cases raise questions about what is meant by a platform for the purposes of platform governance. Parler is an example of a platform that sits somewhere between a fringe, narrow-minded ideological community and a self-proclaimed "digital town square" open to all. In practice, Parler is closer to the former than the latter, but nonetheless occupies a grey area between inclusive and all-encompassing (though far from harmonious) social media platforms like Facebook and Twitter, and extremist-dominated, ideologically homophilic sites like Gab. In terms of app store governance, this has implications with respect to which app store review guidelines are used to judge the (in)appropriateness of the app in question: whether to govern apps such as Parler primarily on the basis of its user-generated content (and thus its ability to moderate this content), or in terms of the risk of physical harm of the platform as a whole, may affect whether it is permitted. Likewise, as we have argued, Babylon also sits in a grey area as a private company providing aspects of a public service, with different governance implications depending on how it is seen. Apps like these and others that occupy various grey areas in terms of their affordances and uses—whether as a quasi-"public sphere", as a quasi-public health provider, or on other dimensions—highlight the limitations of app store guidelines and the need for improved understanding about how classifications such as these are accorded and resulting review decisions made.

In this chapter, we have sought to identify and explore such implications and limitations rather than provide firm theoretical, practical, or legal responses—but each deserves further attention from scholars as a basis for formulating effective, ethical and responsible app store governance.

References

Apple. (2008, December 15). *IPhone to support third-party Web 2.0 applications*. https://web.archive.org/web/20081215230338/; http://www.apple.com/pr/library/2007/06/11iphone.html

Apple. (2021). *Building a trusted ecosystem for millions of apps: A threat analysis of sideloading*. https://www.apple.com/privacy/docs/Building_a_Trusted_Ecosystem_for_Millions_of_Apps_A_Threat_Analysis_of_Sideloading.pdf

Baker, R. (2013). *Before bioethics: A history of American medical ethics from the colonial period to the bioethics revolution.* Oxford University Press. https://doi.org/10.1093/acprof: oso/9780199774111.001.0001

Bloch-Wehba, H. (2019). Global platform governance: Private power in the shadow of the state. *SMU Law Review, 72*(1), 27.

Brodkin, J. (2018, July 18). *EU: Google illegally used Android to dominate search, must pay $5B fine.* Ars Technica. https://arstechnica.com/tech-policy/2018/07/ eu-google-illegally-used-android-to-dominate-search-must-pay-5b-fine/

Bronstein, C. (2021). Deplatforming sexual speech in the age of FOSTA/SESTA. *Porn Studies, 8*(4), 367–380. https://doi.org/10.1080/23268743.2021.1993972

Cath, C. (2021). The technology we choose to create: Human rights advocacy in the Internet Engineering Task Force. *Telecommunications Policy, 45*(6), 102144. https://doi.org/10.1016/j. telpol.2021.102144

DeNardis, L. (2012). Hidden levers of internet control. *Information, Communication & Society, 15*(5), 720–738. https://doi.org/10.1080/1369118X.2012.659199

DeNardis, L. (2014). *The global war for internet governance.* Yale University Press.

Donovan, J. (2019). *Navigating the tech stack: When, where and how should we moderate content?* Centre for International Governance Innovation. https://www.cigionline.org/articles/ navigating-tech-stack-when-where-and-how-should-we-moderate-content/

Douek, E. (2019). Facebook's 'oversight board:' move fast with stable infrastructure and humility. *North Carolina Journal of Law & Technology, 21*(1), 1.

European Commission. (2018, July 18). *Antitrust: Commission fines Google €4.34 billion for abuse of dominance regarding Android devices* [Text]. European Commission. https://ec.europa.eu/ commission/presscorner/detail/en/IP_18_4581

Fraser, H., Coiera, E., & Wong, D. (2018). Safety of patient-facing digital symptom checkers. *The Lancet, 392*(10161), 2263–2264. https://doi.org/10.1016/S0140-6736(18)32819-8

Gilbert, S., Mehl, A., Baluch, A., Cawley, C., Challiner, J., Fraser, H., Millen, E., Montazeri, M., Multmeier, J., Pick, F., Richter, C., Türk, E., Upadhyay, S., Virani, V., Vona, N., Wicks, P., & Novorol, C. (2020). How accurate are digital symptom assessment apps for suggesting conditions and urgency advice? A clinical vignettes comparison to GPs. *BMJ Open, 10*(12), e040269. https://doi.org/10.1136/bmjopen-2020-040269

Gillespie, T. (2018). *Custodians of the Internet: Platforms, content moderation, and the hidden decisions that shape social media.* Yale University Press.

Gorwa, R. (2019). What is platform governance? *Information, Communication & Society, 22*(6), 854–871. https://doi.org/10.1080/1369118X.2019.1573914

Gorwa, R., Binns, R., & Katzenbach, C. (2020). Algorithmic content moderation: Technical and political challenges in the automation of platform governance. *Big Data & Society, 7*(1), 2053951719897945. https://doi.org/10.1177/2053951719897945

Hollister, S. (2021, April 21). *Apple's $64 billion-a-year App Store isn't catching the most egregious scams.* The Verge. https://www.theverge.com/2021/4/21/22385859/ apple-app-store-scams-fraud-review-enforcement-top-grossing-kosta-eleftheriou

Jhaver, S., Birman, I., Gilbert, E., & Bruckman, A. (2019). Human-machine collaboration for content regulation: The case of reddit automoderator. *ACM Transactions on Computer-Human Interaction (TOCHI), 26*(5), 1–35.

Katzenbach, C., & Ulbricht, L. (2019). Algorithmic governance. *Internet Policy Review, 8*(4). https://doi.org/10.14763/2019.4.1424

Kelly, M. (2019, August 4). *Cloudflare to revoke 8chan's service, opening the fringe website up for DDoS attacks.* The Verge. https://www.theverge.com/2019/8/4/20754310/ cloudflare-8chan-fredrick-brennan-ddos-attack

Klonick, K. (2017). The new governors: The people, rules, and processes governing online speech. *Harvard Law Review, 131*, 1598.

Lin, J. (2020, November 25). *How to make $80,000 per month on the apple app store.* Transparency Matters. https://blog.lockdownprivacy.com/2020/11/25/how-to-make-80000.html

Matias, J. N. (2019). The civic labor of volunteer moderators Online. *Social Media + Society, 5*(2), 2056305119836778. https://doi.org/10.1177/2056305119836778

Morley, J., Taddeo, M., & Floridi, L. (2019). Google health and the NHS: Overcoming the trust deficit. *The Lancet Digital Health, 1*(8), e389. https://doi.org/10.1016/S2589-7500(19)30193-1

Morley, J., Cowls, J., Taddeo, M., & Floridi, L. (2020a). Ethical guidelines for COVID-19 tracing apps. *Nature, 582*(7810), 29–31. https://doi.org/10.1038/d41586-020-01578-0

Morley, J., Machado, C. C. V., Burr, C., Cowls, J., Joshi, I., Taddeo, M., & Floridi, L. (2020b). The ethics of AI in health care: A mapping review. *Social Science & Medicine (1982), 260*, 113172. https://doi.org/10.1016/j.socscimed.2020.113172

Musiani, F., Cogburn, D. L., DeNardis, L., & Levinson, N. S. (2016). *The Turn to Infrastructure in Internet Governance*. Springer.

Nickel, P. J. (2020). Disruptive innovation and moral uncertainty. *NanoEthics, 14*(3), 259–269.. Scopus. https://doi.org/10.1007/s11569-020-00375-3

Open Handset Alliance. (2007, July 5). *Industry leaders announce open platform for mobile devices | Open Handset Alliance*. http://www.openhandsetalliance.com/press_110507.html

Roberts, S. T. (2019). *Behind the screen: Content moderation in the shadows of social media*. Yale University Press.

Romano, A. (2017, August 14). *GoDaddy and Google have refused service to a notorious neo-Nazi site*. Vox. https://www.vox.com/policy-and-politics/2017/8/14/16143820/godaddy-and-google-wont-host-daily-stormer-domain

Schulz, W. (2018). Regulating intermediaries to protect privacy online–the case of the German NetzDG. *Personality and Data Protection Rights on the Internet, Forthcoming*.

Sharon, T. (2020). Blind-sided by privacy? Digital contact tracing, the Apple/Google API and big tech's newfound role as global health policy makers. *Ethics and Information Technology*. https://doi.org/10.1007/s10676-020-09547-x

Siegler, M. (2010, April 10). Steve Jobs Reiterates: 'Folks who want porn can buy an Android phone'. *TechCrunch*. https://social.techcrunch.com/2010/04/19/steve-jobs-android-porn/

Taylor, L. (2021). Public actors without public values: legitimacy, domination and the regulation of the technology sector. *Philosophy & Technology*. https://doi.org/10.1007/s13347-020-00441-4

van Dijck, J., Poell, T., & De Waal, M. (2018). *The platform society: Public values in a connective world*. Oxford University Press.

Ziewitz, M., & Pentzold, C. (2014). In search of internet governance: Performing order in digitally networked environments. *New Media & Society, 16*(2), 306–322.

A Legal Principles-Based Framework for AI Liability Regulation

Massimo Durante and Luciano Floridi

Abstract Europe has recently taken the path of regulating artificial intelligence (AI). This is a complex task, in which it is crucial to understand what the purposes of regulation are. In this perspective, it is not enough to identify and set ethical guidelines and legal norms. It is also important to envisage the legal principles that might steer the regulation of AI, which is aimed to reconcile technological innovation, economic development and user trust. Therefore, it may be useful to consider whether some principles may emerge from existing legislation in the AI sector. To this aim, we review the work of the *Expert Group on Product Liability in the field of AI and Emerging Technologies* (2019) as a case study. We show how their work has started to lay the basis for a set of legal principles for AI liability regime. An initial and open list of legal principles can serve as a benchmark for future work on a principles-based AI regulation.

Keywords Artificial intelligence · Governance · Legal principles · Regulation · Trust

1 Introduction

One of the main challenges in regulating artificial intelligence (AI) is finding the right balance between

M. Durante (✉)
Department of Law, University of Turin, Turin, Italy
e-mail: massimo.durante@unito.it

L. Floridi
Oxford Internet Institute, University of Oxford, Oxford, UK

Department of Legal Studies, University of Bologna, Bologna, Italy

(a) promoting technological innovation, which includes economic investment and development, and
(b) fostering users trust, which includes citizens and consumers trust.

This balancing act is clear in many of the recent regulatory proposals from the European Union (EU) relating to AI – including the White Paper on AI, 2020a, the Data Governance Act, 2020b, and the AI Act 2021 – where the relationship between (a) and (b) also refers to *business* and *individuals* as the agents involved in (a) and (b) respectivley. Philipp Hacker (2020, III), for example, has remarked that:

> The latter [the White Paper on AI] is characterized by a clear dichotomy of promotion and regulation. The novel labels for these different perspectives are those of the 'ecosystem of excellence' (promotion of AI) and the 'ecosystem of trust' (regulation of AI). Overall, the White Paper attempts a difficult, but generally worthwhile, balancing act of making Europe a center of AI development and application, while at same time adequately addressing the risks of this technology and ensuring that European fundamental rights and values are adequately enforced.

The balancing act is challenging because, in (a), technological and economic innovation is usually supported by a liberal approach to regulation, in terms of deregulation, self-regulation or co-regulation. However, the need to foster users trust, as indicated in (b), makes implementing (a) more difficult because users trust is required not only for the proper functioning of the digital ecosystem – which is made up of interactions among users and with the widespread diffusion of digital services and products in users' everyday lives – but also for the acquisition of data, on which the digital ecosystem feeds, and this seems to push in the opposite direction from a liberal approach to regulation. As a result, the need for AI regulation is currently and increasingly introduced through reference to the idea of trust or "trustworthy AI" (an expression that has been the subject of an autonomous communication by the High-Level Expert Group on AI, 2019) through mechanisms of *distrust*,[1] that is, through regulatory instruments set up to protect users' rights and prerogatives. In the end, the twofold challenge is to promote technological innovation also by increasing (or at least maintaining) users trust, without relying solely, or primarily, on traditional forms of top-down regulation based on overly prescriptive rules, strict forms of liability, and sanctions, which tend to slow down and discourage technological innovation.

A traditional response to this challenge is to turn to principles-based regulation. This has led to a proliferation of *ethical* AI principles (Floridi & Cowls, 2019; Morley et al., 2020). In this chapter, we argue that it would be useful to envisage the possible design of *legal* principles for AI regulation to deal with the challenge described above.

The chapter is structured as follows. In section two, we outline our methodology. In section three, we address the distinction between ethical and legal principles and discuss the need for legal principles for AI regulation, stressing the main advantages

---

[1] The interplay between trust and distrust is referred to and largely analysed in the literature on trust. See, for instance, Feldman (2018).

of principles-based regulation. In section four, we review the work of the *Expert Group on Product Liability in the field of AI and Emerging Technologies* (2019) (henceforth the Expert Group), as a case study, and show how their work has started to lay the basis for a set of legal principles for AI liability regime. In section five, we propose an initial and open list of legal principles for AI liability, which can serve as a benchmark for future work on principles-based AI regulation. In section six, we examine whether and how a (legal, not just ethical) principled-based AI regulation may be a useful framework for innovation and trust. In section seven, we formulate recommendations for AI policy regulation. We conclude in section eight by summarising the the outcome of the chapter.

2 Methodology

The objective of this chapter is to show the usefulness of an approach to the regulation of AI that is based on legal principles. The fundamental role of legal principles is to make one *see* (understand and evaluate) concisely and more clearly how and why a field of law is structured. Therefore, the chapter has both a normative and an epistemological dimension. Users can trust AI only insofar as they can *see* how and why an AI ecosystem is regulated. They need a clearer understanding of how the technology shapes their lives and how policies and regulatory approaches to AI are consistent with purposes that form an integral part of their everyday forms of life (Durante, 2021). Users trust is based on the perception of shared regulatory goals. In turn, by building users' trust, the principles-based approach to AI regulation enables the creation of an ecosystem favorable to technological development and economic investment.

It is important to note that we are not alone in stressing the need for legal principles for AI.[2] Their formulation can follow a top-down, middle-out, or bottom-up approach. As an example of a more top-down approach, Philip Hacker has recently proposed (2020a) adapting the Lamfalussy process[3] to AI regulation. Ugo Pagallo (Pagallo et al., 2019) has proposed a middle-out approach to AI regulation that combines horizontal and vertical elements. In this chapter, we intend to limit our analysis to the consideration of a case study, examined in section four, to explore the

---

[2] In this regard, it is worth noting the work of the Ad hoc Committee on Artificial Intelligence (Towards AI Regulation, 2021), which assesses the impact of AI on ethical principles, human rights, the rule of law and democracy, and examines key values, rights and principles deriving – in a bottom-up perspective – from sectorial approaches and ethical guidelines; and – in a top-down perspective – from human rights, democracy and the rule of law requirements. It is also worth mentioning the European Declaration on Digital Rights and Principles for the Digital Decade (2022). See also Hacker (2020); Bernitz et al. (2020), notably, Chap. 18; Fjeld et al. (2020). On constitutional and human rights considerations on AI see recently Barfield – Pagallo (2020).

[3] An approach to the development of EU financial services regulation https://ec.europa.eu/info/business-economy-euro/banking-and-finance/regulatory-process-financial-services/regulatory-process-financial-services_en

suitability of a more bottom-up approach. Our reflection on principles-based regulation starts from the Expert Group analysis of an existing sectoral legislation on
product liability in the field of AI. The attempt is to ascertain to what extent the
Expert Group findings can be generalized to elaborate principles for AI liability.
And if so, to what extent these principles can enable us to develop regulatory meta-
principles that can help envisage and design some policy recommendations for AI
regulation. In any case, we shall assume that no regulation of AI is created from
scratch but is always subject to two general constraints: existing legislation, to be
revised and modified, where gaps or shortfalls are shown; and constitutional laws
and principles, particularly fundamental rights and the rule of law, which are essential foundations of democratic life (CAHAI, 2020a, b).

3   Ethical and Legal Principles

A great deal of theoretical work has recently been done on the formulation of ethical
principles for AI. From this there emerges considerable convergence around an ethical approach to AI,[4] even if the picture is not as homogeneous and unproblematic as
it might otherwise appear at first glance. The main challenge concerns the *implementation* of ethical principles, i.e. their translation into mandatory regulation or
practices (Morley et al., 2020). As remarked by (Jobin et al., 2019, 396):

> Although numerical data indicate an emerging convergence around the importance of cer-
> tain ethical principles, an in-depth thematic analysis paints a more complicated picture. Our
> focused coding reveals substantive divergences among all 11 ethical principles in relation
> to four major factors: (1) how ethical principles are interpreted; (2) why they are deemed
> important; (3) what issue, domain or actors they pertain to; and (4) how they should be
> implemented. These conceptual and procedural divergences reveal uncertainty as to which
> ethical principles should be prioritized and how conflicts between ethical principles should
> be resolved, and it may undermine attempts to develop a global agenda for ethical AI. […]
> Such divergences and tensions illustrate a gap at cross-section of principles formulation and
> their implementation into practice.

Even more significantly, Jobin states that:

> These findings have implications for public policy, technology governance and research
> ethics. At the policy level, greater inter-stakeholder cooperation is needed to mutually align
> different AI ethics agendas and to seek procedural convergence not only on ethical princi-
> ples but also their implementation. […] *Furthermore, it should be clarified how AI ethics
> guidelines relate to existing national and international regulation. Translating principles
> into practice and seeking harmonization between AI ethics codes (soft law) and legislation
> (hard law) are important next steps for the global community.* (Jobin et al., 2019, 396, our
> italics)

---

[4] Fjeld et al. (2020); Hagendorff (2020); Jobin et al. (2019).

Let us briefly consider the role of ethical principles, a fuller discussion of which lies outside the scope of the present chapter. It is possible to affirm that, in the context of AI, ethical principles play a crucial role in at least three ways:

(a) epistemological: they reveal the main issues concerning the development of AI which are likely to produce deleterious consequences for individuals or society;
(b) axiological: they designate the values that must be protected, pursued and implemented in the development of AI;
(c) normative: they set forth (as guidelines or soft law) the constraints that the development of AI should satisfy.

Legal principles differ from ethical principles in that they are *meta-regulatory*, in the following sense. Ethical principles concern the purposes of AI, namely the impact of AI on individuals, society, and the environment. Legal principles concern the purposes of AI *regulation*, namely the impact of AI regulation on individuals, society and the environment. Any regulation will have an impact on society, and that impact should be subject to specific analysis concerning its *side effects*. Furthermore, in the current complex framework of digital governance,[5] legal principles affect (AI regulation that concerns) the behaviour not only of individuals, courts of law, officials, and public authorities, but also of private corporations.

There is a large literature on the role of principles in law, and the difference between legal norms and principles (Hart, 1961; Dworkin, 1967; Raz, 1972). Here, suffice it to remark that legal principles are legal standards (and not merely extra-legal factors), which play three essential roles in:

1. guiding behavior in the absence of existing law;
2. interpreting and applying existing law; and
3. creating new law.

Note that, strictly speaking, (2) and (3) are meta-regulatory (about existing or future law), but (1) is not, because it is only about the *absence* of law. With this distinction in the background, let us briefly comment on each role.

In (1), legal principles allow people to adopt measures or implement practices that are not identified in specific terms, but must comply with the general requirements of the principles. In business contexts, the responsibility for applying legal principles typically falls on senior management of companies. Examples of this sort of principles-based regulation can be found in data protection regulation, which sets out high-level data protection principles such as 'accountability', 'purpose limitation' or 'accuracy', and financial regulation, which sets out high-level principles for the conduct of financial companies, such as the requirement to 'treat customers fairly'.

---

[5] See in this respect Mazzini (2019), who stated significantly (note 7, p. 3): "Another very relevant and perhaps more complicated question to address for policymakers in the context of AI governance is the future oriented question of identifying the direction we want to move towards as a society." In this regard, see also crucially Floridi (2020). In the field of digital technologies, as has been remarked, "the 'race to AI' is also bringing forth a 'race to AI regulation'" (Smuha, 2019, 4).

In (2), legal principles help to interpret and apply existing law by elucidating the rationale behind the law. This (a) facilitates the derivation of legal norms, in order to fill regulatory gaps that emerge due to rapid socio-technological change; (b) allows for a fair balancing of hard cases; and (c) creates a more harmonized regulatory framework. From this perspective, legal principles emerge from the overall study of existing law, from case-based analysis, and from generalization of solutions offered in legal disputes. And in this sense, legal principles are closely related to the notion of law as integrity (Dworkin, 1986), which aims to coordinate the interpretation and application of the law in a consistent fashion.

In (3), legal principles contribute to the creation of new law. They highlight whether a given area requires a revision of existing law or the development of more detailed legislation due, for example, to the impossibility of integrating the law by means of analogy or due to the power or information asymmetry that exists among the actors operating in this area. This crucial role is linked to the idea that regulation must evolve with society and ensure that imbalances that affect the society are are resolved into acceptable equilibria.

Because of the previous features, the potential benefits claimed for legal principles are that "they provide flexibility, are more likely to produce behaviour which fulfils the regulatory objectives, and are easier to comply with". (Black et al., 2007, 193)

These normative elements are relevant, but there is another even more important reason for adopting a regulation based on legal principles. To cite Black once again:

> Principles can provide a *basis for open dialogue* between regulator and regulated firm, facilitating a co-operative and educative approach to supervision, particularly with respect to firms who are well intentioned, but either ill informed, or simply confused as to what the regulatory provisions require. (2007, 195)

This open dialogue appears essential today, but it must find an institutional context. Legal principles can define the basic rules of the game, build a level playing field, and display what is at stake, because their elaboration and application provides a basis for a critical, open and institutional dialogue. Legal principles often emerge through time from a large base of legal experience. Once they are formulated the show more clearly, to the actors involved, a general and unifying guidance that was previously hard to perceive. In this sense, legal principles also play a crucial epistemological role, since they show the direction for steering and regulating the digital ecosystem and constitute an institutional forum for an open and critical discussion of AI regulation. At the core of legal principles lies also perspective, that is, a choice of a Level of Abstraction (Floridi, 2008), not only a prescription, that is, a decision about a preferred state of the world.

We have now completed outlining the nature and importance of legal principles in relation to AI legislation. The time has come to turn to a concrete case study to see how general principles can be formulated.

4 A Case Study: How to Derive Legal Principles for AI Regulation

At the explicit behest of the EU Commission, the *Expert Group on Product Liability in the field of AI and Emerging Technologies* (2019) worked precisely within the framework of innovation and trust that we have analysed in the previous two sections:

> The experts were asked to examine whether the current liability regimes are still 'adequate to facilitate the uptake of … new technologies by fostering investment stability and users' trust'. (henceforth: Report, 2019, 13)

The Expert Group started by reviewing existing liability regimes under private law at the national and European levels (for a critical assessment of the Expert Group work, which exceeds the scope of our paper, see Bertolini & Episcopo, 2021). From a methodological standpoint, it is important to assess the state of the art and to establish which legal issues can be dealt with and resolved on the basis of the existing law, and which instead require adjustments to fill gaps in the law. Their initial assessment was that:

> the liability regimes in force in the Member States ensure at least basic protection of victims whose damage is caused by the operation of such new technologies. However, the specific characteristics of these technologies and their applications – including complexity, modification through updates or self-learning during operations, limited predictability, and vulnerability to cybersecurity threats – may make it more difficult to offer these victims a claim for compensation in all cases where this seems justified. It may also be the case that the allocation of liability is unfair or inefficient. To rectify this, certain adjustments need to be made to EU and national liability regimes. (Report 2019, 3)

The Expert Group also laid out ten key findings relating to how liability regimes should be designed – and, where necessary, changed – in order to rise to the challenges emerging digital technologies bring with them. (Expert Group, 2019, 3).

Let us take a closer look at these findings, which are the result of a detailed analysis and evaluation of the existing rules (Expert Group, 2019, 3–4):

1. *"A person operating a permissible technology that nevertheless carries an increased risk of harm to others, for example AI-driven robots in public spaces, should be subject to strict liability for damage resulting from its operation."*
2. *"In situations where a service provider ensuring the necessary technical framework has a higher degree of control than the owner or user of an actual product or service equipped with AI, this should be taken into account in determining who primarily operates the technology."*
3. *"A person using a technology that does not pose an increased risk of harm to others should still be required to abide by duties to properly select, operate, monitor and maintain the technology in use and – failing that – should be liable for breach of such duties if at fault."*

4. *"A person using a technology which has a certain degree of autonomy should not be less accountable for ensuing harm than if said harm had been caused by a human auxiliary."*

5. *"Manufacturers of products or digital content incorporating emerging digital technology should be liable for damage caused by defects in their products, even if the defect was caused by changes made to the product under the producer's control after it had been placed on the market."*

6. *"For situations exposing third parties to an increased risk of harm, compulsory liability insurance could give victims better access to compensation and protect potential tortfeasors against the risk of liability."*

7. *"Where a particular technology increases the difficulties of proving the existence of an element of liability beyond what can be reasonably expected, victims should be entitled to facilitation of proof."*

8. *"Emerging digital technologies should come with logging features, where appropriate in the circumstances, and failure to log, or to provide with reasonable access to logged data, should result in the reversal of the burden of proof in order not to be to the detriment of the victim."*

9. *"The destruction of the victim's data should be regarded as damage, compensable under specific conditions."*

10. *"It is not necessary to give devices or autonomous systems legal personality, as the harm these may cause can and should be attributable to existing persons or bodies."*

As the reader may see, the ten findings just listed can be easily generalised and turned into more general legal principles. We shall do so and discuss them below (see Table 1 for a summary). The ten principles draw upon, but are distinct from, the ten key findings of the Expert Group listed above. Further work should be done to assess the extent to which these principles could apply to other areas of law; that is, the extent to which these principles could converge with other principles deduced from the analysis of other areas of law involved in the regulation of AI. Here, we wish to limit ourselves to showing how the principles can be derived bottom-up from the findings, and the extent to which, once derived and formulated, the principles could be exported to other contexts.

Table 1 The ten legal principles of AI regulation

1. *The level of risk as the attribution of liability regime*
2. *The principle of accountability*
3. *The principle of best information position*
4. *The principle of control as a norm of attribution*
5. *The principle of traceability*
6. *The principle of no advantage from risk*
7. *The principle of functional equivalence*
8. *The principle of equal protection by the law*
9. *The principle of legal unification*
10. *The principle of market placement risk*

In particular, it seems evident that the AI Act satisfies all the principles as formulated below, thus showing a high degree of consistency across different EU regulations, at least when it comes to fundamental, guiding lines of formulation of legislation regulating digital technologies and services (Floridi, 2021). Finally, all the principles below have been based on, and derived from the Expert Group findings, by making the latter more abstract and general, with the only exception of principle 3, which distills a line of reasoning permeating the entire work of the Expert Group, and principle 4, which is based on findings 3 and 4. Further work could be done to verify the extent to which these principles could converge with other principles derived from the analysis of other areas of law involved in the regulation of AI sectors.

1. *The level of risk as the attribution of liability regime*

The diversity of AI systems implies a diverse range of risks, for which it is impossible to devise a single solution. Hence, different levels of risk should be addressed by different liability regimes. This yields *a legal principle for AI regulation according to which the liability regime is set according to the level of risk.* According to this principle, AI systems that pose a high level of risk should be subject to strict liability, while technologies that do not pose a high level of a risk should be subject to fault-based liability. The level of risk – which depends on the nature, scope, context and purposes of the activity, and the rights and freedoms of natural persons involved – is to be preliminary assessed by the operator (defined bt the Expert Group as "the person who is in control of the risk connected with the operation of emerging digital technologies and who benefits from their operation") of the technology according to the principle of accountability, analogously to article 35 of the GDPR, which requires a data protection impact assessment to be carried out in cases of increased risk.

2. *The principle of accountability*

The diversity of AI systems and the rapid pace of technological change tend to discourage the adoption of overly prescriptive rules for AI regulation. Instead, the regulatory framework should establish high-level outcomes that AI systems must deliver, and the system operator should be responsible for establishing the specific measures that are needed for delivering those outcomes, and complying with AI regulation (an outcome-based approach already applies to different sectors, as underlied, for instance, by the UK approach to financial regulation [Black et al., 2007]). This establishes *a legal principle of accountability for AI regulation according to which the operator is called upon to account for, and to be able to demonstrate how, risks inherent to AI systems are envisaged, monitored and managed.* This principle is meant to lie at the heart of AI regulation since, although it does not provide for hard rules, it embodies the spirit of the regulatory framework as the natural complement to the risk-based approach. According to this principle, operators are responsible for complying with AI regulation (i.e., safety and security rules) and for demonstrating compliance. By way of analogy, one can think of the different liability regimes for those who have to set up a given number of fire extinguishers in a building in relation to its size: they are responsible for non-compliance with a safety rule if they do not meet the minimum fire extinguishers requirement and for any resulting harm if the lack of fire extinguishers causes or aggravates the harm caused by an eventual fire.

A key rationale underpinning the principle of accountability, and the outcomes-based approach to AI regulation, is also to be found in the intention of the European regulator to hold responsible and empower those who are (or should be) in the best information situation to predict and control the risks deriving from the use of emerging digital technologies i.e. the companies operating AI systems. The principle of accountability also functions as a meta-principle, by requiring the operator to account for and prove the correct enforcement of the other legal principles and rules that govern the implementation of AI systems.

3. *The principle of best information position*

As observed, this principle cannot be immediately derived from the findings of the Expert Group, although it implicitly underlies many of their considerations. There are several possible criteria or policies for allocating legal responsibility. One possible criterion is that the costs of liability (for non-compliance) should be distributed fairly and efficiently so as to support the ecosystem of trust. This could imply the allocation of liability costs to those,

> whose objectionable behaviour caused the damage; or who benefitted from the activity that caused the damage; or who were in control of the risk that materialized; or who were cheapest cost avoiders or cheapest takers of insurances. (Report, 2019, 5)

A different criterion or policy may be to allocate responsibility to those who are (or are meant to be) in the best information position to predict risks or to decide how to design the network of AI systems and to coordinate the interaction between the different components of this network.[6] As many risks arise from the openness, interoperation and unpredictability of AI systems, which are generally data-intensive, *the legal principle of best information position* can prompt, for instance, operators to deal with and reduce the (algorithmic) uncertainty that might be engendered by the lack of "data accuracy" (which is also a legal principle defined in article 5(1)(f) of the GDPR), where their choices are at the origin of such uncertainty. The legal principle of best information position dovetails with the process- and outcome-based approach to regulation embodied in the principle of accountability, outlined above.

4. *The principle of control as a norm of attribution*

AI systems are typically operated in the context of a multi-agent system, where "the plurality of actors in digital ecosystem makes it increasingly difficult to find out who might be liable for the damage caused". (Report, 2019, 33).

---

[6] See in this regard remarks in the Expert Group Report (2019, 56): "It is also more efficient to hold all potential injurers liable in such cases, as the different providers are in the best position to control risks of interaction and interoperability and to agree upfront on the distribution of the costs of accidents".

This raises the issue of the norms of attribution: namely, how to attribute liability to a specific source of harm. When two or more operators are involved – e.g., a *frontend operator* who primarily decides on, and benefits from, the use of the relevant technology, and a *backend operator* who continuously defines the features of the relevant technology and provides essential and ongoing backend support [Report 2019, 39] – strict liability should lie with the one who has more control[7] over the risks of the operation. This provides (at least for strict liability) *a legal principle of control as a norm of attribution*. This principle implies more generally that strict liability should lie with the person who is in control of the risk connected with the operation of an AI system. Thus, this principle is also coordinated with the first, second, and third legal principle, since one may argue that, in many relevant cases, whoever is in the best information situation to know and foresee the risks involved in the operation of an emerging technology is also the one who has greatest control over the risks themselves.

5. *The principle of traceability*

AI systems and emerging digital technologies should be configured by design in such a way that they leave a record (a *trace*) of their own operations. Legal adjudication is impossible without finding traces that enable one to reconstruct an event as the basis of the attribution of responsibility. This establishes *the principle of traceability*, according to which

> there should be a duty on producers to equip technology with means of recording information about the operation of the technology (logging by design), if such information is typically essential for establishing whether a risk of the technology materialized, and if logging is appropriate and proportionate, taking into account in particular the technical feasibility and the costs of logging, the availability of alternative means of gathering such information, the type and magnitude of the risks posed by the technology, and any adverse implications logging may have on the rights of others. (Report 2019, 47)

This principle identifies a legal standard, the violation of which produces legally relevant consequences in procedural terms. The lack of means for recording and logging information, or the failure to give the victim reasonable access to the information, should entitle the victim to a facilitation of proof, which can result in a reversal of the burden of proof or trigger a

> rebuttable presumption that the condition of liability to be proven by the missing information is fulfilled. (Report, 2019, 47)

This legal principle is closely linked to the next, which has even further reach.

---

[7]Control is a key term for the Expert Group, that has been defined as follows: "'Control' is a variable concept, though, ranging from merely activating the technology, thus exposing third parties to its potential risks, to determining the output or result (…), and may include further steps in between, which affect the details of the operation from start to stop" (Report, 2019, 41).

6. *The principle of no advantage from risk*

It becomes necessary to consider whether there are specific features of AI systems and their functioning that call into question the principle of causality on which the law largely relies.

> "Features of emerging digital technologies, such as opacity, openness, autonomy and limited predictability, may often result in unreasonable difficulties or costs for the victim to establish both what safety an average user is entitled to expect, and the failure to achieve this level of safety. At the same time, it may be significantly easier for the producer to prove relevant facts. This asymmetry justifies the reversal of the burden of proof". (Report, 2019, 43–44)

In more general terms, where the victim is in a weaker position to establish causation, due to the specific features of the AI system (which include the degree of ex-post traceability and intelligibility of processes within the system and the degree of ex-post accessibility and comprehensibility of data collected and generated by the system), the burden of proof can be alleviated (as opposed to the general principle for which *onus probandi incumbit ei qui dicit*, i.e., the burden of proof rests with the claimant), to the point of being reversed.

These considerations bring out a legal principle of general scope, which will be essential for future AI regulation. This principle can be stated as follows: those who produce, operate or make use of AI systems must not gain an advantage (in procedural legal terms) from the degree, nature, or type of risk that the operated system involves, where it is predictable that such a situation of risk (assessed in terms of uncertainty) might occur as a result of the operated system and the existing information asymmetry between the defendant and the victim i.e. the *legal principle of no advantage from risk*.

7. *The principle of functional equivalence*

The *principle of functional equivalence*, which is explicitly referenced by the Expert Group (2019, 25), asserts that assistance by an AI system (or, in their case, any emerging digital technology) should be treated no differently from human assistance in case of harm to a third party. Similar cases must be dealt with in similar ways unless there is a reason to make a distinction. In this regard, a remark by the Expert Group is very relevant: "[It is] challenging to identify the benchmark against which the operation of non-human helpers will be assessed in order to mirror the misconduct element of human auxiliaries". (Report, 2019, 25).

This raises the question of the standard for assessing the operation of AI systems (as non-human helpers). On the one hand, recourse to AI systems should not be discouraged in areas where their use has proved safer than human auxiliaries; on the other hand, it would be desirable and legitimate to expect a higher standard from AI systems (see also Abbott, 2018):

> [...] once autonomous technology outperforms human auxiliaries in terms of preventing harm, the benchmark should be determined by the performance of comparable technology that is available on the market. (Report, 2019, 46)

This principle is complemented by the following one.

8. *The principle of equal protection under the law*

According to this *principle, equal protection under the law* must be ensured for all those who are the receivers of the effects of AI systems or the performance of emerging digital technologies. It clearly emerges as a central tenet of the Expert Group's analysis. It has a high degree of generality and applies both from a substantive and a procedural standpoint. From a substantive standpoint, "victims should be treated alike if they are exposed to and ultimately harmed by similar dangers". (Report, 2019, 40).

Rational distinctions may be made due to AI specificity only where the distinguishing factors are not foreseeable from the start. However, the reasonableness of such distinctions is understood in terms of the predictability of the factors (that are the origin) of uncertainty:

> In view of the need to share benefits and risks efficiently and fairly, the development risk defence, which allows the producer to avoid liability for unforeseeable defects, should not be available in cases where it was predictable that unforeseen developments might occur. (Report, 2019, 43)

From a procedural standpoint, as remarked with reference to the principle of traceability and the principle of no advantage from risk, the possibility of being compensated for loss or injury cannot be reduced where it was predictable that an unforeseen development might occur as a result of the specific artificial nature of the agent producing the harm (in a sense, much as in the case in which a skilled driver starts driving while intoxicated, accepting the risks deriving from his/her diminished driving ability; a specific accident might then be unforeseen and yet quite predictable in general). This suggests a general tenet: AI regulation should always be adopted in such a way as to ensure equal substantive and procedural protection by the law.

9. *The principle of legal unification*

The Expert Group does not support the need to endow artificial autonomous agents with legal personality. What emerges instead is the need to build an ecosystem based on safety and security rules, as a condition for systemic trust in the digital world. To achieve this, a key factor is the implementation of rules that coordinate the actors and other elements making up the ecosystem. This requires the adoption of a legal fiction, which takes the form of a *principle of legal unification*, according to which the diverse elements of the same digital ecosystem are meant to form a single commercial and technological unit. This has practical consequences from a legal point of view, since this principle is likely to create several and possibly joint liabilities towards the victim ("where the victim can demonstrate that at least one element has caused the damage in a way triggering liability, but not which element" [Report, 2019, 56]) of all the potential tortfeasors who have produced or operated the diverse elements of the digital ecosystem. This principle takes on central importance in the light of the progressive construction of an increasingly interconnected digital ecosystem (Internet of Things).

10. *The principle of market placement risk*

In line with these observations, an AI system evolves not only in space, by interacting with other elements of the digital ecosystem, but also over time, by virtue of its openness. One must therefore abandon the concept of a product that is completed at a certain point in time: putting a product on the market no longer marks a necessary limit to the producer's liability. This requires the adoption of a *principle of market placement risk* to mitigate the effects of the contrasting principle of the development risk defense, which allows the producer to avoid liability for unforeseeable defects. According to the principle of market placement risk, a producer can be held liable for unforeseeable consequences in the use of a product, when it was predictable at the time of placing the product on the market that unforeseen consequences could have occurred. This principle is important, because, together with the other principles mentioned before, it helps one realize that, in AI regulation, the actor responsible for the adoption of an AI system that has uncertain effects is responsible for that uncertainty. Or, to put it more precisely, the unpredictability of AI systems (in particular, "black box" ML systems) cannot be invoked as a legal justification if it is the predictable result of the system.

5 General Aims or Meta-principles of AI Regulation

Having identified ten general legal principles for AI liability regulation, which may have a cross-sectoral and broader scope, we can ask ourselves whether this set of principles can suggest some further considerations regarding the main policies that the regulation of AI should pursue. We believe that the overall consideration of such principles may suggest three regulatory meta-principles, or general aims, which illustrate the main policies of AI regulation. They are a third-order kind of regulations, insofar as they orient the legal principles which are meta-regualatory with respect to legislation about AI. Let us examine them.

1. *Answerability*

The European model of AI regulation (notably the *Artificial Intelligence Act* [Floridi, 2021]) is built on a *risk-based approach*. The risk-based approach has been coupled with the principle of *accountability*, remarkably in the area of personal data protection, with reference to the implementation of the GDPR.[8] As with the GDPR, the principle of accountability also functions as a meta-principle, since it governs the application of the other legal principles and of itself. This can be confusing, so we shall use a different terminology here and speak of *answerability* when accountability is used to refer not to a meta-regulatory legal principle, but to a legal meta-principle (strictly speaking, in logical terms, accountability as answerability is a reflective meta-meta-principle). Two basic features describe the principle of

---

[8] See recently, for instance, Quelle (2017), Ivanova (2020). For a comparative analysis between US and Europe, see also Winn (2019).

answerability in the context of the GDPR. First, the data controller, who determines the ends and means of the processing of personal data, is required to adopt adequate technical and organizational measures with regard to the level of risk that the data processing entails. And second, the data controller shall always be able to be transparent about the measures adopted to comply with the GDPR. A data controller is expected to apply all legal principles regarding data protection in accordance with this two-step requirement: identification of security measures based on the level of risk and duty of transparency. By way of analogy, in the context of AI, the European regulator should require, as a general policy, the actor who determines the ends and means of an AI system to apply the legal principles of AI based on a meta-principle of *answerability*, by detecting the level of risk; by adopting the adequate safety measures; by creating a safe AI ecosystem; and by being able to be transparent about the measures adopted.

2. *The reduction of information and power asymmetries*

The field of AI is characterized by information and power asymmetries between the actors involved in an AI ecosystem, because of the uncertainty generated by the use of AI systems (which is mostly a byproduct of the openness, autonomy and unpredictability of AI systems) and since AI systems draw on large volumes of data from the users of those systems, who in turn usually lack access to, control over, and knowledge of the use of their (personal) data. This uncertainty should not be exploited by those in a better informational position. Hence, the European regulator should aim, as a general policy, at reducing these information asymmetries, which are likely to turn into real power asymmetries as, notably, in the case of legal adjudication. More generally, the reduction of information and power asymmetries may encourage the creation of a level playing field and a fairer and clearer AI regulatory framework for innovation and trust. End-users trust, especially, is grounded in the fact that they perceive themselves not only as stake-holders, whose expectations are somehow taken into account, but also as "real interlocutors" (Durante, 2015), who have a say in the construction of the digital ecosystem.

3. *The normative coordination of agency*

The AI ecosystem is a multi-agent system, in which different forms of human and artificial agency interact. AI operations are often the result of this complex interaction, which makes it necessary to combine and regulate these different levels of agency, primarily in order to attribute the results and different actions to their respective sources. Against this backdrop, the EU regulator should envisage and support – as a general policy – the adoption of legal principles and norms for the coordination of agency in the AI ecosystem. This regulatory meta-principle is fundamental for the development of technological innovation. The only way to keep the legal complexity of the multi-agent AI ecosystem from hampering innovation is by adopting clear legal criteria to identify *ex ante* the consequences of AI operations; to attribute them to specific sources; and to share the costs of such consequences among those specific sources.

6 Innovation and Trust

There are several reasons why AI regulation based on legal principles, meta-principles, and general aims can foster innovation and trust. So far, we have argued that the implementation of legal principles for AI can help reduce legal fragmentation, providing greater harmonization of the law and a means for dealing with gray areas and filling gaps in the law. This can create a framework of greater legal certainty that is indispensable for technological innovation, without necessarily implying an increase in overly prescriptive rules. Furthermore, the implementation of legal meta-principles, in terms of answerability, reduction of information and power asymmetries, and normative coordination for agency, can strengthen user trust in AI ecosystems, which is the second prerequisite for the development of AI, along with the drive for technological innovation.

One particular aspect has emerged at several points in this chapter, i.e., the relevance of uncertainty. We are used to considering uncertainty as a feature of the environment that we need to reduce or as an emerging property of complex societies that we need to manage. Today, we face the fact that we are able to produce AI systems that can generate uncertainty by themselves as a by-product of their openness, autonomy, and unpredictability. We refer to this as *artificial uncertainty*.

The management of this artificial uncertainty plays a significant role in the establishment of a regulatory framework of innovation and trust, which may also entail a shift from trustworthy AI to trustworthy regulation procedures and institutions, which are called upon to protect rights, address power imbalances and put in place fair remedies, in order to create stronger conditions for users trust. In this perspective, it should be remarked that technological innovation always brings with it some uncertainty, which must be compensated for by strengthening users' trust in the application of technology. If uncertainty implies, as Floridi points out (2015), some lack of information and, by consequence, a misalignment between questions and answers, then it is essential to understand how to fill this information gap. Against this backdrop, the EU model of AI regulation has encouraged the adoption of *"Ethics Guidelines for Trustworthy AI"* (2019), which has been recently supported by the new *Proposal for a regulation of the European Parliament and the Council laying down harmonised rules on Artificial Intelligence (Artificial Intelligence Act, 2021)*, which aspires to systematize the previous recommendations, guidelines and regulatory instruments inherent to the governance and regulation of AI. "According to the European Data Protection Supervisor website, the AIA is 'the first initiative, worldwide, that provides a legal framework for Artificial Intelligence (AI)'. Regardless of whether this may be true [...], the AIA is one of the most influential regulatory steps taken so far internationally. On the whole, it is a good starting point to ensure that the development of AI in the EU is ethically sound, legally acceptable, socially equitable, and environmentally sustainable, with a vision of AI that seeks to support the economy, society, and the environment". (Floridi, 2021, 215)

Let us go back to the idea of trustworthy AI. This expression has a twofold meaning since it includes a technical and a normative aspect. AI is technically trustworthy, when it holds a set of technical features and standards that make it reliable (Taddeo et al., 2019). In this case, one may also speak of reliance and reliability. AI is normatively trustworthy, when the deployment of an AI system comes along with the provision of legal measures and remedies against the possible negative effects of AI. In this case, one may speak of distrust rather than trust (Durante, 2021). In both cases, we deal with human expectations set under conditions of uncertainty, which are understood, in Floridi's terms, as incomplete information.

We have learned from the analysis carried out so far that both the level and the origin of artificial uncertainty are relevant to the law. The higher the degree of uncertainty produced by the use of an AI system is, the higher the degree of distrust and the related need to provide adequate legal measures and remedies for the risk of uncertainty needs to be. However, when it is possible to envisage and identify the source of uncertainty (i.e., whether an unforeseen result or development might occur), one can fine-tune the legal response and talk about reliance on AI technology. While unforeseeability is usually considered a limitation to legal responsibility, that is no longer the case in AI ecosystems, where it is predictable that unforeseen results or developments might occur (as already pointed out, just as in the case in which some, although unable to foresee certain events, puts themselves in the position to let them happen [like someone who starts driving while intoxicated] and are therefore called upon to accept their relative risks).

How to manage artificial uncertainty in AI ecosystems is the first policy issue that any form of AI regulation needs to address. In every age and society, we institutionalize the principles and standards by which we intend to manage uncertainty in relation to the involved values. In order to decide who will kick-off a football match, we are satisfied with tossing a coin. In order to decide who is guilty or innocent, we require proving guilt beyond any reasonable doubt. In AI ecosystems, it is of crucial importance that we decide on the institutional principles and standards by which we aim to manage the uncertainty produced by the AI systems we produce.

7   Policy Recommendations

Considerations made in this chapter translate into a series of recommendations, which together outline an innovative approach to AI regulation. These recommendations can be divided into three progressive steps: (1) reviewing the existing law in various sectors of AI; (2) formulating legal principles; and (3) formulating regulatory meta-principles.

1. *Reviewing the existing law in various sectors of AI*
The ubiquitous use of AI can generate problems of a new kind, making it necessary to review the existing law, to fill legal gaps, to balance opposing interests and to resolve conflicts arising from the need to adapt the law to new needs. This review process has mostly been entrusted to groups of experts. One must not loose track of their work, or treat it as a mere communication of guidelines or as a description of

problems. It has highlighted that solutions offered to specific problems are likely to reveal areas of convergence that may result in the elaboration of legal principles, understood as real normative standards, useful for the interpretation and enforcement of existing law, the resolution of controversial cases, and the elaboration of a more harmonized regulatory framework. Principles-based regulation cannot be entirely separated from the reference to a body of existing legal norms, which it necessarily accompanies.

2. *Formulating legal principles*
AI sectors need legal principles together with ethical principles of a general nature and local sector regulations as a means of mediating between the generality of ethical requests and the particularity of legal requests. Legal principles can function as a tool for translating principles into practice, because they enable the generalization of legal cases bottom-up, The analysis of several sectors of AI is likely to reveal different sets of legal principles, which can be further compared in order to verify second degree convergences between principles: this may reveal legal principles with a cross-sectoral dimension. Legal principles thus play an important if not crucial role of mediation in creating and delimiting a level playing field, and in showing the dynamics of the game, by making one grasp the rationale behind rules. Above all, legal principles force to question the very aims of a regulation. Indeed, a series of cross-sectoral legal principles can reveal the purposes or meta-principles of AI regulation.

3. *Formulating regulatory meta-principles*
The overall consideration of a series of legal principles at an intra-sectoral level has suggested three general aims or meta-principles that AI regulation can and perhaps must have. This point is crucial: an innovative and mature approach to AI regulation must question not only the purposes of AI but more importantly the purposes of AI regulation. In fact, from the implementation of these regulatory aims comes the true potential for people to enjoy the benefits deriving from the use of AI systems. Therefore, the regulation of AI must pursue these three aims or meta-principles: (1) the formulation of a general principle of answerability for the field of AI as a necessary complement to the EU risk-based approach to AI and a tool for articulating top-down and bottom-up regulations; (2) the progressive reduction of information and power asymmetries, which are capable of undermining the development of AI for the common good; (3) the elaboration of normative coordination between all the actors, stakeholders, and real interlocutors involved in the governance of AI.

8 Conclusions

In this chapter, we have highlighted the need of what we see as a fundamental regulatory tool for the development of policies and the implementation of practices within the current digital ecosystem: legal principles. The current regulation of AI is characterized, on the one hand, by (too) many ethical principles, and, on the other, by sometimes forgotten or neglected legal limits provided by the rule of law, fundamental rights, and mandatory existing law. There is an increasing need for something that mediates between these two regulatory poles and that can help us understand how we want to design and regulate the digital ecosystem.

For a digital ecosystem to be based on a framework of innovation and trust, it is necessary to develop and share a set of meta-regulatory legal principles and general aims that can show clearly, concisely and intelligibly to all the concerned stakeholders on which basic pillars this ecosystem should be built and in which direction AI regulation should be steered. Our analysis suggested a set of legal principles, which may have a cross-sectoral dimension, and three general aims, which are fundamental for the development and design of AI regulation. In line with this approach, we have suggested a series of recommendations useful for the development and design of such legal principles and regulatory meta-principles. We have highlighted what may be the main issue that AI regulation will need to address in every sector of application: i.e., the spread of artificial uncertainty generated as a by-product of the openness, autonomy and unpredictability of AI systems. In an up-to-date legal framework of innovation and trust, AI regulation must define how to reduce and manage this uncertainty along the path drawn by the proposed legal principles and regulatory meta-principles.

Funding Massimo Durante's research was supported by a fellowship funded by Google EU.

References

Abbott, R. (2018). The reasonable computer: Disrupting the paradigm of tort liability (November 29, 2016). *George Washington Law Review, 86*(1) Available at SSRN: https://ssrn.com/abstract=2877380

Ad hoc Committee On Artificial Intelligence (CAHAI). (2020a). *Feasibility Study*, CAHAI, 23.

Ad hoc Committee on Artificial Intelligence (CAHAI). (2020b). *Towards Regulation of AI Systems. Global perspectives on the development of a legal framework on Artificial Intelligence systems based on the Council of Europe's standards on human rights, democracy and the rule of law*, Council of Europe, DGI, 16.

Barfield, W., & Pagallo, U. (2020). *Advanced introduction to law and artificial intelligence*. Edward Elgar.

Bernitz, U., Groussot, X., Paiu, J., & De Vries, S. (2020). *General principles of EU law and the EU digital order*. Wolters Kluwer.

Bertolini, A. – Episcopo, F., The expert Group's report on liability for artificial Intelligence and other emerging digital technologies: A critical assessment, in European Journal of Risk Regulation, 2021, pp. 1–16.

Black, J., Hopper, M., & Band, C. (2007, March). Making a success of Principles-based regulation. *Law and Financial Markets Review*, 191–206.

Durante, M. (2015). *The democratic governance of information societies. A critique to the theory of stakeholders*. Philosophy and Technology, 28, 11–32.

Durante, M. (2021). Computational power. The Impact of ICT on Law. In *Society and knowledge*. Routledge.

Dworkin, R. (1967). The model of rules. *University of Chicago Law Review, 35*(1), 14–46. Article 3.

Dworkin, R. (1986). *Law's empire*. Fontana Press.

European Commission. (2020a). On artificial intelligence – A European approach to excellence and trust. *White Paper, COM*, 65 final.

European Commission. (2020b). *Proposal for a regulation of the European Parliament and of the council on European data governance*, COM, 767 final.

European Commission. (2021). *Proposal for a regulation of the European Parliament and the Council laying down harmonised rules on Artificial Intelligence (AI Act)*, COM, 206 final.

Expert Group On Liability And New Technologies. (2019). *Liability for artificial intelligence and other emerging digital technologies*.

Feldman, R.-C. (2018). Artificial intelligence. The importance of trust and distrust. *Green Bag, 21*(3), 1–13. Available at: https://ssrn.com/abstract=3118523

Fjeld, J., Nele, A., Hilligoss, H., Nagy, A., & Srikumar, M. (2020, January 15). *Principled artificial intelligence: Mapping consensus in ethical and rights-based approaches to principles for AI*. Berkman Klein Center Research Publication No. 2020–1. Available at SSRN: https://ssrn.com/abstract=3518482

Floridi, L. (2008). The method of levels of abstraction. *Minds and Machines, 18*(3), 303–329.

Floridi, L. (2015). The politics of uncertainty. *Philosophy and Technology, 28*, 1–4.

Floridi, L. (2020). The fight for digital sovereignty: What it is, and why it matters, especially for the EU. *Philosophy and Technology, 33*, 369–378.

Floridi, L. (2021). The European legislation on AI: A brief analysis of its philosophical approach. *Philosophy and Technology, 34*, 215–222.

Floridi, L., & Cowls, J. (2019). A unified framework of five principles for AI in society. *Harvard Data Science Review, 1*(1), 1–15.

Hacker, P. (2020, May 7). *AI regulation in Europe*. Available at SSRN: https://ssrn.com/abstract=3556532

Hagendorff, D.-T. (2020). The ethics of AI ethics — An evaluation of guidelines. *Minds & Machines, 30*, 99–120.

Hart, H. (1961). *The concept of law*. OUP.

High Level Expert Group on AI. (2019, April 8). *Ethics guidelines for trustworthy AI*.

Ivanova, Y. (2020). Data controller, processor or a joint controller: Towards reaching GDPR compliance in the data and technology driven world. In M. Tzanou (Ed.), *Personal data protection and legal developments in the European Union*. IGI Global.

Jobin, A., Ienca, M., & Vayena, E. (2019). The global landscape of AI ethics guidelines. *Nature Machine Intelligence, 1*, 389–399.

Mazzini, G. (2019). A system of governance for artificial intelligence through the lens of emerging intersections between AI and EU law. In A. De Franceschi & R. Schulze (Eds.), *Digital revolution – New challenges for law*. Available at SSRN: https://ssrn.com/abstract=3369266

Morley, J., Floridi, L., Kinsey, L., et al. (2020). From what to how: An initial review of publicly available AI ethics tools, methods and research to translate principles into practices. *Science and Engineering Ethics, 26*, 2141–2168.

Pagallo, U., Casanovas, P., & Madelin, R. (2019). The middle-out approach: Assessing models of legal governance in data protection, artificial intelligence, and the Web of Data. *The Theory and Practice of Legislation, 7*(1), 1–25.

Quelle, C. (2017, April 7). Privacy, proceduralism and self-regulation in data protection law. In *Teoria Critica della Regolazione Sociale*.

Raz, J. (1972). Legal principles and the limits of law. *Yale Law Journal, 81*, 823.

Smuha, N.-A. (2019, November). *From a 'Race to AI' to a 'Race to AI Regulation' – Regulatory competition for artificial intelligence*. Available at SSRN: https://ssrn.com/abstract=3501410

Taddeo M., McCutcheon, T., & Floridi, L. (2019, December 01). *Trusting artificial intelligence in cybersecurity is a double-edged sword*. Available at SSRN: https://ssrn.com/abstract=3831285

Winn, J. (2019, July 11). *The governance turn in information privacy law*. Available at SSRN: https://ssrn.com/abstract=3418286

The New Morality of Debt

Nikita Aggarwal

Abstract This chapter examines the new morality of debt in light of the rise of machine learning and increasing datafication of lending. It argues that existing regulatory frameworks governing consumer lending are inadequate as they fail to sufficiently alleviate the privacy, autonomy and dignity harms of datafied lending. The chapter recommends ways to close the privacy gap in consumer credit markets through substantive and institutional regulatory reforms. *Note that an earlier version of this chapter first appeared in the IMF's Finance and Development magazine.*

Keywords Fintech · Privacy · Financial regulation · Artificial intelligence · Big data · Consumer finance

Throughout history, society has debated the morality of debt. In ancient times, debt—borrowing from another on the promise of repayment—was viewed in many cultures as sinful, with lending at interest especially repugnant. The concern that borrowers would become overindebted and enslaved to lenders meant that debts were routinely forgiven (Graeber, 2011). These concerns continue to influence perceptions of lending and the regulation of credit markets today. Consider the prohibition against charging interest in Islamic finance and interest rate caps on payday lenders—companies that offer high-cost, short-term loans. Likewise, proponents of debt forgiveness appeal in part to morality when they advocate relieving hard-up debtors of the burden of unsustainable debt.

N. Aggarwal (✉)
Faculty of Law, University of Oxford, Oxford, UK

Oxford Internet Institute, University of Oxford, Oxford, UK
e-mail: nikita.aggarwal@law.ox.ac.uk

© The Author(s), under exclusive license to Springer Nature Switzerland AG 2022
J. Mökander, M. Ziosi (eds.), *The 2021 Yearbook of the Digital Ethics Lab*, Digital Ethics Lab Yearbook, https://doi.org/10.1007/978-3-031-09846-8_8

1 "Datafied" Lending

In much of this debate, the principal moral value at play is fairness; specifically, distributional fairness. Debt is deemed to be unfair and thus immoral because of the inequality of knowledge, wealth, and power between borrowers and lenders, which lenders can and often do exploit. Recent technological advances in lending are adding new dimensions to debt's morality. Notably, the datafication of consumer lending is amplifying moral concerns about harm to individual privacy, autonomy, identity, and dignity (Aggarwal, 2021). Datafication in this context describes the rapidly growing use of personal data for consumer credit decision-making—particularly "alternative" social and behavioral data, such as a person's social media activity and mobile phone data—together with more sophisticated data-driven machine learning algorithms to analyze those data (Hurley & Adebayo, 2017).

These techniques enable lenders to predict the behavior of consumers and shape their financial identities in much more granular ways than in the past. For example, it has been shown that borrowers who use iOS devices (Berg et al., 2020), have larger and more stable social networks (Björkegren & Grissen, 2020), or make less spelling mistakes in filling out a loan application form (Lee & Singh, 2021) are more likely to be creditworthy and repay debt on time (of course, many of these variables proxy for fundamental credit life-cycle variables, such as income). Innovation in datafied lending has been driven largely by fintech start-ups, particularly peer-to-peer lending platforms such as LendingClub and Zopa and Big Tech companies like Alibaba/Ant Group. However, alternative data and machine-learning techniques are also increasingly being adopted by traditional bank lenders, as highlighted by recent surveys from the Bank of England (2019) and the Cambridge Centre for Alternative Finance (Ryll et al., 2020).

These practices diminish consumers' ability to craft their own identity as they become increasingly chained to their "data self" (Cheney-Lippold, 2017), or algorithmic identity. Moreover, the ubiquitous collection of data and surveillance that fuels datafied lending constrains consumers from acting freely lest their actions negatively affect their creditworthiness. Additionally, the commodification of certain types of personal data for lending decisions raises moral concern about harm to individual dignity. Is it moral for lenders to use highly intimate health and relationship data—for example, captured from social media and dating apps—to determine consumer creditworthiness? Consumers may willingly share their data in specific contexts and for specific purposes, such as to facilitate online dating and social interaction. However, this does not imply that they consent to the use of that information in new contexts and for different purposes, particularly commercial purposes such as credit scoring and marketing (Nissenbaum, 2014).

Datafication also amplifies existing concerns about fairness and inequality in consumer lending (Baradaran, 2019). Lenders are prone to abuse data-driven insights, for example, to target the most vulnerable consumers with unfavorable credit offers (Engel & McCoy, 2002). Data-driven profiling of borrowers also facilitates more aggressive and intrusive debt-collection practices against the poor

(Roussi, 2020). And more accurate screening and price discrimination using alternative data and machine learning increase the cost of borrowing for consumers previously subsidized by hidden information (Fuster et al., 2020).

In addition, increasingly data-driven, algorithmic lending could amplify unfairness as a result of racial and gender-based discrimination, as highlighted by the recent Apple Card debacle, when women were offered smaller lines of credit than men (New York Department of Financial Services, 2021). In particular, biases and proxy variables in the data used to train machine-learning models could exacerbate indirect discrimination in lending against minority groups—particularly where the data reflect long-standing structural discrimination (Barocas and Selbst, 2016). Alternative data, such as social media data, are typically more feature-rich than financial credit data and thus embed more proxy variables for protected characteristics, such as race and gender. Furthermore, the limited interpretability of certain machine-learning methods (such as deep neural networks) could impede efforts to detect discrimination by proxy. Deploying these machine learning models without rigorously testing their results, and without meaningful human oversight, therefore risks reinforcing social biases and historical patterns of unlawful discrimination, perpetuating the exclusion of less-advantaged and minority groups from consumer lending markets.

Yet the datafication of consumer lending could also uphold the morality of debt by improving other dimensions of distributional fairness in consumer credit markets. Notably, more accurate credit assessment thanks to machine learning and alternative data in algorithmic credit scoring will improve access to credit, particularly for (creditworthy) "thin-file" and "no-file" consumers previously locked out of mainstream credit markets because of insufficient credit data, such as a credit history (Aggarwal, 2018). Estimates suggest that nearly 10% of the UK population,[1] and nearly 20% of the US population (Brevoort et al., 2015), have thin files or no files and lack access to affordable credit. In developing economies, this figure is several times greater. According to the World Bank Global Financial Inclusion Index, more than 90% of people living in south Asia and sub-Saharan Africa lack access to formal credit (Demirgüç-Kunt et al., 2017), in part due to the lack of formal credit data (Agarwal et al., 2019).

Given that these consumers are often the least-advantaged members of society, typically from ethnic minority and lower-income groups, improving their access to credit supports financial inclusion and enhances fairness—as well as efficiency—in consumer lending markets. Datafied, algorithmic lending also stands to support fairness by reducing more visceral forms of direct discrimination in lending—for example, stemming from the sexist or racist preferences of a (human) loan officer (Bartlett et al., 2017). Moreover, better access to credit and the accompanying opportunities can itself enhance the autonomy and dignity of consumers.

More broadly, the digitalization and automation of lending stand to increase financial inclusion by reducing transaction costs and making it more feasible for

---

[1] https://www.experian.com/blogs/news/2020/03/16/uk-invisible-challenge/

lenders to extend small-value loans and reach consumers traditionally excluded from borrowing by their remote physical location (for example, a lack of bank branches in "banking deserts"). Data-driven technology also can support financial inclusion by improving consumer financial literacy and personal debt management. For example, automated saving and debt pay-down features of many fintech credit apps can help overcome some of the more common behavioral biases that undermine consumer financial decision-making (Campbell et al., 2011).

2 Recasting Regulation

The rise of machine learning and datafied lending renders the morality of debt much more nuanced. The Goldilocks challenge for regulators is to find the right balance between the benefits and harms of datafied lending. They must protect consumers from its greatest harms—in terms of privacy, unfair discrimination, and exploitation—while still capturing the key benefits, particularly improved access to credit and financial inclusion. However, existing regulatory frameworks governing consumer credit markets and datafied lending in places such as the United Kingdom, United States, and European Union do not strike the right balance. In particular, they do not sufficiently alleviate the privacy, autonomy, and dignity harms of datafied lending.

The prevailing approach to regulating consumer privacy in these jurisdictions is distinctly individualistic. It relies on consumers to consent to all aspects of data processing and to self-manage their privacy (Solove, 2013)—for example, by exercising their right to access, correct, and erase their own data. However, this approach cannot protect consumers in ever-more-datafied consumer credit markets. These markets display steep asymmetries of information and power between borrowers and lenders, negative externalities related to data processing, and behavioral biases that impede consumer decision-making, such that individuals cannot on their own safeguard their privacy and autonomy.

In a recent article in the Cambridge Law Journal, I recommend ways to address these inadequacies and close the privacy gap in consumer credit markets through substantive and institutional regulatory reforms (Aggarwal, 2021). To begin with, a more top-down regulatory approach is needed. Firms should be subject to more rigorous obligations to justify the processing of personal data under the paradigm of datafied lending. This should include stricter ex ante restrictions on the types and granularity of (personal) data that can be used for credit decision-making. For example, the use of intimate, feature-rich data, such as social media data, should be explicitly prohibited, and anonymization of personal data should be the default.

Firms should, moreover, bear a higher burden of proof regarding the necessity and proportionality of processing personal data and thus their encroachment on consumer privacy. This should include stricter, ongoing model validation and data quality verification obligations, particularly for nonbank fintech lenders. For example, in the context of algorithmic credit scoring, lenders should be required to demonstrate that the processing of alternative data yields a sufficiently significant improvement in the accuracy of creditworthiness assessment to be justified.

These reforms should be accompanied by changes to the regulatory architecture to improve the enforcement of consumer privacy protection in consumer credit markets. In particular, regulatory agencies responsible for consumer financial protection, such as the UK Financial Conduct Authority, should have expanded authority to enforce privacy and data protection in consumer credit markets. I argue that data protection is consumer financial protection. Given their expertise and experience working with consumer credit firms, sectoral agencies are in many ways better positioned than cross-sectoral data protection and consumer protection agencies to enforce data protection in consumer financial markets. However, they should continue to collaborate with cross-sectoral regulators, such as the UK Information Commissioner's Office, that have expertise in data protection regulation.

Of course, these reforms are not needed only for datafied consumer lending and its regulation. To truly safeguard the privacy of (credit) consumers, stricter limits on the processing of personal data are called for in all contexts, not only consumer credit markets, and on all actors in the development life cycle of consumer-facing information systems. Likewise, in an increasingly datafied economy, the optimal institutional arrangement for data protection regulation entails a greater role for sectoral regulators and deeper collaboration between sectoral and cross-sectoral regulators everywhere—not just in consumer credit markets.

References

Agarwal, S., et al. (2019). *Financial inclusion and alternate credit scoring: Role of big data and machine learning in Fintech.* https://doi.org/10.2139/ssrn.3507827

Aggarwal, N. (2018). Machine learning, big data and the regulation of consumer credit markets: The case of algorithmic credit scoring. In *Autonomous Systems and the Law.* Beck. https://papers.ssrn.com/abstract=3309244

Aggarwal, N. (2021). The norms of algorithmic credit scoring. *The Cambridge Law Journal, 80*(1), 42–73. https://doi.org/10.1017/S0008197321000015

Bank of England and Financial Conduct Authority. (2019). *Machine learning in UK financial services.* https://bit.ly/3l3YITa

Baradaran, M. (2019). Jim Crow credit. *U.C. Irvine Law Review, 9*(4), 887. https://scholarship.law.uci.edu/ucilr/vol9/iss4/4

Bartlett, R. P., Morse, A., Stanton, R., & Wallace, N. (2017). *Consumer lending discrimination in the FinTech era* (SSRN Scholarly Paper ID 3063448). Social Science Research Network. https://papers.ssrn.com/abstract=3063448

Barocas, S., & Selbst, A. D. (2016). Big data's disparate impact. *California Law Review, 104*, 671.

Berg, T., et al. (2020). *On the rise of FinTechs—Credit scoring using digital footprints. 33*(7) The *Review of Financial Studies 2845–2897.*

Björkegren, B., & Grissen, D. (2020). Behaviour revealed in mobile phone usage predicts credit repayment. *The World Bank Economic Review, 34*(3), 618.

Brevoort, K., Grimm, P., & Kambara, M. (2015). *Data point: Credit invisibles.* U.S. Consumer Financial Protection Bureau. https://files.consumerfinance.gov/f/201505_cfpb_data-point-credit-invisibles.pdf

Campbell, J. Y., et al. (2011). Consumer financial protection. *Journal of Economic Perspectives, 25*(1), 91.

Cheney-Lippold, J. (2017). *We are data: Algorithms and the making of our digital selves*. New York University Press.

Demirgüç-Kunt, A., et al. (2017). *The global Findex database*. https://globalfindex.worldbank.org/

Engel, K. C., & McCoy, P. A. (2002). A tale of three markets: The law and economics of predatory lending. *Texas Law Review, 80*(2), 1255.

Fuster, A., Goldsmith-Pinkham, P., Ramadorai, T., & Walther, A. (2020). Predictably unequal? *The Effects of Machine Learning on Credit Markets*. https://doi.org/10.2139/ssrn.3072038

Graeber, D. (2011). *Debt: The first 5,000 years*. Melville House.

Hurley, M., & Adebayo, J. (2017). Credit scoring in the era of big data. *Yale Journal of Law and Technology, 18*(1) https://digitalcommons.law.yale.edu/yjolt/vol18/iss1/5

Lee, M. S. A., & Singh, J. (2021). *Spelling errors and non-standard language in peer-to-peer loan applications and the borrower's probability of default*. https://papers.ssrn.com/abstract=3609834

New York Department of Financial Services. (2021). *Report on Apple Card Investigation*. https://www.dfs.ny.gov/reports_and_publications/press_releases/pr202103231

Nissenbaum, H. (2014). Privacy as contextual integrity. *Washington Law Review, 79*(Part 1), 119–158.

Roussi, A. (2020). Kenyan borrowers shamed by debt collectors chasing Silicon Valley loans. *Financial Times*. https://on.ft.com/2FtPY95

Ryll, L., et al. (2020). *Transforming paradigms: A global AI in financial services survey*. https://www.jbs.cam.ac.uk/faculty-research/centres/alternative-finance/publications/transforming-paradigms/

Solove, D. J. (2013). Privacy self-management and the consent dilemma. *Harvard Law Review, 126*, 1880.

Site of the Living Dead: Clarifying Our Moral Obligations Towards Digital Remains

Mira Pijselman

Abstract As internet users, we leave behind both a physical and informational corpse when we die. Practices regarding the disposition of physical remains have been developed over millennia and subjected to considerable ethical scrutiny. However, the same cannot be said for digital remains, for which there is no unified roadmap regarding their appropriate management. This chapter clarifies the conceptual relationship between the human dignity of the online dead, duties to preserve our digital cultural heritage, and our associated moral obligations. Using the levels of abstraction method, I draw upon the perspectives of archaeology, grave re-use, and archival curation to guide my analysis. In doing so, I argue that decisions surrounding the use, preservation, and destruction of post-humous data are a complex negotiation between the directives of the dead, the interests of families, cultural norms, and their scientific value. Further, I highlight that this negotiation is dynamic, as our moral obligations towards the online dead shift over time due to an ethically significant difference between the *near* and *forgotten dead*. For the *forgotten dead,* certain *post-humous harms,* such as data deletion, are morally permissible in the interests of humanity, a stakeholder group comprised of past, present, and future generations. In response to this macro-ethical analysis, I advance pragmatism and selectivity as core infraethics that ought to structure the future of digital remains management.

Keywords Online death studies · Ethics of information · Levels of abstraction · Digital remains · Data ethics

M. Pijselman (✉)
Oxford Internet Institute, University of Oxford, Oxford, UK
e-mail: mira.pijselman@mail.utoronto.ca

© The Author(s), under exclusive license to Springer Nature Switzerland AG 2022
J. Mökander, M. Ziosi (eds.), *The 2021 Yearbook of the Digital Ethics Lab*, Digital Ethics Lab Yearbook, https://doi.org/10.1007/978-3-031-09846-8_9

1   Introduction

When we die, the living must dispense with our corporeal remains. Regardless of the method of disposition, corpses are central objects in end-of-life rituals, as embodiments of the deceased. Despite the fact that a person has died, the extension of personhood through one's physical remains create moral obligations for the living regarding their ethical treatment. It is for this reason that we intuitively recoil at the thought of corpses being desecrated after death or being tossed haplessly into mass graves. Well put by the Advisory Panel on the Archeology of Burials in England, 2017: while "a corpse has no more eternal significance than an empty shell [...] the material body is invested with meaning as the visible manifestation of one with whom we lived, laughed, and conversed" (p. 26).

Those who have died or will die following the dawn of information and communication technologies (ICTs) will not just leave behind a physical corpse, but an informational corpse – their *digital remains* (Lingel, 2013). There is a distinction between heritable assets like money and our data, as a result of data's encapsulation of personal identity; just as we *are* our bodies, we *are* our data (Floridi, 2016), and we *are* our digital remains (Stokes, 2012, 2015). However, unlike our bodies, there is no unified roadmap for how we ought to manage digital remains – the subject of this chapter.

Facebook and Instagram alone will accumulate 9.5 billion informational corpses by the year 2100 if usership continues to grow at its present rate (Öhman & Watson, 2019, 2020). Rigorous ethical investigation is thus required to regulate companies that will control our digital remains and, by extension, our *digital cultural heritage* (Cameron & Kenderdine, 2007). In this chapter, I build upon macro-ethical analyses[1] of online death to balance existing literature, which has historically catered to ethnographic studies of online grief (Ohman, 2020). Previous, foundational work in this space and in the ethics of information, more broadly, have identified a denial or manipulation of control over our constitutive information to be a violation of human dignity (Floridi, 2016) and that the co-presence of the living and the dead afforded by ICTs entails a persistence of human dignity after biological death that must be protected (Ohman, 2020; Stokes, 2012, 2015, 2019). From these foundations, I consider how human dignity and a duty to preserve our digital cultural heritage informs our moral obligations towards the online dead. Further, I consider how the conceptual relationship between human dignity, digital cultural heritage, and our moral obligations towards the online dead informs an infraethical[2] framework (Floridi, 2013) for digital remains management.

---

[1] Macro-level ethical analyses are analyses that focus on broader "societal processes," as opposed to individual-level experiences (Ohman, 2020, p. 25).

[2] Infraethics, an ethics of information term, refers to values that "support the flourishing of moral actions" though "are not good in themselves, nor are they sufficient to determine morally good outcomes" (Taddeo, 2016, p. 7).

In this chapter, I employ the levels of abstraction (LoA) method[3] to contemplate the human dignity of the online dead from the perspectives of archaeology, grave re-use, and archival curation – practices that either manage physical human remains or preserve physical cultural heritage. In doing so, I locate an ethically significant difference between the *near* and the *forgotten* dead that shapes our corresponding moral obligations. In managing the digital remains of the near dead, we must privilege the inherent human dignity of the informational corpse and their sentimental value to the living over their potential contributions to our digital cultural heritage. However, in managing the digital remains of the forgotten dead, we must privilege the meaningful curation of our digital cultural heritage, wherein certain data is preserved and other data is selectively deleted in the interests of humanity: a stakeholder group composed of past, present, and future generations (Ohman, 2020). Therein, I argue that pragmatism and selectivity are core infraethics that must be embraced in the regulation of digital remains. This chapter makes two contributions to online death studies. First, I enable conceptual clarity regarding the relationship between the human dignity of the dead, our duty to preserve our digital cultural heritage, and our moral obligations towards digital remains. Second, I make a policy contribution regarding the values that should be incorporated into the designs of platforms that currently or will house informational corpses.

2   Contextualizing Online Death

Online death studies have been dominated by two research agendas. First, there has been a great deal of research conducted regarding ICTs and experiences of grief (see Brubaker et al., 2013; Mitchell et al., 2012; Stillman, 2014; Walter et al., 2012). Walter et al., 2012, for example, argue that the internet has altered how we experience and interact with death by providing pathways for the living to continue their social connection to the online dead and their identities. Therein, they differentiate between physical and social death. The notion of two deaths aligns well with aforecited literature regarding informational versus physical corpses. ICTs do not furnish humankind with digital immortality, as they do not prevent biological death. Instead, they permit for an extension of personal identity with which the living can engage as a source of comfort, remembrance, and, in a limited form, continued sociality (Walter et al., 2012). Collectively, this research suggests that online grief has de-sequestered death from day-to-day life by unifying the realms of the living and the dead (Walter et al., 2012).

Understanding how the affordances of digital platforms facilitate grief is important research. However, such studies saturate the literature and, in focusing on the level of individuals, do not address the broader ethical concerns raised by digital remains, such as how they should be managed. In light of this gap, a small but

---

[3] A detailed description of this method is provided in Sect. 3.

emerging corpus of normative scholarship has sought to ascertain the moral status of digital remains and establish ethical principles that are relevant to their treatment. A key author in this vein of scholarship is Carl Öhman, whose doctoral work has been fundamental in advancing macro-ethical analyses of the online dead and the principles that ought to guide regulation in this space. Öhman's research is unified by a central theoretical concept, the *post-mortal condition*, which he uses to describe the collapse between the domains of the dead and living due to the continued "information[al] presence" of the deceased after biological death (Ohman, 2020, p. 160). The idea of a post-mortal condition reaffirms the aforementioned de-sequestering hypotheses of online grief researchers, while refocusing the importance of this domain-collapse towards an examination of our associated moral obligations.

Öhman's work adopts a Floridian understanding of personal identity and privacy. Floridi (2014a, b) notes that "human life is quickly becoming a matter of *onlife* experience," wherein ICTs are rendered "technologies of the construction of the self" (p. 210). In this way, we do not own information in the same way that we own material possessions. Instead, "information plays a constitutive role" in one's personal identity (Floridi, 2016, p. 308). Floridian ethics positions the source of harm in privacy violations to be connected to a denial of human dignity, wherein dignity is equated to a denial of the power to "[master] our own journeys" via "control over our own constitutive information" (Floridi, 2016, p. 310). Öhman's work also expands upon Patrick Stokes' research, which views biological and informational remains to be similar in moral status precisely because of their constitutive properties (see Stokes, 2012, 2015, 2019).

Macro-ethical analyses of online death consider both the human dignity of the online dead and their contribution to our digital cultural heritage: the "intellectual capital" captured by historical data (Cameron & Kenderdine, 2007, p. 1). Öhman 2020 speaks of the value of informational corpses to the "self-understanding of future generations" (p. 106) in a way that is evocative of how we view the value of physical cultural heritage objects that we might find in museums or archives. However, in contrast to physical cultural heritage objects, cheaper data storage capabilities afforded by advances in ICTs have made it easier to collect, compile, and analyze digital cultural heritage assets, in what Mayer-Schöenberger 2009 claims is a societal shift from forgetting to remembering. His overarching thesis is that forgetting is a virtue that should be protected in design. However, he does not advocate for an end to remembering. Instead, he finds that we must acknowledge the value of forgetting to societal progress via information expiration, as forgetting removes "the shackles of the past" and allows us "to live in the present" (p. 20). Thus, while authors have advanced the view that the preservation of our digital cultural heritage is a form of public good (see Waters, 2002, for example), there are concerns regarding what should be preserved, by whom, for how long, and for what purposes (Galloway, 2005).

This chapter thus adopts an ethics of information perspective and starts from the assumptions outlined in this formative scholarship. Specifically, I assume that one's information is constitutive of one's identity and that a deprivation or manipulation of control over one's personal identity is a violation of human dignity. I also take it to be true that our informational corpses are worthy of human dignity, as our data is still constitutive information after biological death. However, while an informational corpse can be said to be deserving of human dignity, it is still unclear how the human dignity of the online dead shapes our moral obligations towards them and how these moral obligations are further shaped by a duty to preserve our digital cultural heritage – the focus of my contributions.

3 Methodology

This research leverages the levels of abstraction (LoA) method, a core method within the ethics of information (Floridi, 2014a, b). A LoA is "a finite but non-empty set of observables accompanied by a statement of what feature of the system under consideration such a LoA stands for" (Taddeo & Floridi, 2016, p. 1577). LoAs assist philosophers in examining systems from different points of reference, which is accomplished by narrowing analyses to a limited set of observables, "interpreted typed variable[s]", relevant to a LoA (Floridi, 2014a, b, p. 31). A collection of more than one LoA is referred to as an interface and is used to holistically evaluate a phenomenon (Floridi, 2014a, b), such as digital remains management. In this research, my interface is comprised of three, non-hierarchical[4] LoAs, each of which take a stance on the appropriate treatment of *physical* human remains or *physical* cultural heritage. At each LoA, I consider how digital remains might be treated based on relevant observables.

The first LoA that I employ is that of an archeologist (LoA_{Arch}). Archeology, a discipline that partakes in the discovery, preservation, and exhibition of human physical remains for the purposes of conserving our physical cultural heritage, was adopted as a LoA because of Öhman's 2020 suggestion that the field may provide a fruitful ethical framework with which to adapt to digital remains management (see Öhman & Floridi, 2018 also). To build out the observables relevant to the LoA_{Arch}, I focus on two guidelines published by the World Archeological Congress (WAC), the Vermillion Accord on Human Remains (1989) and the Tamaki Makau-rau Accord on the Display of Human Remains and Sacred Objects (2005). The second LoA that I employ is that of grave re-use practices (LoA_{GR}). This LoA was included in my interface to unpack the sustainability dimensions of physical death and its subsequent translation online. To build out the observables relevant to the LoA_{GR}, I leverage the case study of Greek burial practices. The final LoA that I employ is that of archival curation (LoA_{Cur}). This LoA was included in my interface to understand how archivists maintain the integrity of archives, which entails decision-making about the preservation and destruction of records over time. To build out the observables relevant to the LoA_{Cur}, I refer to the International Council on Archives' Code of Ethics (1996).

---

[4] My interface is non-hierarchical because each LoA stands alone; there are no parent or daughter LoAs.

4 Analysis

There are few stories that better encapsulate the elevated moral status of human physical remains than Sophocles' *Antigone*. In the play, Antigone grieves the deaths of her brothers, Eteocles and Polyneices, following a civil war for control over the city of Thebes. While Eteocles is granted a burial because he had fought on the winning side of the war, King Creon, the new leader of Thebes, prevents Polyneices' burial by law and leaves his corpse to rot on the battlefield, as a traitor. In defiance of King Creon, Antigone honors her brother through a proper burial out of a belief that "there are honors due all the dead" (Sophocles, par. 410). While *Antigone* was first performed in 441 BCE, the pain that the heroine experiences at the indignant treatment of her brother's corpse still resonates today as a useful metaphor surrounding the ethical treatment of the dead, both offline and online. As such, this story will serve as the unifying thread in my analysis.

4.1 LoA_{Arch} – A Balancing Act

Ethical treatment of the dead figures heavily in archaeological ethics. In the Vermillion Accord, which pertains to the acquisition of human physical remains via archaeological excavation, the dead, the community of the dead (where relevant), and the scientific interests of broader society are all referenced as objects of archaeological duty. Similarly, while norms regarding the exhibition of human physical remains captured by the Tamaki Makau-rau Accord recognize the scientific value of displaying human remains in a museum context, this value is only permissibly extracted upon negotiation and acceptance with the community of the deceased. The Tamaki Makau-rau Accord also emphasizes the importance of cultural respect when exhibiting human physical remains. Thus, to practice cultural respect, archaeologists must engage in ongoing dialogue with living descendants of the deceased in order to define how cultural respect is practiced.

  While both archaeological accords emphasize respect for the body of the dead and the interests of living descendants in the excavation and display of human physical remains, there is no outright prohibition on such activities. The fact that no prohibition exists to prevent the disturbance of ancient remains for archaeological study acknowledges that an ability to understand our past through physical cultural heritage objects is of vital societal importance. In considering the treatment of human physical remains from the LoA_{Arch}, a balancing thus emerges between the inherent human dignity of the deceased, the interests of living descendants, and the scientific interests of broader society.

  There is a comparable balancing of interests that must take place regarding the management of the online dead. Digital traces on social media networks, such as Facebook, act as relics of the past that provide a comprehensive record of how humans lived at a particular point in time. As a result of the breadth and depth of human behavioral data that exists on online social networks, there is immense sociological value in preserving this information for the benefit of future generations and their self-understanding (Ohman, 2020). These traces also have inherent value as extensions of the identities of the deceased, which may be of particular sentimental value to living kin, in agreement with afore-cited literature on online grief. One can imagine a modern retelling of *Antigone*, where our heroine is not defying King Creon's mistreatment of Polyneices' physical corpse, but instead, defying Facebook's potential misuse of his digital remains, which may include premature deletion.

The notion of cultural reflexivity also resonates with digital remains management. Öhman 2020 notes that the majority of users who will become part of the online dead will be from non-Western nations, which means that the regulations that we introduce in this space must be sensitive to differences in cultural practices and beliefs surrounding death (see also Öhman et al., 2019; Öhman & Watson, 2019, 2020). For example, Indigenous peoples in Australia consider the physical remains of their ancestors to be sacred even thousands of years after death and disturbances of their remains are viewed as highly disruptive to living descendants (Blake, 2007). Developing policies for digital remains management must be culturally reflexive and take into account non-Western belief systems to be representative of the projected population of the online dead. The present task is to advance an infraethics that enables the ethical balancing of the interests presented in the LoA$_{Arch}$.

For archaeologists, it is normally impossible to know if the individual that they are exhuming would assent to their remains being used for scientific research. However, when it comes to our digital remains, users are well-positioned to leave directives regarding the preservation of their data after death and the conditions of its use. Facebook, for example, provides an option for users to set up a legacy contact that can delete your account or transition your page into a "memorialized account" that friends and family can virtually visit (Facebook, 2021). As per the LoA$_{Arch}$, which emphasizes the importance of the disposition of human remains in accordance with the wishes of the dead (or their living descendants), when final data directives are known, they must be adhered to in order to preserve the human dignity of the dead. Companies such as Facebook, which will control the social media data of the deceased, should undertake serious efforts to understand users' data directives *before* they die to ensure that they are not denying dignity to the dead. Where data directives are not available, companies that manage digital remains ought to defer to the wishes of their closest living kin, just as King Creon should have privileged Antigone's demands for Polyneices' burial over his own interests in painting Polyneices as a traitor.

4.2 LoA$_{GR}$ – The Near and Forgotten Dead

Considering digital remains management from the LoA$_{Arch}$ has shown that decisions surrounding data use after death is a complex negotiation between the directives of the dead (where they exist), the interests of the family (where they are expressed), and their inherent scientific value to society. Therein, upholding the human dignity of the dead by carrying out their final digital wishes, whether deletion or preservation, is our primary moral obligation. However, are our moral obligations towards the online dead static or dynamic?

Around the world, countries have been running out of grave space to bury their dead. As a result, grave re-use, which has been practiced in Europe since the eighteenth century, has become a practical solution for cemeteries in the face of increasing urban density and population growth (Rugg & Holland, 2017). For my purposes, I define grave re-use as any permissible disturbance of human physical remains, whether it be due to the overcrowding of burial sites or for a desired change in land use. In contemporary Greece, the site of Sophocles' tragedy, the Orthodox religion states that physical remains have to be "buried whole to make resurrection possible" (Mar 2011). A combination of religious prohibitions regarding cremation and a growing lack of grave space has made death a difficult process for Greek families. If they choose cremation, many are forced to travel to surrounding countries to access cremation services (Smith, 2019). If they choose burial, they are compelled to purchase a temporary gravesite, billed annually, where their loved one can remain buried for a period of 3 years, after which their bodies are exhumed and placed into an ossuary (Hadjimatheou, 2015). If there is no longer living family to pay for the storage of the bones or the family decides to cease payments, the bones are put into a mass grave referred to as a "digestive pit" (Hadjimatheou, 2015). While families with more financial resources can keep their loved ones buried for longer, Greece's population growth compared to the availability of burial plots has made grave re-use a necessary practice.

While Greek grave re-use practices may make the reader recoil in a similar fashion to corpse desecration, evaluating the management of digital remains at the LoA$_{GR}$ highlights an important tension between pragmatism, sustainability, and upholding the dignity of the dead (Rugg & Holland, 2017). Imagine Polyneices had been buried in a Theban cemetery that was struggling to bury the newly dead. Hundreds of years have passed and there are no living kin left to visit his gravesite or pay for its upkeep. His physical remains, which now consist mainly of bone fragments, are exhumed, cremated, and scattered in a mass grave. Is this scenario morally equivalent to a Greek woman being forced to exhume her mother just a few years after she has died? I would argue that it is not, as a result of an ethically significant difference between the *near dead* and the *forgotten dead* that alters our corresponding moral obligations towards the deceased.

While the near dead refers to those who have died that still have living kin, the forgotten dead refers to individuals that have been dead for a long enough period that everyone who would have known them have also died. Does grave re-use of the forgotten dead still constitute a violation of the human dignity of the deceased? Under Floridian ethics, the answer would be yes. Human dignity is not organic matter that degrades over time. When we exhume ancient graves to make room for the near dead or to put in a new rail line to meet the demands of growing populations, we are still compromising constitutive information about a person. However, when an individual shifts from being the near dead to the forgotten dead, grave re-use becomes a *morally permissible posthumous harm* in the interests of humanity. On this view, Antigone's authority over Polyneices' treatment is privileged when he is a member of the near dead. However, when Polyneices becomes a member of the forgotten dead, which assumes Antigone's demise, decisions regarding the treatment of his remains must be made in accordance with what is best for Thebans, past, present, and future.

The LoA_{GR} is a useful lens with which to examine the management of digital human remains because it asks us to consider how preserving the past affects our present and future. Unlike human physical remains, our digital remains do not decay, as long as the format in which they are preserved remains accessible and no data corruption occurs. A lack of data decay combined with cheaper data storage capabilities has led to companies retaining "inordinate amounts of information" rooted in economic calculations, wherein holistic data preservation is easier and more cost-effective than selective deletion (Mayer-Schönberger, 2009, p. 47). However, remembering everything requires the maintenance of an extensive physical infrastructure to store data that has a corresponding environmental impact. While the private nature of social media companies makes it difficult to ascertain the precise environmental footprint of their data centers, Facebook's highest business expense is their electricity expenditure (Hogan, 2013). In a New York Times investigation of the environmental impacts of large-scale data storage, it was uncovered that "online companies […] run their facilities at maximum capacity […] whatever the demand" and that this attitude of instantaneous, holistic access entails a wastage of "90 percent or more of the electricity [data centers] pull off the grid" (Glanz, 2012). Further, high energy demands have made data centers key offenders when it comes to excessive water usage and greenhouse gas emissions (Gmach et al., 2010), with data center maintenance accounting for 1% of all global electricity use (Mytton, 2020). The scale of this infrastructure and the demands that it makes on our environment will only continue to grow as the number of informational corpses we passively choose to preserve rises.

The idea that we can persist in the minds of future generations and be of use to others after our deaths through our data may serve as a source of comfort (Mayer-Schönberger, 2009). We would like to believe that someone, sometime in the future, might be curious about our lives even after we have ceased living them. But, just as the Greek stance against cremation is not a viable policy, the maintenance of every informational corpse that does exist or will exist in perpetuity is not a viable policy. The status quo surrounding online death makes a false assumption that we can or should try to withstand a growing environmental footprint that our earth cannot foreseeably support (Hogan, 2013, p. 9). However, even if we could withstand the size of this footprint does not mean that we should.

4.3 LoA_{Cur} – Selective Memory

When you acquire a dataset to conduct research, the first step in your analysis is to clean the data. You take the parts of the dataset that are relevant to your research questions, set them aside for future use, and remove the parts that are not relevant.

Data cleaning is closely related to the much older practice of archival curation. The ICA code of archival ethics states that "archivists should protect the integrity of archival material and thus guarantee that it continues to be reliable evidence of the past" (1996, p. 2). Therein, archivists are tasked with making essential decisions regarding what records to acquire, maintain, and destroy based upon the needs and purposes of the archive. Decisions surrounding the selective preservation of records in archival curation are based upon the value of a record to society (Shilton, 2012). The ICA acknowledges that curatorial decisions in archives "must have regard to the legitimate, but sometimes conflicting, rights, and interests of employers, owners, data subjects and users, past, present, and future" and that their "objectivity and impartiality" pertaining to what is best for the archive "is the measure of their professionalism" (1996, p. 1).

Like the curation of a physical archive, a duty to preserve our digital cultural heritage is less of a technical issue and more of a societal issue pertaining to how our society "chooses to preserve the record of its existence" (Galloway, 2005, p. 552). Of course, this is no simple task. The vast array of data that we presently collect provides a highly detailed look into the past that risks being burdensome to future generations without selective curation. Pitsillides et al. (2012) note that, as humans, we accumulate a vast amount of material objects over the course of our lives. While most of these objects are common and not essential to our identities, there may be a few items that we come to associate with a person. When they die, it is those items that we reflect on as memory objects, while the others are discarded. Returning to a modern retelling of *Antigone*, a time will come when all the social connections of a hypothetical, Facebook-using Polyneices will pass away. When they do, it will be up to the digital archivists of that time to determine what parts of Polyneices, if any, to preserve as additions to our digital cultural heritage, and which to delete.

We are still in the early stages of the phenomenon of digital death. At present, we have an opportunity to establish infraethics rooted in pragmatism and selectivity that will foster the ethical balancing of the human dignity of the dead, their sentimental value, and their scientific value. Pragmatism will ensure that our technical designs regarding the storage of digital remains are sustainable and do not overstep practical limitations, such as the health of our environment. Selectivity will ensure that data deletion driven by pragmatism will amplify the value of our digital cultural heritage to humanity by crystallizing the essential aspects of our digital remains and forgetting the rest. Without the pragmatic and selective curation of our digital remains, we will not vest future generations with a corpus of data that enables their self-understanding, but rather, contributes to their collective confusion.

5 Conclusion

As stated by Öhman and Floridi 2017: "the Internet will continue to be an integral part of everyday life and [...] humans will (at least in the organic sense) continue to die" (p. 657). This chapter has identified an urgent need for infraethics that balance the inherent, sentimental, and scientific value of the online dead. Using three levels of abstraction, I have proposed an infraethical framework, composed of pragmatism and selectivity, to guide the regulation of digital remains management. The LoA$_{Arch}$ identified that human remains may be permissibly disturbed to advance our self-understanding, as long as we take into account the dignity of the dead and the culturally-specific interests of descendants. The LoA$_{GR}$ identified that pragmatism must play a role in ensuring that digital afterlife practices remain sustainable for future generations, wherein a key distinction emerges regarding the near and forgotten dead. In recognizing that holistic, perpetual memory is not pragmatic, the LoA$_{Cur}$ identified the importance of selectivity in curating our digital archive to maximize its value to society.

Outside of my findings, this chapter presents several avenues for future research. As I only consider a limited selection of LoAs, future studies may seek to evaluate digital remains management from different perspectives. It will be especially useful to include LoAs that represent non-Western cultural beliefs surrounding death. Second, there are a dearth of suggestions pertaining to the appropriate data controller of informational corpses and the economic viability of different models of data ownership. Being in a position of deciding what to keep and what to delete from the contents of our informational corpses is one of great power that demands greater academic attention in subsequent studies.

At *Antigone's* tragic end, the head of the chorus states: "there is no happiness where there is no wisdom; No wisdom but in submission to the gods. Big words are always punished, And proud men in old age learn to be wise" (Sophocles, par. 1040). King Creon's hamartia was believing that his convictions should take precedence over god's law. Our hamartia, perhaps, is believing that every data point that we produce throughout our lives is worthy of remembrance. While Creon was a victim of his hubris, the infraethics proposed in this chapter present an opportune way for us to learn from his mistakes.

References

Advisory Panel on the Archeology of Burials in England. (2017). *Guidance for best practice for the treatment of human remains excavated from Christian burial grounds in England* (Vol. 2, pp. 1–51). Advisory Panel on the Archeology of Burials in England. https://www.archaeologyuk.org/apabe/pdf/APABE_ToHREfCBG_FINAL_WEB.pdf

Blake, J. (2007, July). Beyond death: The treatment of indigenous human remains—A human rights perspective. *Islam and Christian–Muslim Relations, 18*(3), 367–375. https://doi.org/10.1080/09596410701396113

Brubaker, J. R., et al. (2013). Beyond the grave: Facebook as a site for the expansion of death and mourning. *The Information Society, 29*(3), 152–163. https://doi.org/10.1080/01972243.2013.777300

Cameron, F., & Kenderdine, S. (2007). *Theorizing digital cultural heritage a critical discourse.* MIT Press. *Open WorldCat,* http://proxy.library.carleton.ca/login?url=http://site.ebrary.com/lib/oculcarleton/Top?id=10190483

Facebook. (2021). Choose a legacy account. *Facebook*. https://www.facebook.com/hel p/103897939701143?helpref=faq_content

Floridi, L. (2013). Infraethics. *The Philosophers' Magazine, 60*, 26–27. https://doi.org/10.5840/ tpm20136010

Floridi, L. (2016). On human dignity as a foundation for the right to privacy. *Philosophy & Technology, 29*(4), 307. https://doi.org/10.1007/s13347-016-0220-8

Floridi, L. (2014a). *The ethics of information*. Oxford University Press.

Floridi, L. (2014b). *The onlife manifesto : Being human in a hyperconnected era*. Springer Open.

Galloway, P. (2005, September). Preservation of digital objects. *Annual Review of Information Science and Technology, 38*(1), 549–590. https://doi.org/10.1002/aris.1440380112

Glanz, J. (2012, September). Power, pollution, and the internet. *The New York Times, 22*. https:// www.nytimes.com/2012/09/23/technology/data-centers-waste-vast-amounts-of-energy-belying-industry-image.html

Gmach, D., et al. (2010). Profiling sustainability of data centers. In *Proceedings of the 2010 IEEE international symposium on sustainable systems and technology* (pp. 1–6). IEEE. https://doi. org/10.1109/ISSST.2010.5507750

Hadjimatheou, C. (2015). "Why Greeks Are Exhuming Their Parents." *BBC News, 26*, https:// www.bbc.co.uk/news/magazine-34920068

Hogan, M. (2013, November). Facebook data storage centers as the archive's underbelly. *Television & New Media, 16*(1), 3–18. SAGE. https://doi.org/10.1177/1527476413509415.

International Council on Archives. (1996). *Code of ethics* (pp. 1–3). International Council on Archives. https://www.ica.org/sites/default/files/ICA_1996-09-06_code%20of%20 ethics_EN.pdf

Lingel, J. (2013, May). The digital remains: Social media and practices of online grief. *The Information Society, 29*(3), 190–195. https://doi.org/10.1080/01972243.2013.777311

Mayer-Schönberger, V. (2009). *Delete: The virtue of forgetting in the digital age*. Princeton University Press.

Mitchell, L. M., et al. (2012). Death & grief on-line: Virtual memorialization and changing concepts of childhood death and parental bereavement on the internet. *Health Sociology Review*, 2125–2164. https://doi.org/10.5172/hesr.2012.2125

Mytton, D. (2020, August). Hiding greenhouse gas emissions in the cloud. *Nature Climate Change, 10*(8), 701–701. https://doi.org/10.1038/s41558-020-0837-6

Öhman, C., et al. (2019, June). Prayer-bots and religious worship on twitter: A call for a wider research agenda. *Minds and Machines, 29*(2), 331–338. https://doi.org/10.1007/ s11023-019-09498-3

Ohman, C. (2020). *The post-mortal condition : Being with the dead in the age of digital media*. ProQuest Dissertations Publishing.

Öhman, C., & Floridi, L. (2018). An ethical framework for the digital afterlife industry. *Nature Human Behaviour, 2*(5), 318. https://doi.org/10.1038/s41562-018-0335-2

Öhman, C. (2017). The political economy of death in the age of information: A critical approach to the digital afterlife industry. In *Minds and machines* (Vol. 27). Springe. https://doi.org/10.1007/ s11023-017-9445-2

Öhman, C., & Luciano F. (2017). "The Political Economy of Death in the Age of Information: A Critical Approach to the Digital Afterlife Industry." *Minds and Machines*, vol. 27. https://doi. org/10.1007/s11023-017-9445-2.

Öhman, C. J., & Watson, D. (2019). Are the dead taking over Facebook? A big data approach to the future of death online. *Big Data & Society, 6*(1). https://doi.org/10.1177/2053951719842540

Öhman, C., & Watson, D. (2020). Are the dead taking over Instagram? A follow-up to Öhman and Watson 2019. In J. Cowls & J. Morely (Eds.), *The 2020 yearbook of the digital ethics lab*. Springer.

Pitsillides, S., et al. (2012). Museum of the self and digital death: An emerging curatorial dilemma for digital heritage. In E. Giaccardi (Ed.), *Heritage and social media: Understanding heritage in a participatory culture* (1st ed.). Routledge.

Rugg, J., & Holland, S. (2017, January). Respecting corpses: The ethics of grave re-use. *Mortality, 22*(1), 1–14. https://doi.org/10.1080/13576275.2016.1192591

Smith, H. (2019, Mar). "Greece Defies Church with Step towards First Crematorium." *The Guardian, 12.* https://www.theguardian.com/world/2019/mar/12/greecedefies-church-with-step-towards-first-crematorium.

Sophocles. *Antigone*. Edited by Dudley Fitts and Robert Fitzgerald. https://mthoyibi.files.wordpress.com/2011/05/antigone_2.pdf.

Stillman, A. (2014). Virtual graveyard: Facebook, death, and existentialist critique. In C. M. Moreman & A. D. Lewis (Eds.), *Digital death : Mortality and beyond in the online age.* Praeger.

Stokes, P. (2015). Deletion as second death: The moral status of digital remains. *Ethics and Information Technology, 17*(4), 237–248. https://doi.org/10.1007/s10676-015-9379-4

Stokes, P. (2019). *Digital remains: Ethical preservation, disposal and reuse of online artefacts of the dead* (pp. 1–6). Submission, New South Wales Law Reform Commission. http://172.105.175.196/wp-content/uploads/NSWLRC-Submission.pdf

Stokes, P. (2012). Ghosts in the machine: Do the dead live on in Facebook? *Philosophy & Technology, 25*(3), 363–379. https://doi.org/10.1007/s13347-011-0050-7

Taddeo, M. (2016, December). Data philanthropy and the Design of the infraethics for information societies. *Philosophical Transactions of the Royal Society A: Mathematical, Physical and Engineering Sciences, 374*(2083), 20160113. https://doi.org/10.1098/rsta.2016.0113

Taddeo, M., & Floridi, L. (2016). The debate on the moral responsibilities of online service providers. *Science and Engineering Ethics, 22*(6), 1575–1603. https://doi.org/10.1007/s11948-015-9734-1

Walter, T., et al. (2012, June). Does the internet change how we die and mourn? Overview and analysis. *OMEGA - Journal of Death and Dying, 64*(4), 275–302. https://doi.org/10.2190/OM.64.4.a

Waters, D. (2002). Good archives make good scholars: Reflections on recent steps toward the archiving of digital information. In *The state of digital preservation: An international perspective* (pp. 78–95). Council on Library and Information Resources.

World Archaeological Congress. (2006). The Tamaki Makau-Rau accord on the display of human remains and sacred objects. *World Archaeological Congress*. https://worldarch.org/code-of-ethics/

World Archaeological Congress. (1989). Vermillion accord on human remains. *World Archaeological Congress*. https://worldarch.org/code-of-ethics/

The Statistics of Interpretable Machine Learning

David S. Watson

Abstract Statisticians, especially those who do applied work in the natural and social sciences, have long been interested in understanding model parameters and predictions. However, the distinct subfield of interpretable machine learning (iML), with its focus on general purpose explanations at varying degrees of resolution, is a much more recent development. In this chapter, I go beyond mere taxonomies and survey some of the most influential iML proposals at length. My goal in so doing is not to undertake the Sisyphean task of constructing a comprehensive review – numerous articles and at least one monograph have attempted this feat, only to be rendered obsolete within a few months – but rather to catalogue the assumptions and methods that characterise the most influential directions of research in iML. This overview is an essential step to ground future analysis, introducing a range of technical concepts and notation designed to enable broader discussion across disciplines. The examples from this chapter give a sense of the methodological breadth of iML techniques, which draw on literature in statistics, computer science, and game theory in the effort to make ML models and predictions more interpretable.

Keywords Machine learning · Interpretability · Explainability · Algorithmic transparency

1 Introduction

Explainability is a relatively young subfield in ML, yet already the area is booming with active research on multiple fronts. In just the last few years – roughly concurrent to the dramatic uptick in academic publications on the topic (see Fig. 1) – iML has gone mainstream. It has been the subject of lengthy articles in both *The New Yorker* (Mukherjee, 2017) and *The New York Times Magazine* (Kuang, 2017),

D. S. Watson (✉)
Department of Statistical Science, University College London, London, UK

© The Author(s), under exclusive license to Springer Nature
Switzerland AG 2022
J. Mökander, M. Ziosi (eds.), *The 2021 Yearbook of the Digital Ethics Lab*,
Digital Ethics Lab Yearbook, https://doi.org/10.1007/978-3-031-09846-8_10

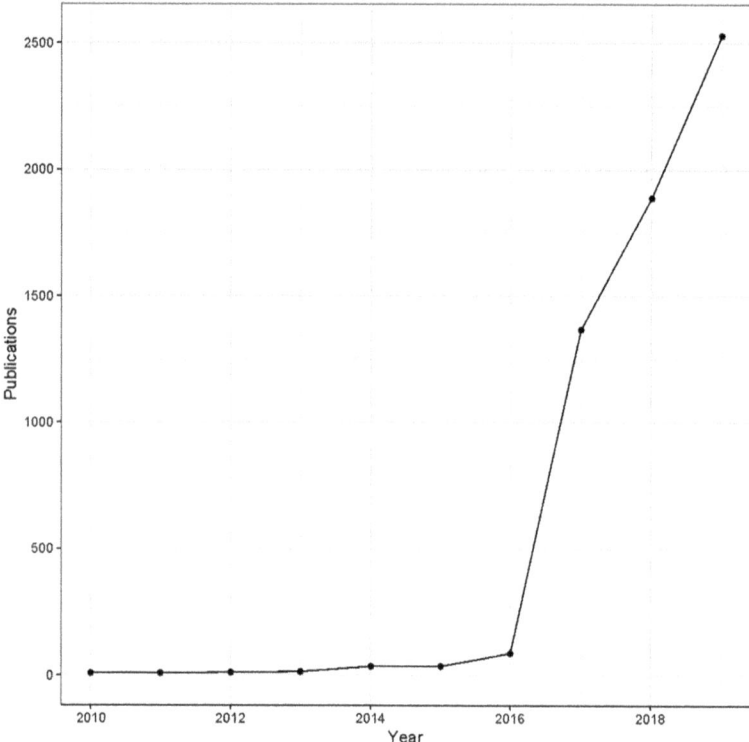

Fig. 1 Surge in research interest. The plot depicts the number of academic publications with "interpretable machine learning" or "explainable artificial intelligence" in the title, abstract, or keywords published between 2010 and 2019. (Source: Google Scholar)

as well as numerous TED talks (Hasani, 2019; Holzinger, 2019) and other public lectures that have collectively gathered hundreds of thousands of views online (Doshi-Velez, 2017; Hall, 2018; Lundberg, 2019). Meanwhile, major tech companies have begun to take notice. Google,[1] Microsoft,[2] and IBM[3] have all released open-source algorithmic explainability toolkits. Cloud computing services from Amazon,[4] Microsoft,[5] and Google[6] now include native implementations of various model interpretability methods.

In this chapter, I review some of the most prominent and promising proposals to have emerged from this burgeoning discourse. Building on the typology originally

---

[1] See https://pair-code.github.io/what-if-tool/

[2] See https://github.com/interpretml/interpret

[3] See http://aix360.mybluemix.net/

[4] See https://aws.amazon.com/blogs/machine-learning/ml-explainability-with-amazon-sagemaker-debugger/

[5] See https://docs.microsoft.com/en-us/azure/machine-learning/how-to-machine-learning-interpretability

[6] See https://cloud.google.com/explainable-ai/

introduced by Molnar (2020), I will explore both intrinsic and post-hoc methods operating at global and local resolutions. However, I will generally focus on model-agnostic approaches to the exclusion of model-specific alternatives, as the latter are simply too large in number and narrow in scope to warrant comprehensive coverage in this literature review. For good survey articles, see (Adadi & Berrada, 2018; Arrieta et al., 2020; Guidotti et al., 2018a, b; Murdoch et al., 2019); for book-length works, see (Molnar, 2020; Samek et al., 2019).

A number of commentators contend that the technical goals of iML are under-specified. Whereas formal criteria for algorithmic fairness abound (Narayanan, 2018), explainability is generally harder to quantify or optimise. Lipton identifies a vague and heterogeneous collection of concepts that jointly form "the mythos of model interpretability" (2016). He argues that without greater clarity on what exactly interpretability means and why it is important, efforts to build more explainable models will continue to be dragged down by implicit presuppositions that make it difficult or impossible to compare proposals. Doshi-Velez & Kim (2017) likewise emphasise the lack of consensus regarding the definition or operationalisation of algorithmic explainability. They provide a thorough taxonomy of iML, ultimately advocating a pragmatic focus on empirically demonstrating a model's ability to promote greater understanding among human subjects. In the years since their influential preprint was posted to *arXiv*, the research agenda outlined by Doshi-Velez & Kim has attracted enormous interest and helped inspire a wide array of new statistical methodologies.

The remainder of this chapter is structured as follows. I consider local linear approximators in Sect. 2, focusing especially on the popular iML tools LIME and SHAP. A critical discussion of rule lists follows in Sect. 3, examining both global and local recursive partitioning schemes. I review a number of case-based methods in Sect. 4, including algorithms that identify or construct prototypes and counterfactuals. I analyse feature importance measures in Sect. 5, including marginal and conditional metrics. (Note that, in an effort to maintain consistency, I will adapt notation throughout to conform to my own formalisms.) I conclude in Sect. 6 with a summary, and some future directions for research.

2   Local Linear Approximators

Many people believe, rightly or wrongly, that linear models are somehow inherently interpretable (Lipton, 2016). Several features of this function class probably contribute to this dogma, but I will focus on three assumptions in particular: that effects are *additive*, *monotonic*, and *constant*.[7] The additivity assumption means that predictions can be expressed as a weighted sum of input features. This is a convenient decomposition that allows users to quickly scan model parameters and potentially

---

[7] In generalised linear models (GLMs), it should be noted that these properties hold only in a transformed space defined by the link function. For instance, they are true in logit space (not probability space) for logistic regression.

even grasp the relative importance of variables (presuming they have been properly standardised and are not too numerous). The monotonicity constraint means that the association between a predictor and an outcome is either positive, negative, or zero. No further internal variation is permitted. For instance, say we want to evaluate the impact of alcohol consumption on sociability via linear regression, which assigns a coefficient of 1.5 to the predictor variable. Then, according to this model, more alcohol *always* equates with greater sociability. The regression does not care whether you just arrived at a party or just woke up from a hangover. Moreover, because linear effects are constant, the increase in sociability is directly proportional to the increase in alcohol intake. For every unit of alcohol imbibed, sociability goes up exactly 1.5 units, on average.

Of course, these traits that make linear models so easy to interpret are precisely what can make them so inaccurate in practice. Many systems of interest are not additive, monotonic, or constant, and imposing these assumptions where they do not apply results not only in poor predictive performance but in model parameters with no clear interpretation. To continue with the example above, say we add a binary predictor to our model that indicates whether or not data subjects have a history of alcoholism. In this case, sociability is not a simple additive function of the inputs; instead, we need an interaction term that assigns one (presumably positive) coefficient to alcohol intake for non-alcoholics, and another (presumably negative) coefficient for alcoholics. Linear models can adapt to this setting when interactions or mixed effects are made explicit, but cannot detect such subtleties on their own the way many ML algorithms do. Moreover, the monotonicity constraint is unreasonable here. A bit of alcohol may tend to make people chattier, but there are quantities of alcohol that would make even the world's most gregarious individual decidedly unsociable. In any event, the relationship is surely not constant. The effect of another drink on one's sociability is largely determined by how many drinks one has already had.

In full awareness of these considerations, many researchers in iML have promoted a solution I shall term *local linear approximation*. The idea is not to reconstruct the global behaviour of some complex regressor or classifier with a linear model – for this is likely impossible – but rather to use linear techniques merely to approximate the regression surface or decision boundary near a datapoint of interest. If you zoom in close enough to any point on a differentiable function, you will eventually find a tangent that can be expressed as a linear combination of input variables (see Fig. 2). By analysing the formula for this approximation, we may gain some intuition for the target function's behaviour in a particular region of the feature space.

This is the logic behind the locally interpretable model explanations (LIME) algorithm, an early and influential iML tool (Ribeiro et al., 2016). Specifically citing issues of algorithmic trust as a motivation for their work, the authors propose a novel explainability technique based on random sampling and regularised linear models. Given an input datapoint x and some target function f, LIME simulates a synthetic dataset by perturbing the coordinates of x, thereby creating a collection of counterfactual observations x'. By querying f at each point, LIME generates a

Fig. 2 A complex decision boundary (the pink blob/blue background) separates red crosses from blue circles. This function cannot be well-approximated by a linear model. But the boundary near the large red cross is roughly linear, as indicated by the dashed line. (From Ribeiro et al., 2016, p. 1138)

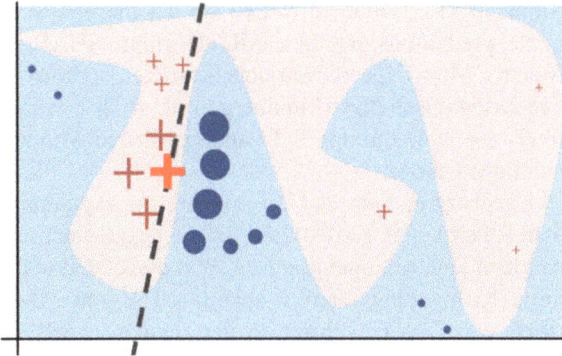

training dataset of $(x', y' = f(x'))$ pairs. The goal now is to learn an explanation model $g : X \rightarrow Y$. Points are weighted by their distance from the original target x using some appropriate similarity kernel k – the authors recommend the radial basis $\exp(-d(x, x')^2/\sigma^2)$, where d is a user-selected distance function and σ is the kernel bandwidth – and a sparse linear regression is fit to the data via weighted least squares (WLS). The lasso penalty that Ribeiro et al. apply to compute g is a sparsity-inducing regularisation technique that automatically assigns zero weight to uninformative predictors through a tuneable Lagrangian penalty λ on the L_1-norm of model coefficients (Tibshirani, 1996). Rather than selecting λ via cross-validation, as one typically would in a lasso prediction problem (Friedman et al., 2010), the parameter is selected to ensure at most m nonzero coefficients, where m is a user-selected hyperparameter. Ribeiro et al. demonstrate LIME's utility on text classification and image recognition problems, empirically validating their method through a human subject study in which participants were able to correctly identify which of two models was superior based on explanations extracted by LIME. The algorithm has been implemented in a popular Python library, available through the Python package index (PyPI).

A number of other local linear approximators debuted around the same time as LIME, each relying on different assumptions and optimised for different use cases. Examples include quantitative input influence (Datta et al., 2016); layer-wise relevance propagation (Bach et al., 2015); DeepLIFT (Shrikumar et al., 2017); and Shapley sampling values (Štrumbelj & Kononenko, 2014). In an award-winning NeurIPS paper, Lundberg & Lee (2017) show that these methods are all formally equivalent in the limit, up to some variation in their choice of kernel. The authors advocate for a particular kernel inspired by Shapley values, a foundational concept in game theory originally derived to solve *the attribution problem*, which asks how to fairly distribute surplus across a coalition of players in cooperative games (Shapley, 1953). It can be shown that Shapley values are the unique solution to the attribution problem satisfying certain desirable properties (see below). In this reframing, players are replaced by input features and Shapley values measure their contribution to a given prediction. Directly computing classical Shapley values is NP-hard, however numerous approximations have been proposed (Sundararajan &

Najmi, 2019). Lundberg & Lee are not the first to use this framework to explain model predictions, but their iML algorithm, SHAP, is especially efficient and user-friendly. Model-specific variants have been optimised for deep neural networks and tree-based ensembles (Lundberg et al., 2020), while a model-agnostic version is freely available through PyPI and distributed with all of the explainability toolkits mentioned above.

Lundberg & Lee (2017) formulate the explainability problem a bit differently than Ribeiro et al. (2016). Let f be the target function and g a corresponding explanation model. An input point $x \in R^p$ is associated with a simplified input $x' \in \{0, 1\}^p$ through a mapping function, $x = h_x(x')$. The goal is then to ensure that $g(z') \approx f(h_x(z'))$ whenever $z' \approx x'$, subject to certain constraints on the explanation model g. Specifically, Lundberg & Lee (2017) identify three desirable properties:

1. **Local Accuracy**

$$f(x) = g(x') = \phi_0 + \sum_{j=1}^{p} \phi_j x_j'$$

The explanation model matches the original model when $x = h_x(x')$, where $\phi_0 = f(h_x(0))$ represents the model output with all simplified inputs set to zero.

2. **Missingness**

$$x_j' = 0 \Rightarrow \phi_j = 0$$

Features with simplified values of zero have no attributed impact.

3. **Consistency**

Let $f_x(z') = f(h_x(z'))$ and $z' \backslash j$ denote setting $z_j' = 0$. For any two models f and f', if

$$f_x'(z') - f_x'(z' \backslash j) \ge f_x(z') - f_x(z' \backslash j)$$

for all inputs $z' \in \{0, 1\}^p$, then $\phi_j(f', x) \ge \phi_j(f, x)$.

Adapting Shapley's (1953) original theorem, Lundberg & Lee (2017) show that only one possible linear function g satisfies these properties, with coefficients given by:

$$\phi_j(f,x) = \sum_{z' \subseteq x'} \frac{m!(p-m-1)!}{p!} \left[f_x(z') - f_x(z' \backslash j) \right]$$

where m is the number of nonzero entries in z', and $z' \subseteq x'$ represents all z' vectors where the nonzero entries are a subset of the nonzero entries in x'. This result implies a so-called "Shapley kernel", which can be substituted for Ribeiro et al.'s exponential kernel to recover Shapley values through WLS regression.[8] For details, see Theorem 2 in (Lundberg & Lee, 2017).

Local linear approximators offer a fast and principled method for generating algorithmic explanations. However, they are inherently limited in several respects. First, as noted above, linear regression methods rely on strong assumptions that are often violated in practice. The restriction to local settings ameliorates these concerns somewhat but does not remove them altogether. Both LIME and SHAP are bound to produce unstable estimates for highly nonlinear regions of the model space. The best linear approximation can vary wildly depending on the range of application and/or associated kernel weights. It is unclear if and how classical methods for quantifying uncertainty in linear parameters may apply in this setting. Neither iML algorithm offers any method for evaluating the quality of the underlying approximation or its probable scope of applicability.

A second issue with these methods is that they provide no option for users to specify a contrast class of interest. The default behaviour of both LIME and SHAP is to explain why and how an outcome deviates from the mean response for the entire dataset, real or simulated. In many contexts, this makes sense. For instance, if a patient receives a rare and unexpected diagnosis, then she may want to know what differentiates her from the majority of patients. However, it seems strange to suggest, as these algorithms implicitly do, that "normal" predictions are somehow inexplicable. There is nothing confusing or improper about someone wondering, for instance, why they received an average credit score instead of a better-than-average one. Yet in their current form, neither LIME nor SHAP can accommodate such inquiries. To address these issues, we need more flexible iML tools.

3   Rule Lists

A growing body of psychological research suggests that humans are especially adept at generating and interpreting explanations in the form of *rule lists* – i.e., sequences of if-then statements (Lage et al., 2018). This accords with the privileged position of material implication in propositional logic, where its symbol → is typically regarded as a primitive relation, along with conjunction ($\wedge$), disjunction ($\vee$), and negation ($\neg$). These logical connectives form a functionally complete class, capable of expressing all possible Boolean operations.

In statistical contexts, rule lists are generally learned through some process of recursive partitioning. For instance, the pioneering classification and regression tree (CART) algorithm (Breiman et al., 1984) predicts outcomes by finding the optimal split c in a continuous predictor X such that error is minimised by segregating those samples with X-values greater than c from those with X-values less than c. For

---

[8] This relies on a strong but not uncommon assumption in the local linear approximation literature, namely that features are independent. See Equation 11 in (Lundberg & Lee, 2017), as well the critical discussion in (Sundararajan & Najmi, 2019).

categorical predictors, samples are separated by class label. A collection of such rules can be visually depicted as a tree, with branches corresponding to split points. Trees violate all of the aforementioned assumptions of linear models. They naturally detect interaction effects – the depth of a tree (i.e., number of recursive partitions) corresponds to the degree of interactions the model can capture – and easily adapt to nonmonotonic and even discontinuous functions. Computing optimal decision trees is NP-complete (Hyafil & Rivest, 1976), but CART uses greedy heuristics that typically work well in practice. Because individual trees tend to be unstable predictors, they are frequently combined through ensemble methods such as bagging (Breiman, 2001), in which predictions are averaged across trees trained on random bootstrap samples, and boosting (Friedman, 2001), in which predictions are summed over a series of trees, each sequentially optimised to improve upon the last. The resulting algorithms are extremely fast and flexible, putting recursive partitioning at the core of some of the most popular and powerful techniques in all of supervised learning (Biau & Scornet, 2016; Chen & Guestrin, 2016).

While combining basis functions tends to improve predictions, it unfortunately makes it difficult if not impossible to extract individual rules for better model interpretation. However, some regularisation schemes have been developed to postprocess complex learning forests for precisely this purpose. For instance, Friedman & Popescu (2008) propose the RuleFit algorithm, which mines a collection of Boolean variables by extracting splits from a gradient boosted forest. These engineered features are then combined with the original predictors in a lasso regression, producing a sparse linear combination of splits and inputs. Nalenz & Villani (2018) develop a similar procedure using a Bayesian horseshoe prior (Carvalho et al., 2010) instead of an L_1 penalty to encourage shrinkage. They also add splits extracted from a random forest with those learned via gradient boosting to promote greater diversity.

Another strand of research in this area has focussed on *falling* rule lists, which create monotonically ordered decision trees such that the probability of the outcome $Y = 1$ strictly decreases as one moves down the list. These models were originally designed for medical contexts, where doctors must evaluate patients quickly and accurately. For instance, Letham et al. (2015) design a Bayesian rule list to predict stroke risk, resulting in a model that outperforms leading clinical diagnostic methods while being small enough to fit on an index card (see Fig. 3). Falling rule lists can be very challenging to compute – see the note above about NP-completeness – and subsequent work has largely focussed on efficient optimisation strategies. Specifically, researchers have developed fast branch-and-bound techniques to prune the search space and reduce training time (Chen & Rudin, 2018; Yang et al., 2017), culminating in several tree-learning methods that are provably optimal under some restrictions on the input data (Angelino et al., 2018; Hu et al., 2019). For instance, Fig. 4 depicts the output of one such algorithm on the ProPublica dataset, for which the goal is to predict two-year criminal recidivism based on demographic and personal information. This simple tree outperforms the notorious COMPAS algorithm, a proprietary system used to assist judges evaluating pretrial risk in nine US states.

---

if hemiplegia **and** age > 60 **then** *stroke risk* 58.9% (53.8%–63.8%)
else if cerebrovascular disorder **then** *stroke risk* 47.8% (44.8%–50.7%)
else if transient ischaemic attack **then** *stroke risk* 23.8% (19.5%–28.4%)
else if occlusion and stenosis of carotid artery without infarction **then** *stroke risk* 15.8% (12.2%–19.6%)
else if altered state of consciousness **and** age > 60 **then** *stroke risk* 16.0% (12.2%–20.2%)
else if age ≤ 70 **then** *stroke risk* 4.6% (3.9%–5.4%)
else *stroke risk* 8.7% (7.9%–9.6%)

Fig. 3 Decision list for determining 1-year stroke risk following diagnosis of atrial fibrillation from patient medical history. Risk is given by the posterior mean, with 95% credible interval in parentheses. (From Letham et al., 2015, p. 1361)

if $(age = 18 - 20)$ **and** $(sex = male)$ **then predict** *yes*
else if $(age = 21 - 23)$ **and** $(priors = 2 - 3)$ **then predict** *yes*
else if $(priors > 3)$ **then predict** *yes*
else predict *no*

Fig. 4 Output of the certifiably optimal rule list (CORELS) algorithm. This rule list predicts two-year recidivism in the ProPublica dataset. (From Angelino et al., 2018, p. 2)

More customisable solutions are proposed by Lakkaraju et al. (2016, 2019), who implement a number of methods for computing interpretable decision sets simultaneously optimised for accuracy and sparsity. Their model understanding through subspace explanations (MUSE) algorithm allows users to specify a fixed number of features through which to explain the behaviour of an underlying target function, effectively modulating between global and local resolutions with user-specified granularity. The resulting objective function is non-normal, non-negative, non-monotone, submodular, and constrained by matroids – a class of budgeted coverage problems known to be NP-hard (Khuller et al., 1999). However, approximate optimality can be guaranteed under mild assumptions, which Lakkaraju et al. exploit to compute efficient and interpretable decision sets that perform favourably against alternative rule list approaches in a number of user studies.

For all the advantages of these fast and occasionally optimal algorithms, rule lists remain prohibitively expensive to compute on data with more than a few dozen variables. But recall that our goal in iML is often not to learn a globally explainable model, but just to understand particular algorithmic prediction(s). Locally interpretable decision trees are not nearly as common as their global counterparts, but there have been some recent advances in this direction. Guidotti et al. (2018) introduce local rule-based explanations (LORE), which simulate a balanced dataset of cases using a genetic algorithm designed to sample heavily from points near the decision boundary. A decision tree g is then fit to the synthetic dataset, with special emphasis on both the input point of interest and the nearest counterfactual cases on the opposite side of the boundary. Explanations extracted from g then take a conjunctive form, providing short rule lists to explain both why $f(x) = 1$ and why $f(x) \neq 0$. Sokol & Flach (2020) introduce LIMEtree, a rule list variant of the aforementioned *locus classicus* of explainability methods, which allows users to interrogate particular predictions through a series of "What-if?" questions about possible perturbations of feature values. The method comes with local fidelity guarantees and is more adaptable than its linear forebear.

More recently, the authors of LIME have proposed a follow up method called *anchors* (Ribeiro, Singh, & Guestrin, 2018). Anchors are specifically designed to address the scoping problem raised in Sect. 2. Given an explanandum $f(x)$, the goal is to find a set of Boolean conditions A (the eponymous anchor) such that $A(x) = 1$ and

$$E_{D_x(z|A)}\left[I\big(f(x) = f(z)\big)\right] \geq \tau,$$

where the expectation is taken with respect to a conditional perturbation distribution $D_x(\bullet \mid A)$, which represents a density centred at x where the conditions in A hold. $I(\bullet)$ denotes the indicator function and τ a tuneable threshold parameter that Ribeiro et al. (2018) call *precision*. Once τ is fixed, the goal is to maximise *coverage*, formally defined as $E_{D_x(z)}\left[A(z) = 1\right]$, i.e. the proportion of datapoints to which the anchor applies. This generalises the notion of rule lists to include both global and local explanations, as the former can simply be expressed as anchors with high (ideally unit) coverage. The authors reframe recursive partitioning as a reinforcement learning problem, combining graph search with a multi-armed bandit algorithm to compute anchors. This iML approach is original and rigorous. It is a rare and welcome departure from standard work in this literature, where authors hardly ever bother to quantify the uncertainty of explanations or evaluate expected error rates. That said, anchors face some major hurdles in practice. First, continuous predictors often need to be discretised to avoid low-coverage anchors. Second, results depend heavily on tuning parameters buried in the method's subroutines. Finally, anchors do not scale well with data dimensionality.

Setting aside computational concerns, rule lists face a very different set of statistical challenges than linear models. Whereas the latter start from the assumption that all effects must be monotonic and constant, decision trees struggle to detect smooth, linear functions. Recursive partitioning naturally produces jagged regression surfaces that can only be smoothed out by increasing complexity, usually by incorporating more basis functions (see Fig. 5). The resulting models are generally not differentiable, which means parameters cannot be learned using popular optimisation techniques like gradient descent or the Newton-Raphson algorithm. An exception to this rule is posed by so-called "soft trees" (Kontschieder et al., 2015), which treat splits probabilistically rather than categorically. By parametrising the probability of splitting a given direction as a logistic function of a linear combination of inputs – a common formulation in neural networks – model parameters can be learned through

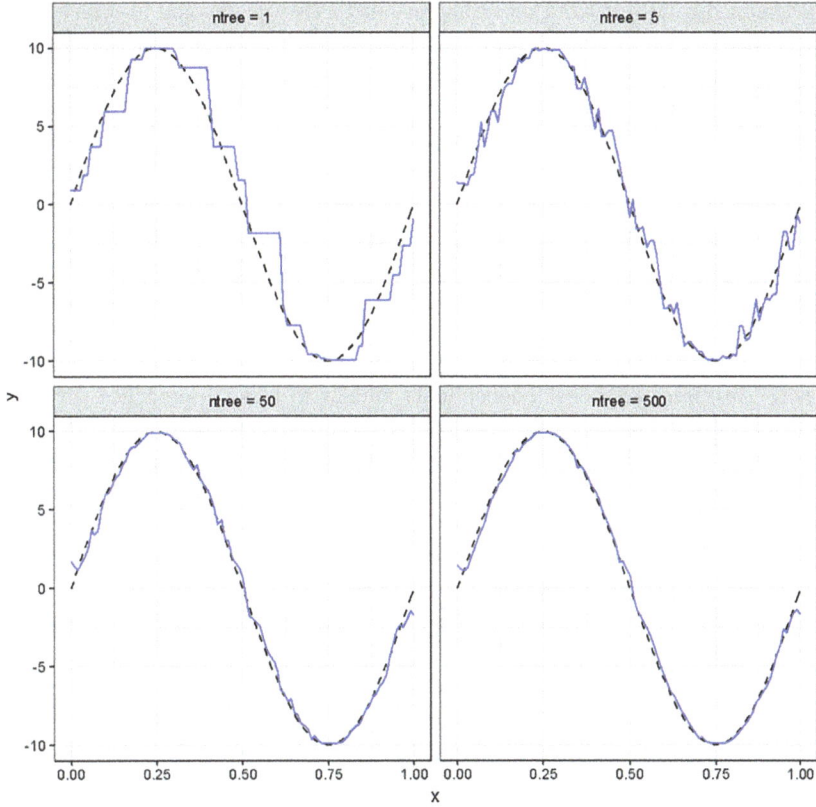

Fig. 5 Predictions from a random forest regression converging on a sine function as the number of trees in the ensemble increases. (From Watson, 2019, p. 429)

back-propagation. Some have even proposed this as a method for distilling deep networks into shallow trees for greater interpretability (Frosst & Hinton, 2017).

4   Case-Based Methods

Whatever the differences between linear models and rule lists, they share a certain formal similarity in the context of iML, in that both attempt to approximate some complex functional relationship with an alternative method considered more readily interpretable. The proposals considered in this subsection take an entirely different approach. Rather than building a new model g to explain a target model f, example-based methods opt for a strategy that might be summed up as "show, don't tell." They offer explanations in the form of particular cases intended either to exemplify a given class or highlight subtle, decisive differences between samples.

Some of the earliest work on exemplary methods was devoted to *prototypes*. A prototypical example of a given class might be thought of as a sort of Platonic ideal, a central theme upon which all other instances are merely a variation. In statistical learning, the concept is perhaps most familiar in the context of clustering, where the classic k-means algorithm (Forgy, 1965) identifies k separate centroids, one for each cluster. A centroid is a kind of prototype where coordinates for each variable are set to the group-wide mean. This may create impossible datapoints, however, for example when some variables are categorical or integer-valued. An alternative method more appropriate in such settings is the k-medoids algorithm (Kaufman & Rousseeuw, 1990), which replaces centroids with medoids, where coordinates for each variable are set to the group-wide median. By construction, these coordinates must exist somewhere in the training data. The prototype selection (PS) method of Bien & Tibshirani (2011) is conceptually similar to these approaches. Unlike in unsupervised learning, PS uses labels to partition the data. The prototype for class $Y = 1$ can then be understood in geometric terms as whichever sample is closest to other $Y = 1$ cases and farthest from all $Y \neq 1$ cases on average. Thus the output of PS is neither a centroid (containing coordinates found nowhere in the data) nor a medoid (containing coordinates cobbled together from various samples), but rather some actual training case that best represents its label.

Another approach in this area is the Bayesian case model (BCM), a generative algorithm for identifying prototypes (Kim et al., 2014). BCM treats observations as the result of a mixture of k classes, where k is supplied by the user. The method works by learning the subspace of features most strongly associated with a particular label. The sample that maximises the posterior probability of class membership conditional on the cluster subspace is the prototype. This inference relies on a complex set of hierarchical priors and collapsed Gibbs sampling procedures. Kim et al. (2016) modify the original BCM procedure by finding not just a prototype for each class, but a *criticism* as well – i.e., the datapoint least well represented by its prototype. Their MMD-critic algorithm embeds the data in a reproducing kernel Hilbert space (RKHS), where distances between distributions can be efficiently computed with the maximum mean discrepancy (MMD) statistic (Gretton et al., 2007). Prototypes and criticisms are found via greedy search in the RKHS, by maximising and minimising, respectively, the distance between their distributions and the rest of the data (see Fig. 6).

One especially promising work in this area is the ProtoPNet, which Chen et al. (2019) implement using state of the art image classification techniques based on deep convolutional neural networks. Their model architecture includes a so-called "prototype layer" in between the convolutional layers and the fully connected layer, tasked with learning prototypes of each class label based on high-level features (e.g., classifying bird species by the shape of beaks and wings) and computing distances between such prototypes and input image segments. The resulting model can be said to reason by analogy – the title of the paper is "*This* looks like *that*" – which, the authors persuasively argue, is more intelligible to humans than complex optimisation procedures over some high-dimensional parameter space (see Fig. 7).

Fig. 6 Learned prototypes and criticisms from the ImageNet dataset (two types of dog breeds). (From Kim et al., 2016, p. 8)

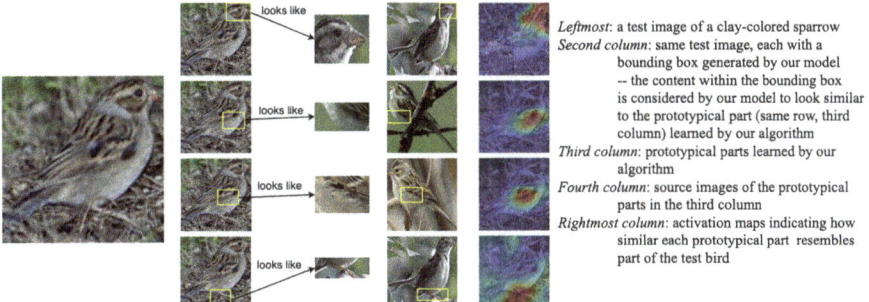

Leftmost: a test image of a clay-colored sparrow
Second column: same test image, each with a bounding box generated by our model -- the content within the bounding box is considered by our model to look similar to the prototypical part (same row, third column) learned by our algorithm
Third column: prototypical parts learned by our algorithm
Fourth column: source images of the prototypical parts in the third column
Rightmost column: activation maps indicating how similar each prototypical part resembles part of the test bird

Fig. 7 Example output from the ProtoPNet model on a clay-coloured sparrow. (From Chen et al., 2019, p. 2)

More recent work in case-based reasoning has focused on *counterfactuals*, as opposed to prototypes. Wachter et al. (2018) suggest explaining the algorithmic decision $f(x) = y$ by finding a set of nearest counterfactual neighbours – say, all simulated datapoints x' within an ε-ball of x such that $f(x') = y' \neq y$. Differences between these x' and x are explanatory to the extent that they indicate a minimal set of perturbations sufficient to change a prediction by some prespecified threshold. Building on seminal results in deep learning, Wachter et al. use generative adversarial networks (GANs) to optimise for x'. GANs were originally developed in machine vision, where Goodfellow et al. (2014) first observed that minor perturbations of input pixels can result in new synthetic images indistinguishable from the original – at least to the human eye – yet capable of fooling otherwise high-performing machine vision algorithms into strange and consistent misclassifications. Using a similar approach, Wachter et al. propose the following objective function:

$$\lambda\left(f\left(x'\right) - y'\right)^2 + d\left(x, x'\right)$$

where d is a distance measure between points and λ controls the trade-off between squared error and distance. The authors recommend a standardised L_1 metric for d, as this tends to promote sparse solutions that minimise the number of axes along which x and x' are likely to differ. A simplified version of this algorithm that avoids optimisation altogether is implemented in Google's What-If Tool (Wexler et al., 2020), which selects counterfactual cases by finding the observed case(s) x' nearest to x such that $f(x') = y'$.

The counterfactual method is attractive in several respects. Not only does it elegantly tie together disparate disciplinary influences, but it does so in a manner that is sophisticated and novel. Wachter et al.'s (2018) paper has shifted the iML discourse away from the approximation methods chronicled in Sects. 2 and 3 and towards case-based reasoning with greater success than the prototype methods that preceded it (Artelt & Hammer, 2019). It has spawned an especially active and promising subfield known as "algorithmic recourse," which studies automatic methods for explaining unfavourable outcomes and recommending actions to alter them. See (Karimi et al., 2020a, b) for a recent survey.

However, counterfactual explanations are not without their difficulties. First, the method is liable to produce a wide range of plausible explanations x' within some ε-ball of x. Some authors consider this a virtue, since different explanations can provide different details about a given outcome. But without some further method for evaluating the quality of these candidate x', this pluralism may become confusing or worse – it could give bad actors an easy way to avoid accountability should some explanations reflect more favourably than others upon algorithmic decisions. A second, arguably more important concern regards the unrestricted nature of adversarial attacks, which offer no guidance on how to limit the search space to genuine possibilities. The problem is especially acute in high dimensions, where data are often presumed to lie on a low-dimensional manifold from which the GAN is likely to deviate. This could potentially be mitigated through pre-processing steps that reduce the dimensionality of the input data, but this introduces whole new sources of potential confusion and error. Subsequent work has focused on restricting the search space to counterfactuals that are "actionable" (Ustun et al., 2019) or "coherent" (Russell, 2019) using mixed integer programming; or else those that are "feasible" in some causal sense (Karimi et al., 2020a, b; Mahajan et al., 2019), which is crucial to guarantee actionable recourse (Barocas et al., 2020). However, these methods only work at the expense of parametric constraints on f and/or strong assumptions about the data generating process.

5   Variable Importance

A final class of iML techniques I will consider in this literature review are variable importance (VI) measures. A helpful way to think about these methods is as different sorts of *interventions*, in the technical sense of the term common in computer science and statistics (Imbens & Rubin, 2015; Pearl, 2000). Let $Z = (X, Y)$ denote a dataset of $z_i = (x_i, y_i)$ pairs drawn i.i.d. from some fixed but unknown joint probability distribution, $P(Z) = P(X, Y)$. If our goal is to estimate the importance of feature subset $X^S \subseteq (X_1, \ldots, X_p)$ on learner $f : X \to Y$, then we can simply compare model performance before and after various perturbations of X^S. Let z_i denote the ith data-point in Z following an intervention on subset X^S (e.g., permutation or deletion). Let $L(f, z_i)$ denote the loss function for model f, evaluated at point z_i. Then we may define a random variable

$$\Delta_i = L\left(f, \tilde{z}_i\right) - L\left(f, z_i\right)$$

as the sample-wise difference in loss between perturbed and original data. This measure of local VI is formally equivalent to an individual treatment effect (ITE) for the intervention on X^S. The global measure can be estimated by taking the mean of this variable across the complete dataset, which in this causal reframing could be considered an average treatment effect (ATE).

This high-level overview glosses over many important details discussed below, such as how particular interventions are designed, whether $P(Z)$ factorises into a (known) structural model such that causal effects can be propagated throughout the associated graph, and whether f is held fixed or retrained following the perturbation of X^S. But hopefully this brief exposition provides some intuition for what unifies these heterogeneous VI methods. Whatever their assumptions or strategies, they all represent different attempts to quantify just how much predictive information is encoded in some feature subset – typically, just a single variable X_j.

VI methods can be categorised by three dichotomies: local/global, model-specific/model-agnostic, and marginal/conditional. The first contrast is formally outlined above. The second is a familiar distinction in iML. As noted previously, I will restrict my focus here to model-agnostic methods. The third dichotomy is argu-ably the most fundamental. To evaluate response variable Y's marginal dependence on predictor X_j, we test against the following hypothesis:

$$H_0^m : X_j \perp Y, X_{-j},$$

where X_{-j} denotes a set of covariates. A measure of conditional dependence, on the other hand, tests against a different null hypothesis:

$$H_0^c : X_j \perp Y | X_{-j}.$$

Observe that the former entails the latter, as conditional independence is just one possible form of marginal independence. Since H_0^c is more restrictive, we may find instances in which it holds but H_0^m does not. Specifically, this will be the case whenever X_j's marginal VI is high due to its association with X_{-j} rather than Y. This is why measures of marginal importance tend to favour correlated predictors. Often, however, our goal is to determine whether X_j adds any *new* information – in other words, whether Y is dependent on X_j even after conditioning on X_{-j}. This becomes especially important when the assumption of feature independence is violated.

A number of popular marginal VI methods are what I will call permute and predict (PaP) procedures. The first and most famous of these is Breiman's (2001) permutation importance, originally designed for random forests. The approach has since been generalised to other function classes, including most recently by Fisher et al. (2019), who introduce a number of "reliance" statistics that can be computed for any supervised learning algorithm. The basic idea with PaP procedures is to perturb some feature subset X^S by permuting its rows and observe the impact on performance for some fixed model f. Fisher et al. study the behaviour of this VI measure in individual models and diverse Rashomon sets, establishing uniform error bounds with the theory of U-statistics and drawing some unexpected connections between their measure and several causal estimands of interest.

The permutation approach has also led to a number of popular graphical tools that can help visualise complex associations in data. For instance, Friedman's (2001) partial dependence plots (PDPs) are widely used to evaluate the shape of relationships between predictors and outcomes. The partial dependence function is closely related to Breiman and Fisher et al.'s permutation measures, and can be formally defined as follows:

$$PD\left(x_i^S\right) = E_{X^R}\left[f\left(x_i^S, X^R\right)\right] = \int f\left(x_i^S, X^R\right) dP\left(X^R\right)$$

In other words, we integrate predicted values for f over the marginal distribution of covariates $X^R = X \backslash X^S$ while holding the point x_i^S constant. Zhao & Hastie (2019) observe that this is formally identical to Pearl's (2000) famous backdoor adjustment for estimating causal effects in graphical models, thereby enabling a causal interpretation of PD when the complementary subset X^R satisfies the backdoor criterion. Plotting empirical PD-estimates against X^S provides a visual summary of the (potentially causal) effect of the feature subset on model predictions $f(X^S, X^R)$. PDPs can also be adapted to visualise feature interactions and visually check for additive effects (Friedman, 2001; Friedman & Popescu, 2008). Goldstein et al. (2015) extend the method by decomposing partial dependence into n unique curves, each calculated by ranging over the empirical distribution of X^S while holding x_i^R constant. The mean of these curves at any given point is the corresponding PD-value. Assuming once again that the backdoor criterion holds, this helps visualise not just average but individual treatment effects (see Fig. 8).

Despite the success and conceptual appeal of PaP approaches, they face several practical and theoretical obstacles. First, the computational expense of permutations can be enormous for large datasets. Sampling procedures can offset the cost of such operations, but also introduce new sources of error and compromise any theoretical guarantees that hold in the full sample setting. More troubling, numerous commentators have pointed out that these methods tend to be badly biased in favour of correlated predictors (Gregorutti et al., 2015; Nicodemus et al., 2010; Toloşi & Lengauer, 2011). Hooker & Mentch (2019) explain the issue as one of extrapolation. Permuting features that are strongly associated with covariates results in improbable or even impossible samples very far from any in the algorithm's training

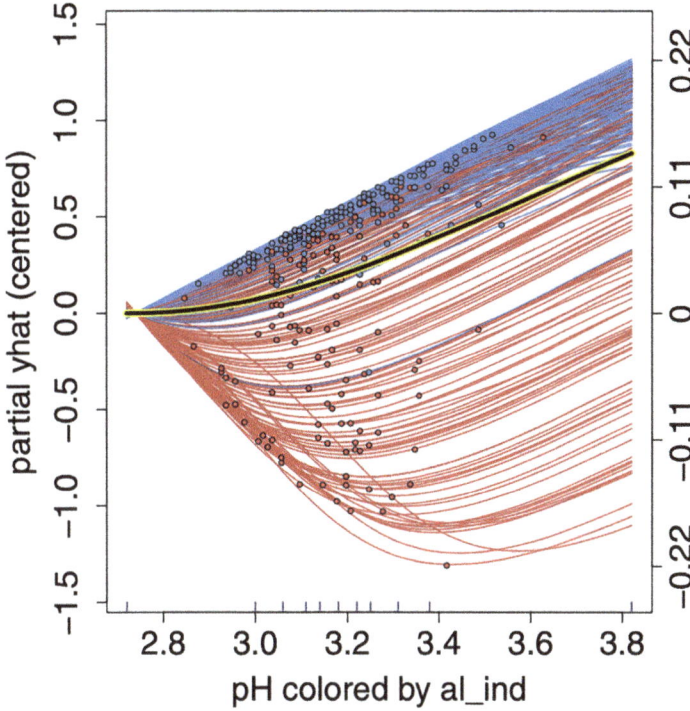

Fig. 8 An individual conditional expectation (ICE) plot depicting the partial dependence of wine ratings on pH level, with colour indicating whether the alcohol content is high (blue) or low (red). The black curve depicts Friedman's (2001) partial dependence function. (From Goldstein et al., 2015, p. 58)

data. For example, they note that a PaP procedure evaluating the importance of pregnancy status in a model f that also includes gender would force f to predict outcomes for pregnant men as often as pregnant women. Should f perform poorly on such datapoints – as we might expect – then pregnancy will receive high VI, even if it is independent of the response Y. Without further analysis, PaP methods cannot distinguish between variables that are truly predictive and those that merely appear so due to their association with covariates.

Broadly, there are two (model-agnostic) ways to evaluate VI that avoid the errors of PaP methods. One is to design a more targeted intervention on X^S that breaks its tie with the outcome Y while preserving the general covariance structure of the predictors. This requires some model of $P(X^S | X^R)$. Options in this direction include structural causal models (Pearl, 2000; Peters et al., 2017), which can rule out impossible coordinate combinations; semiparametric inference (van der Laan, 2006; van der Laan & Rose, 2011; van der Laan & Rose, 2018), which maximises likelihood for a single estimand while regressing out the effect of nuisance parameters; conditional permutation schemes (Berrett et al., 2020; Doran et al., 2014; Strobl et al., 2008), which constrain the set of allowable permutations in a data-adaptive manner;

or knockoff approaches (Barber & Candès, 2015; Candès et al., 2018), which compare observed data with synthetic data designed to minimise dependence between features and outcomes, conditional on covariates. Another way to estimate conditional VI is to simply relearn the model after permuting or deleting X^S. That is, we train null and alternative models from the same function class F on perturbed and original data, respectively:

$$f_0 = \frac{1}{n}\sum_{i=1}^{n} L(f,\tilde{Z}), \; f_1 = \frac{1}{n}\sum_{i=1}^{n} L(f,Z).$$

We then redefine the Δ variable accordingly:

$$\Delta_i = L(f_0,\tilde{z}_i) - L(f_1,z_i)$$

and use it to compute unbiased conditional estimates of global and local VI. In the parametric setting, the remove-and-relearn approach corresponds to the well-known likelihood ratio test in classical statistics (Lehmann & Romano, 2005). A more general leave one covariate out (LOCO) procedure is proposed by Lei et al. (2018) and studied further by Rinaldo et al. (2019). Mentch & Hooker (2016) advocate the permute-and-relearn approach, which they point out has the advantage of maintaining the same dimensionality for perturbed and original data. This helps ensure fair comparison of f_0 and f_1, especially when testing large feature subsets.

Of course, there are drawbacks to all of these methods. For example, it is not always clear how or if one can model $P(X^S| X^R)$ while maintaining independence from Y without assuming a great deal of domain knowledge or incurring an impractical computational burden. These are major hurdles for the structural modelling framework and targeted maximum likelihood estimation. Conditional permutation methods are limited to low-dimensional settings or particular model architectures, and generally require binning strategies to discretise continuous predictors. The original knockoff method relies on strong parametric assumptions. More general alternatives have been proposed (see, e.g., Bates et al., 2020; Berrett et al., 2020; Romano et al., 2019), but their applicability remains a matter of some dispute. Meanwhile, relearning f for each input variable – let alone the powerset of such variables – can be infeasible with even moderately large datasets and complex learning algorithms.

6 Conclusion

This chapter has introduced a number of different iML approaches, including methods based on local linear approximation, rule lists, case-based reasoning, and variable importance measures. There are advantages and disadvantages to all approaches chronicled herein. Notably, each places certain restrictions on the realm of possible functions or distributions considered. Such assumptions are necessary to get even minimally explanatory results. Imposing few or no assumptions may boost flexibility – this is the basic idea behind universal approximators such as neural networks and Gaussian processes – but the complexity of the resulting models is what led researchers to begin developing iML algorithms in the first place. Drawing inspiration from human psychology – attempting to adapt the limitations of an explanatory function class to the cognitive limitations of humans – has led to a diverse and expanding array of iML methodologies.

It is unlikely that iML will converge on any one class of explanations in the near future, at least not for arbitrary prediction tasks. There are at least two good reasons for this. First, there is no consensus about what sorts of explanations are most successful in general, although some common trends have been identified (Miller, 2019). With no unanimously agreed upon standard for explanations in any realm of science or human endeavour, we should not expect any greater consensus to emerge from AI research. Second, it is not obvious that any single explanatory method *should* reign supreme. Different agents require different explanations in different contexts, a perfectly rational instance of epistemological pluralism (Watson & Floridi, 2020). Thus, there is no single paradigm of iML that currently dominates – and that is a good thing. Rather than converge on any particular research agenda, a more likely and productive strategy for iML moving forward would be a sustained proliferation of new methods. This process is already evident in the last few years and likely to continue for some time.

Just because various iML algorithms are simultaneously in use, however, does not mean that practitioners should avoid alignment on certain standards and heuristics. For instance, there is already some evidence that counterfactual explanations are especially appropriate in legal settings (Artelt & Hammer, 2019), while feature attributions are more popular in bioinformatics (Watson, 2021). This sort of gradual standardisation is common in mature disciplines with complex workflows. The important point is simply for practitioners to stay apprised of ongoing research in iML, and not allow their explanatory preferences to ossify. This is no different in principle from encouraging researchers used to working with one set of familiar tools (e.g., linear regression) from maintaining an open mind to more novel alternatives (e.g., kernel methods).

In summary, there is a large and growing body of high-quality research in iML. The horizon is wide and bright for this fast-moving subdiscipline. Though statisticians and computer scientists are at the forefront of this burgeoning field, their work is of general interest to a wide range of practitioners – from doctors to bankers, policymakers to end users. It is incumbent upon anyone who relies on algorithms to aid in real-world decision-making to familiarise themselves with the challenges and opportunities presented by iML. I hope this literature review goes some way toward helping them fulfil that task.

References

Adadi, A., & Berrada, M. (2018). Peeking inside the black-box: A survey on explainable artificial intelligence (XAI). *IEEE Access, 6*, 52138–52160.

Angelino, E., Larus-Stone, N., Alabi, D., Seltzer, M., & Rudin, C. (2018). Learning certifiably optimal rule lists for categorical data. *Journal of Machine Learning Research, 18*(234), 1–78.

Artelt, A., & Hammer, B. (2019). On the computation of counterfactual explanations: A survey.

Bach, S., Binder, A., Montavon, G., Klauschen, F., Müller, K. R., & Samek, W. (2015). On pixel-wise explanations for non-linear classifier decisions by layer-wise relevance propagation. *PLoS One, 10*(7), 1–46.

Barber, R. F., & Candès, E. J. (2015). Controlling the false discovery rate via knockoffs. *Ann. Statist., 43*(5), 2055–2085.

Barocas, S., Selbst, A. D., & Raghavan, M. (2020). The hidden assumptions behind counterfactual explanations and principal reasons. In *Proceedings of the 2020 conference on fairness, accountability, and transparency* (pp. 80–89).

Barredo Arrieta, A., Díaz-Rodríguez, N., Del Ser, J., Bennetot, A., Tabik, S., Barbado, A., et al. (2020). Explainable Artificial Intelligence (XAI): Concepts, taxonomies, opportunities and challenges toward responsible AI. *Information Fusion, 58*, 82–115.

Bates, S., Candès, E., Janson, L., & Wang, W. (2020). Metropolized knockoff sampling. *Journal of the American Statistical Association*, 1–15.

Berrett, T. B., Wang, Y., Barber, R. F., & Samworth, R. J. (2020). The conditional permutation test for independence while controlling for confounders. *Journal of the Royal Statistical Society: Series B (Statistical Methodology), 82*(1), 175–197.

Biau, G., & Scornet, E. (2016). A random forest guided tour. *TEST, 25*(2), 197–227.

Bien, J., & Tibshirani, R. (2011). Prototype selection for interpretable classification. *The Annals of Applied Statistics, 5*(4), 2403–2424. https://doi.org/10.1214/11-AOAS495

Breiman, L. (2001). Random forests. *Machine Learning, 45*(1), 1–33.

Breiman, L., Friedman, J., Stone, C. J., & Olshen, R. A. (1984). *Classification and regression trees*. Taylor & Francis.

Candès, E., Fan, Y., Janson, L., & Lv, J. (2018). Panning for gold: 'Model-X' knockoffs for high dimensional controlled variable selection. *Journal of the Royal Statistical Society: Series B (Statistical Methodology), 80*(3), 551–577.

Carvalho, C. M., Polson, N. G., & Scott, J. G. (2010). The horseshoe estimator for sparse signals. *Biometrika, 97*(2), 465–480.

Chen, C., Li, O., Tao, D., Barnett, A., Rudin, C., & Su, J. K. (2019). This looks like that: Deep learning for interpretable image recognition. *Advances in Neural Information Processing Systems, 32*, 8930–8941.

Chen, C., & Rudin, C. (2018). An optimization approach to learning falling rule lists. In A. Storkey & F. Perez-Cruz (Eds.), *Proceedings of the twenty-first international conference on artificial intelligence and statistics* (pp. 604–612).

Chen, T., & Guestrin, C. (2016). XGBoost: A scalable tree boosting system. In *Proceedings of the 22Nd ACM SIGKDD international conference on knowledge discovery and data mining* (pp. 785–794).

Datta, A., Sen, S., & Zick, Y. (2016). Algorithmic transparency via quantitative input influence: Theory and experiments with learning systems. In *Proceedings – 2016 IEEE symposium on security and privacy* (pp. 598–617).

Doran, G., Muandet, K., Zhang, K., & Schölkopf, B. (2014). A permutation-based kernel conditional Independence test. In *Proceedings of the Thirtieth Conference on Uncertainty in Artificial Intelligence* (pp. 132–141).

Doshi-Velez, F., & Kim, B. (2017). Towards A Rigorous Science of Interpretable Machine Learning. arXiv preprint, 1702.08608.

Doshi-Velez, F. (2017). *A roadmap for the rigorous science of interpretability*. Retrieved from Talks at Google website: https://www.youtube.com/watch?v=MMxZlr_L6YE

Fisher, A., Rudin, C., & Dominici, F. (2019). All models are wrong, but many are useful: Learning a Variable's importance by studying an entire class of prediction models simultaneously. *Journal of Machine Learning Research, 20*(177), 1–81.

Forgy, E. (1965). Cluster analysis of multivariate data: Efficiency versus interpretability of classification. *Biometrics, 21*(3), 768–769.

Friedman, J. H. (2001). Greedy function approximation: A gradient boosting machine. *The Annals of Statistics, 29*(5), 1189–1232.

Friedman, J. H., & Popescu, B. E. (2008). Predictive Learning via Rule Ensembles. *The Annals of Applied Statistics, 2*(3), 916–954.

Friedman, J., Hastie, T., & Tibshirani, R. (2010). Regularization paths for generalized linear models via coordinate descent. *Journal of Statistical Software, 33*(1), 1–41.

Frosst, N., & Hinton, G. E. (2017). Distilling a neural network into a soft decision tree. In T. R. Besold & O. Kutz (Eds.), *Proceedings of the first international workshop on comprehensibility and explanation in AI and ML*.

Goldstein, A., Kapelner, A., Bleich, J., & Pitkin, E. (2015). Peeking inside the black box: Visualizing statistical learning with plots of individual conditional expectation. *Journal of Computational and Graphical Statistics, 24*(1), 44–65.

Goodfellow, I., Pouget-Abadie, J., Mirza, M., Xu, B., Warde-Farley, D., Ozair, S., et al. (2014). Generative adversarial nets. *Advances in Neural Information Processing Systems, 27*, 2672–2680.

Gregorutti, B., Michel, B., & Saint-Pierre, P. (2015). Grouped variable importance with random forests and application to multiple functional data analysis. *Computational Statistics & Data Analysis, 90*, 15–35.

Gretton, A., Borgwardt, K., Rasch, M., Schölkopf, B., & Smola, A. J. (2007). A kernel method for the two-sample-problem. *Advances in Neural Information Processing Systems, 19*, 513–520.

Guidotti, R., Monreale, A., Ruggieri, S., Pedreschi, D., Turini, F., & Giannotti, F. (2018a). Local rule-based explanations of Black Box decision systems. *arXiv* preprint, 1805.10820.

Guidotti, R., Monreale, A., Ruggieri, S., Turini, F., Giannotti, F., & Pedreschi, D. (2018b). A survey of methods for explaining black box models. *ACM Computing Surveys, 51*(5), 1–42.

Hall, P. (2018). *Building explainable machine learning systems: The good, the bad, and the ugly.* Retrieved from H20.ai website: https://www.youtube.com/watch?v=Q8rTrmqUQsU

Hasani, R. (2019). *A journey inside a neural network.* TED Talk. Retrieved from https://www.ted.com/talks/ramin_hasani_a_journey_inside_a_neural_network.

Holzinger, A. (2019). *From explainable AI to human-centered AI.* Ted Talk. Retrieved from https://www.ted.com/talks/andreas_holzinger_from_explainable_ai_to_human_centered_ai.

Hooker, G., & Mentch, L. (2019). Please Stop Permuting Features: An Explanation and Alternatives. *arXiv* preprint, 1905.03151.

Hu, X., Rudin, C., & Seltzer, M. (2019). Optimal sparse decision trees. *Advances in Neural Information Processing Systems, 32*, 7267–7275.

Hyafil, L., & Rivest, R. L. (1976). Constructing optimal binary decision trees is NP-complete. *Information Processing Letters, 5*(1), 15–17.

Imbens, G. W., & Rubin, D. B. (2015). *Causal inference for statistics, social, and biomedical sciences: An introduction.* Cambridge University Press.

Karimi, A.-H., Barthe, G., Schölkopf, B., & Valera, I. (2020a). A survey of algorithmic recourse: definitions, formulations, solutions, and prospects. *arXiv* preprint, 2010.04050.

Karimi, A.-H., Schölkopf, B., & Valera, I. (2020b). Algorithmic Recourse: From Counterfactual Explanations to Interventions. *arXiv* preprint, 2002.06278.

Kaufman, L., & Rousseeuw, P. (1990). *Finding groups in data.* Wiley.

Khuller, S., Moss, A., & Naor, J. (Seffi). (1999). The budgeted maximum coverage problem. *Information Processing Letters, 70*(1), 39–45.

Kim, B., Khanna, R., & Koyejo, O. O. (2016). Examples are not enough, learn to criticize! Criticism for interpretability. In *Advances in neural information processing systems 29* (pp. 2280–2288). Curran Associates, Inc.

Kim, B., Rudin, C., & Shah, J. (2014). The Bayesian case model: A generative approach for case-based reasoning and prototype classification. In *Proceedings of the 27th international conference on neural information processing systems – volume 2* (pp. 1952–1960). MIT Press.

Kontschieder, P., Fiterau, M., Criminisi, A., & Bulò, S. R. (2015). Deep neural decision forests. *IEEE International Conference on Computer Vision (ICCV), 2015*, 1467–1475.

Kuang, C. (2017, November). Can AI be taught to explain itself? *The New York Times Magazine.*

Lage, I., Chen, E., He, J., Narayanan, M., Gershman, S., Kim, B., & Doshi-Velez, F. (2018). An evaluation of the human-interpretability of explanation. *Advances in Neural Information Processing Systems.*

Lakkaraju, H., Bach, S. H., & Leskovec, J. (2016). Interpretable Decision Sets: A Joint Framework for Description and Prediction. *Proceedings of the 22Nd ACM SIGKDD International Conference on Knowledge Discovery and Data Mining*, 1675–1684.

Lakkaraju, H., Kamar, E., Caruana, R., & Leskovec, J. (2019). Faithful and Customizable Explanations of Black Box Models. *Proceedings of the 2019 AAAI/ACM Conference on AI, Ethics, and Society*, 131–138.

Lehmann, E. L., & Romano, J. P. (2005). *Testing statistical hypotheses (third edit)*. Springer.

Lei, J., G'Sell, M., Rinaldo, A., Tibshirani, R. J., & Wasserman, L. (2018). Distribution-free predictive inference for regression. *Journal of the American Statistical Association, 113*(523), 1094–1111.

Letham, B., Rudin, C., McCormick, T. H., & Madigan, D. (2015). Interpretable classifiers using rules and Bayesian analysis: Building a better stroke prediction model. *The Annals of Applied Statistics, 9*(3), 1350–1371.

Lipton, Z. (2016). The mythos of model interpretability. *arXiv* preprint, 1606.03490.

Lundberg, S. (2019). *Explainable AI for science and medicine*. Microsoft Research, Retrieved from. https://www.microsoft.com/en-us/research/video/explainable-ai-for-science-and-medicine/

Lundberg, S. M., Erion, G., Chen, H., DeGrave, A., Prutkin, J. M., Nair, B., et al. (2020). From local explanations to global understanding with explainable AI for trees. *Nature Machine Intelligence, 2*(1), 56–67.

Lundberg, S. M., & Lee, S.-I. (2017). A unified approach to interpreting model predictions. *Advances in Neural Information Processing Systems, 30*, 4765–4774.

Mahajan, D., Tan, C., & Sharma, A. (2019). *(2019). Preserving causal constraints in counterfactual explanations for machine learning classifiers.* CausalML.

Mentch, L., & Hooker, G. (2016). Quantifying uncertainty in random forests via confidence intervals and hypothesis tests. *Journal of Machine Learning Research, 17*(1), 841–881.

Miller, T. (2019). Explanation in artificial intelligence: Insights from the social sciences. *Artificial Intelligence, 267*, 1–38.

Molnar, C. (2020). *Interpretable machine learning: A guide for making black box models interpretable*. Christoph Molnar.

Mukherjee, S. (2017, April). A.I. versus M.D. *The New Yorker.*

Murdoch, W. J., Singh, C., Kumbier, K., Abbasi-Asl, R., & Yu, B. (2019). Definitions, methods, and applications in interpretable machine learning. *Proceedings of the National Academy of Sciences, 116*(44), 22071–22080.

Nalenz, M., & Villani, M. (2018). Tree ensembles with rule structured horseshoe regularization. *The Annals of Applied Statistics, 12*(4), 2379–2408.

Narayanan, A. (2018). *Tutorial: 21 fairness definitions and their politics*. Retrieved April 8, 2020, from https://www.youtube.com/watch?v=jIXIuYdnyyk

Nicodemus, K. K., Malley, J. D., Strobl, C., & Ziegler, A. (2010). The behaviour of random forest permutation-based variable importance measures under predictor correlation. *BMC Bioinformatics, 11*(1), 110.

Pearl, J. (2000). *Causality: Models, reasoning, and inference*. Cambridge University Press.

Peters, J., Janzing, D., & Schölkopf, B. (2017). *The elements of causal inference: Foundations and learning algorithms*. The MIT Press.

Ribeiro, M. T., Singh, S., & Guestrin, C. (2016). "Why should I trust you?": Explaining the predictions of any classifier. In *Proceedings of the 22nd ACM SIGKDD international conference on knowledge discovery and data mining* (pp. 1135–1144).

Ribeiro, M. T., Singh, S., & Guestrin, C. (2018). Anchors: High-precision model-agnostic explanations. *AAAI*, 1527–1535.

Rinaldo, A., Wasserman, L., & G'Sell, M. (2019). Bootstrapping and sample splitting for high-dimensional, assumption-lean inference. *Ann. Statist., 47*(6), 3438–3469.

Romano, Y., Sesia, M., & Candès, E. (2019). Deep Knockoffs. *Journal of the American Statistical Association,* 1–12.

Russell, C. (2019). Efficient search for diverse coherent explanations. In *Proceedings of the conference on fairness, accountability, and transparency* (20–28).

Samek, W., Montavon, G., Vedaldi, A., Hansen, L. K., & Müller, K. R. (Eds.). (2019). *Explainable AI: Interpreting, explaining, and visualizing deep learning.* Springer.

Shapley, L. (1953). A value for n-person games. In *Contributions to the theory of games* (pp. 307–317).

Shrikumar, A., Greenside, P., & Kundaje, A. (2017). Learning important features through propagating activation differences. In *Proceedings of the 34th international conference on machine learning.*

Sokol, K., & Flach, P. (2020). LIMEtree: Interactively customisable explanations based on local surrogate multi-output regression trees. *arXiv* preprint, 2005.01427.

Strobl, C., Boulesteix, A.-L., Kneib, T., Augustin, T., & Zeileis, A. (2008). Conditional variable importance for random forests. *BMC Bioinformatics, 9*(1), 307.

Štrumbelj, E., & Kononenko, I. (2014). Explaining prediction models and individual predictions with feature contributions. *Knowledge and Information Systems, 41*(3), 647–665.

Sundararajan, M., & Najmi, A. (2019). The many Shapley values for model explanation. In *Proceedings of the ACM conference.* ACM.

Tibshirani, R. (1996). Regression shrinkage and selection via the lasso. *Journal of the Royal Statistical Society. Series B (Methodological), 58*(1), 267–288.

Toloşi, L., & Lengauer, T. (2011). Classification with correlated features: Unreliability of feature ranking and solutions. *Bioinformatics, 27*(14), 1986–1994.

Ustun, B., Spangher, A., & Liu, Y. (2019). Actionable recourse in linear classification. In *Proceedings of the conference on fairness, accountability, and transparency* (pp. 10–19).

van der Laan, M. J., & Rose, S. (Eds.). (2011). *Targeted learning: Causal inference for observational and experimental data.* Springer.

van der Laan, M. J. (2006). Statistical inference for variable importance. *The. International Journal of Biostatistics, 2*(1).

van der Laan, M. J., & Rose, S. (Eds.). (2018). *Targeted learning in data science: Causal inference for complex longitudinal studies.* Springer.

Wachter, S., Mittelstadt, B., & Russell, C. (2018). Counterfactual Explanations without Opening the Black Box: Automated Decisions and the GDPR. *Harvard Journal of Law and Technology, 31*(2), 841–887.

Watson, D. (2019). The rhetoric and reality of anthropomorphism in artificial intelligence. *Minds and Machines, 29*(3), 417–440.

Watson, D. S. (2021). Interpretable machine learning for genomics. *Human Genetics.* https://doi.org/10.1007/s00439-021-02387-9

Watson, D., & Floridi, L. (2020). The explanation game: A formal framework for interpretable machine learning. *Synthese.* https://doi.org/10.1007/s11229-020-02629-9

Wexler, J., Pushkarna, M., Bolukbasi, T., Wattenberg, M., Viégas, F., & Wilson, J. (2020). The what-if tool: Interactive probing of machine learning models. *IEEE Transactions on Visualization and Computer Graphics, 26*(1), 56–65.

Yang, H., Rudin, C., & Seltzer, M. (2017). Scalable Bayesian rule lists. In *Proceedings of the 34th international conference on machine learning.*

Zhao, Q., & Hastie, T. (2019). Causal interpretations of black-box models. *Journal of Business & Economic Statistics,* 1–10.

Formalising Trade-Offs Beyond Algorithmic Fairness: Lessons from Ethical Philosophy and Welfare Economics

Michelle Seng Ah Lee, Luciano Floridi, and Jatinder Singh

Abstract There is growing concern that decision-making informed by machine learning (ML) algorithms may unfairly discriminate based on personal demographic attributes, such as race and gender. Scholars have responded by introducing numerous mathematical definitions of fairness to test the algorithm, many of which are in conflict with one another. However, these reductionist representations of fairness often bear little resemblance to real-life fairness considerations, which in practice are highly contextual. Moreover, fairness metrics tend to be implemented in narrow and targeted toolkits that are difficult to integrate into an algorithm's broader ethical assessment. In this chapter, we derive lessons from ethical philosophy and welfare economics as they relate to the contextual factors relevant for fairness. In particular we highlight the debate around acceptability of particular inequalities and the inextricable links between fairness, welfare and autonomy. We propose Key Ethics Indicators (KEIs) as a way towards providing a more holistic understanding of whether or not an algorithm is aligned to the decision-maker's ethical values.

Keywords Algorithmic fairness · Algorithmic ethics · Key ethics indicators · Algorithmic trade-offs · Ethical assessment · Fairness in welfare economics · Fairness in ethical philosophy

M. S. A. Lee (✉) · J. Singh
Department of Computer Science & Technology, Compliant & Accountable Systems
Research Group, University of Cambridge, Cambridge, UK
e-mail: michelle.sengah.lee@cl.cam.ac.uk; jatinder.singh@cl.cam.ac.uk

L. Floridi
Oxford Internet Institute, University of Oxford, Oxford, UK

Department of Legal Studies, University of Bologna, Bologna, Italy
e-mail: Luciano.floridi@oii.ox.ac.uk

J. Mökander, M. Ziosi (eds.), *The 2021 Yearbook of the Digital Ethics Lab*,
Digital Ethics Lab Yearbook, https://doi.org/10.1007/978-3-031-09846-8_11

1 Introduction

Algorithms are increasingly used to inform critical decisions across high impact domains, from credit risk evaluation to hiring to criminal justice. These algorithms are using more data from non-traditional sources and employing advanced techniques in machine learning (ML) and deep learning (DL) that are often difficult to interpret. The result is rising concern that these algorithmic predictions may be misaligned to the designer's intent, an organisation's legal obligations, and societal expectations, such as discriminating based on personal demographic attributes. In response, there has been a proliferation of literature on algorithmic fairness aiming to quantify the deviation of their predictions from a formalised metric of equality between groups (e.g. male and female). Dozens of notions of fairness have been proposed, prompting efforts (Verma & Rubin, 2018) to disentangle their differences and rationale.

In line with this, a number of fairness *toolkits*[1] have been introduced to test the algorithm's predictions against various fairness definitions. The fairness toolkit landscape so far reflects the reductionist understanding of fairness as mathematical conditions, as the implementations rely on narrowly defined fairness metrics to provide "pass/fail" reports. These toolkits can sometimes give practitioners conflicting information about an algorithm's fairness, which is unsurprising given that it is mathematically impossible to meet some of the fairness conditions simultaneously (Kleinberg et al., 2016). This is reflective of the conflicting visions of fairness espoused by each mathematical definition and the underlying ethical assumptions (Binns, 2020).

A recent paper surveying the fairness toolkit landscape (Lee & Singh, 2020) found there were significant gaps between ML practitioner needs and the toolkits' features, especially regarding means that helped practitioners account for the contextual specifics of their use case – one practitioner commenting the toolkits "make everything look clear-cut, which it really isn't 'in the wild'" (Lee & Singh, 2020). Other studies involving ML practitioners have similarly emphasised the need for domain-specific and contextual factors to be closely considered to improve algorithmic fairness (Veale et al., 2018). In particular, in many domains, practitioners claim that fairness cannot be understood in terms of well-defined quantitative metrics (Holstein et al., 2019).

This disconnect between real-world needs and axiomatic fairness definitions is not new. Hutchinson and Mitchell (2019) warn of the gap between the unambiguous formalisation of fairness metrics and the contextual and practical needs of society, politics, and law. They compared the recent surge in ML fairness research to literature from the 60s and 70s, which fizzled as "no statistic that could unambiguously indicate whether or not an item is fair was identified. There were no broad technical solutions to the issues involved in fairness" (Cole, 1973). From a legal standpoint,

---

[1] For example: IBM Fairness 360 (Bellamy et al., 2018), UChicago Aequitas tool (Saleiro et al., 2018).

the approach in automating "fairness testing" appears incompatible with the requirements of EU non-discrimination law, which relies heavily on the context-sensitive, intuitive, and ambiguous evidence (Wachter et al., 2020).

Fairness toolkits aim to be widely accessible, drawing attention to fairness considerations, and encouraging and supporting practitioners to consider, assess (and therefore mitigate) their algorithms in leading to unfair outcomes. However, without a consideration of the *relevant context* in the socio-technical system surrounding the algorithm, these tools risk engendering false confidence in flawed algorithms. Different considerations come into play for each use case. That is, organisations should not rely solely on one-dimensional algorithmic fairness metrics to account for its ethical concerns. These narrow applications of fairness could mislead organisational strategy, risk management, and policies.

Towards this, in this chapter we draw from literature in *ethical philosophy* and *welfare economics* to pinpoint the relevant contextual information that should be considered in an understanding of a model's ethical impact. We argue that any future development of fairness toolkits should be framed within a broader view of ethical concerns to ensure their adoption promotes a contextually appropriate assessment of each algorithm.

To this end, we propose a new approach using *Key Ethics Indicators (KEIs)* to provide a more holistic understanding of whether or not an algorithm is aligned to the decision-maker's values. Though resembling some previous work on domain-specific trade-off analyses in fairness metrics vs. public safety (Corbett-Davies & Goel, 2018) and vs. financial inclusion (Lee & Floridi, 2020), our chapter generalises the steps required for a holistic ethical assessment.

Our contribution is two-fold: (1) the identification of *relevant contextual factors* for fairness as drawn from ethical philosophy and welfare economics and (2) the proposal of a "Key Ethics Indicator" approach for a more comprehensive understanding of an algorithm's potential impact.

2 Definitions

We start by defining key terms: ethics, justice, fairness, equality, discrimination, and protected characteristics. This will frame our subsequent discussions on the contextual considerations for algorithmic ethics beyond what can be assessed using fairness metrics. While the dimensions these terms cover do not comprehensively cover all relevant aspects of algorithmic ethics, they clearly demonstrate the limitations of mathematical fairness formalisations in capturing necessary information about the algorithmic system.

As many organisations have launched initiatives to establish ethical principles, "AI ethics" definitions may vary; however, Floridi and Cowls identify the five common themes across these sets of principles: *beneficence, nonmaleficence, autonomy, justice, and explicability* (Floridi & Cowls, 2019). We define algorithmic ethics along these five dimensions.

A study of proposed ethical principles finds that different countries' and organisations' understanding of *justice* varies for each document, from the elimination of discrimination to promoting diversity to shared prosperity (Floridi & Cowls, 2019). For the purpose of this chapter, we distinguish between *justice* and *fairness* in accordance with legal and organisational science literature, with *justice* denoting adherence to the standards agreed upon in society (e.g. based on laws) and *fairness* as a related principle of an evaluative judgement of whether a decision is morally right (Goldman & Cropanzano, 2015).

In line with this definition, fairness is inherently subjective. The concept is based on the egalitarian foundation that humans are fundamentally *equal* and should be treated *equally*. However, how equality should be measured and to what extent it is desirable have been a source of debate in both philosophical ethics from a moral standpoint, and welfare economics from a market efficiency standpoint. What are the relevant criteria based on which limited resources should be distributed? For example, Aristotle wrote that if there are fewer flutes available than people who want to play them, they should be given to the best performers (Aristotle & Sinclair, 1962).

From a legal standpoint, *discrimination* refers to the notion that certain demographic characteristics, such as race and gender, should not result in a relative disadvantage of deprivation. Non-discrimination laws aim to not only prevent ongoing discrimination but also to change societal policies and practices to achieve more substantive equality – an aim which is described as incompatible with some fairness metrics (Wachter et al., 2021). While legal analysis is outside the scope of this chapter, we refer to protected characteristics as those commonly referenced and reflected in non-discrimination laws, such as race and ethnicity, gender, religion, age, disability, and sexual orientation, given these personal demographic features are central to discussions in algorithmic fairness literature. We also refer to 'direct' discrimination, which concerns differential treatment based on a protected characteristic and "indirect" discrimination, which represents an inadvertent negative impact on a protected group (Wachter et al., 2021). Computer science literature

2.1 Fairness Metrics

Existing mathematical definitions of fairness, while loosely derived from a notion of egalitarianism, should be calculated while keeping in mind the nuances and context-specificity present in philosophical discourse. We will walk through a use case: a lender building a model to predict a prospective borrower's risk of default on a loan. In this case, the False Positives (FP) represent lost opportunity (predicted default, but would have repaid), and the False Negatives (FN) represent lost revenue (predicted repayment, but defaulted).

The calculations of error rates used in the metrics are defined below, with some of the most commonly cited fairness definitions in Table 1:

Table 1 Fairness metrics

Fairness metric	Equalising	Intuition (Example)
Maximise total accuracy	N/A	The most accurate model gives people the loan and interest rate they 'deserve' by minimising errors
Demographic parity, group fairness, disparate impact (Feldman et al., 2015)	Outcome	Black and white applicants have the same loan approval rates
Equal opportunity/false negative error rate balance (Hardt et al., 2016)	FNR	Among applicants who are creditworthy and would have repaid their loans, both black and white applicants should have similar rate of their loans being approved
False positive error rate balance/predictive equality (Chouldechova, 2017)	FPR	Among applicants who would default, both black and white applicants should have similar rate of their loans being denied
Equal odds (Hardt et al., 2016)	TPR, TNR	Meets both of above conditions
Positive predictive parity (Chouldechova, 2017)	PPV	Among credit-worthy applicants, the probability of predicting repayment is the same regardless of race
Positive class balance (Kleinberg et al., 2016)	Average probability of positive class	Both credit-worthy white and black applicants who repay their loans have an equal average probability score
Negative class balance (Kleinberg et al., 2016)	Average probability of negative class	Both white and black defaulters have an equal average probability score
Counterfactual fairness (Kusner et al., 2017)	Prediction in a counterfactual scenario in which the person had a different protected feature	For each individual, if he/she were a different race, the prediction would be the same
Individual fairness (Dwork et al., 2012)	Outcome for 'similar' individuals	For each individual, he/she has the same outcome as another 'similar' individual of a different race

- True Positive Rate (TPR) = $TP/(TP + FN)$
- True Negative Rate (TNR) = $TN/(FP + TN)$
- False Positive Rate (FPR) = $FP/(FP + TN) = 1 - TNR$
- False Negative Rate (FNR) = $FN/(FN + TP) = 1 - TPR$
- Positive Predictive Value (PPV) = $TP/(TP + FP)$

There are difficulties in deciding which metric is most appropriate for each use case (Lee & Floridi, 2020). Is a 3% increase in positive predictive parity preferable over a 5% increase in equal odds? Moreover that many of these metrics cannot be satisfied at the same time (Kleinberg et al., 2016), it is not intuitive on which metric best represents the lender's interests. These issues will be further discussed in §3, where we will link each fairness metric to its philosophical origin and address the gaps. In the next section, we challenge the types of inequalities that the fairness metrics assume are acceptable vs. unacceptable.

2.2   Acceptability of Inequalities

First, we challenge the fairness metrics' assumed simplicity and separability of unacceptable bias by discussing the complexity of the debates on equality in ethical philosophy. Note that these metrics are aimed at a class of machine learning algorithms that are supervised, i.e. with a known outcome, and for classification purposes, i.e. for a discrete outcome (e.g. default vs. repayment) rather than a continuous outcome (e.g. amount repaid).

These algorithms aim to identify the features that are associated with the outcome of interest. For example, one with higher income is more likely to be approved for a loan due to its association with higher ability to repay. In this case, differences in socioeconomic status is accepted as an inequality that is important to consider in the loan decision. Previously, scholars have made the distinction between "acceptable" vs. "unacceptable" inequalities based on legal precedents between "explainable" and "non-explainable" discrimination (Kamiran & Žliobaitė, 2013) based on Rawlsian philosophy between "relevant" and "irrelevant" features (Rawls, 1999). For example, income may be considered a "relevant" feature, and gender or race may be considered an "irrelevant" feature. The former should influence the algorithmic decisions, but the latter should not.

Scholars attempting to formalise these criteria into a mathematical definition of fairness have needed to address what type of equality is deemed to be fair. Some assume that any disparity in a given outcome metric is unacceptable (Gajane & Pechenizkiy, 2017): for example, loan approval rate for men and women should be the same. Others assume a level playing field (Gajane & Pechenizkiy, 2017): for example, there is no gender or racial discrimination in the real world that may affect the data. More recent work has taken a more nuanced stance, suggesting that the only features that should contribute to the outcome disparity are those that can be controlled by the individual, emphasising a distinction between the features driven by "effort" vs. "circumstances" (Heidari et al., 2018). This is following the logic of Dworkin's theory of Resource Egalitarianism, no one should end up worse off due to bad luck, but rather, people should be given differentiated economic benefits as a result of their own choices (Dworkin, 1981).

Mathematical fairness formalisations must first determine which inequalities are "unacceptable." Some assume that all disparity in a given outcome metric is unacceptable, while others assume a level playing field (Gajane & Pechenizkiy, 2017), an assumption rarely met in societal challenges. Others emphasise the need to separate "effort" and "circumstances," suggesting that the only features that should contribute to the outcome disparity are those that can be controlled by the individual (Heidari et al., 2018). Another paper distinguishes between "benign" disparities and structural bias that should be corrected (Binns, 2020).

In reality, the layers of inequality between two individuals are intertwined, dynamic, and difficult to disentangle from one another. Consider the layers of inequality in Table 2. Two individuals may be unequal on several levels – in their level and type of talent, parents' socioeconomic status, behaviour, etc. – that may affect the target outcome of interest, whether it is credit-worthiness, predicted performance at a job, or insurance risk. It is possible that the differences in the observed outcome are attributable to one or more of the above inequalities. Building an algorithm to predict the outcome could result in a faithful representation of these inequalities and the resulting replication and perpetuation of the same inequality through decisions informed by its predictions. However, which of the inequalities should be allowed to influence the model's prediction? We present the limitations of proposals thus far on how this question should be addressed.

Legally Protected Characteristics

The open source fairness toolkits often refer specifically to protected attributes in their assessment of fairness. For example, Fairness 360 defines protected attribute one that "partitions a population into groups whose outcomes should have parity. Examples include race, gender, caste, and religion" (Bellamy et al., 2018). While they acknowledge protected attributes may be application-specific, there is limited guidance on under what circumstances two groups should have parity in outcomes. In addition, how much disparity is acceptable in each use case and for each subgroup of interest? In order for the applications to be adopted across domain areas, it is important for practitioners to have a clear idea of what types of demographic features are acceptable vs. unacceptable to consider in an algorithm. However, often, computer science literature uses these demographic features to assess fairness without challenging whether they are relevant to the decision at-hand.

Whether a disparity in fairness metrics between legally protected groups is fair depends on the context. Race and gender may be causally relevant in differential medical diagnosis (e.g. sickle cell anaemia, ovarian cancer) due to the different biological mechanisms in question. If the differences in outcome are causally related to the protected feature, the difference in decisions may be arguably fair. If a man

Table 2 Layers of inequality affecting the ground truth (Partial and Indicative)

Types of inequality	Examples	Variable
Natural inequality	Disability at birth	Inequality 0
Socioeconomic inequality	Parents'/guardians' assets	Inequality 1
Talent inequality	Intelligence, skills, employment prospects	Inequality 2
Preference inequality	Saving behaviour, cultural prioritisation of values associated with economic opportunities	Inequality 3
Treatment inequality/societal discrimination (external)	Discrimination in job market and education system affecting income stability	Inequality 4

has a higher income than a woman, he may receive a higher credit limit given his higher ability to repay.

Effort vs. Circumstances

The suggestion to distinguish between the features driven by "effort" vs. "circumstances" in algorithmic fairness (Heidari et al., 2018) follows the logic of Dworkin's theory of Resource Egalitarianism, no one should end up worse off due to bad luck, but rather, people should be given differentiated economic benefits as a result of their own choices (Dworkin, 1981).

However, in reality, it is difficult to separate out what is within an individual's *genuine control*. For example, a credit market does not exist in a vacuum; while potential borrowers can improve their creditworthiness to a certain extent, e.g. by building employable skills and establishing a responsible payment history, it is difficult to isolate the features from discrimination in other markets, layers of inequality, and the impact of their personal history.

In addition to the challenge of defining what is within our control, some circumstances are necessary to take into account in a decision-making process. For example, one may not be in full control of one's income or education level, but they are crucial indicators of credit risk given they indicate greater job security. Socioeconomic and talent inequalities may be considered relevant in a credit risk evaluation algorithm, but they are not necessarily within our control.

Source of Inequality

Scholars have also proposed that the *source* of inequality should determine which fairness metric is appropriate for each use case, i.e. whether the outcome disparity is explainable, justifiable, or benign or due to structural discrimination (Kamiran & Žliobaitė, 2013; Binns, 2020). Binns (2020) suggests group fairness metrics assumes disparities are benign, e.g. the loan approval difference between white and black applicants is solely due to their differences in ability to repay; statistical parity assumes structural bias that requires correction, e.g. historically, black applicants' risk have been inflated due to past discriminatory practices. However, in reality, there is rarely such a separation. For example, Lee and Floridi review the literature on U.S. mortgage lending and suggest that there are many structural and statistical factors that lead the lenders to both overestimate and under-estimate the risk of black borrowers (Lee & Floridi, 2020).

Any attempt to isolate the impact of discrimination from the impact of "benign" inequality needs to also consider the intersectional discrimination faced by those already marginalised in society (Crenshaw, 1989), e.g. the inter-connectivity of gender and racial discrimination (Collins, 2002). The boundary between what is an acceptable representation of existing inequalities and what is due to systematic discrimination and marginalisation of a group is challenging to ascertain.

Fleurbaey et al. (2008) also cautions that "responsibility-sensitive egalitarianism" in welfare economics could be used to hastily justify inequalities and unfairly chastise the "undeserving poor" (Fleurbaey et al., 2008). He points out that the idea that people should bear the consequences of their choices is not as simple as it seems; it only makes sense when individuals are put in equal conditions of choice. Such an equality is not true in most systems. When one has fewer opportunities than another, one cannot be held fully responsible insofar as one's choice is more constrained.

Takeaways

The assumed clear and intuitive separation between acceptable and unacceptable inequalities, whether based on their source or the role of luck, rarely exists in real-life models. Not only is making the distinction impractical, the boundary itself is more controversially debated than it is often portrayed in algorithmic fairness literature, especially in computer science. The criteria for desirable equality depend on the philosophical perspective, which is ultimately a subjective judgement.

The decision on the target state – the way it ought to be – is an ethical decision with mathematically inevitable trade-offs between objectives of interest. Heidari et al. dismiss the distinction between relevant vs. irrelevant features in practice as out of scope for their paper: "Determining accountability features and effort-based utility is arguably outside the expertise of computer scientists" (Heidari et al., 2018). On the contrary, we argue that computer scientists and model developers must be actively engaged in the discussion on what layers of inequality should and should not be influencing the model's prediction, as this directly influences not only the model design and feature selection but also the selection of performance metrics.

3   Lessons from Ethical Philosophy on (In)Equalities

Ethical philosophers have long debated whether equality is desirable and – if so – what type of equality people should pursue in society. Table 3 gives an example of philosophical perspectives and their perceptions of what types of inequality are acceptable. Formal equality of opportunity (EOP), or procedural fairness, posits that all opportunities should be equally open to all applicants (e.g. jobs, loans, etc.) based on a relevant definition of merit. However, in theory, this can be fully satisfied even if only a minority segment of a population (e.g. those with family wealth and connections) have realistic prospects of accessing the opportunity. In other words, as long as the opportunity is theoretically *available*, it is irrelevant whether it is *practically accessible*.

The Rawlsian fair EOP goes further to propose that any individuals with the same native talent and ambition should have the same prospects of success, requiring that all competitive advantage (e.g. parental efforts) be offset (Rawls, 1999).

Table 3 Key philosophical perspectives on inequality

Philosophical perspective	Acceptable inequalities	Unacceptable inequalities
Formal equality of opportunity/procedural fairness (Greenberg, 1987)	Any inequality as long as the opportunity was open to all	Treatment inequality
"Fair equality of opportunity" (Rawls, 1999, 2001)	Natural, talent, and preference inequalities	Socioeconomic, treatment inequalities
Rawlsian EOP + Difference principle (Rawls, 1999)	Natural, talent, and preference inequalities, plus any inequality benefiting the most disadvantaged society members in long-term impact	Socioeconomic, treatment inequalities
Equality of outcome/ condition/welfare (Greenberg, 1987)	None – all members should get the exact same outcome	All
Luck egalitarianism (Dworkin, 1981)	Effort-based inequalities (e.g. preference)	Circumstances (e.g. natural inequality)
Equality of freedom/ autonomy (Sen, 1992)	Inequality resulting in" genuinely free" choices	Any inequality hindering freedom
Sufficiency/Equality of capability (Walzer, 1983)	Any inequality as long as everyone is above the level of sufficiency	Any resulting in people falling below sufficiency levels
Prioritarianism (Scheffler, 1994; Parfit, 1991)	Any inequality reduction should prioritise resource allocation to those who are worst off	None as long as the worst off are prioritised
Desert (Kagan, 1999, 2014)	Any inequality based on what he/she" deserves"	Any inequality that does not equate to the person's deservingness

This is at odds with Lockean and libertarian ideals that assert the value of each person's freedom insofar as there is no harm to another (Nozick, 1974), which naturally extends to the right to ownership and capital. Rawls also proposes the Difference Principle as an exception: economic and social inequalities can only be justified if they benefit the most disadvantaged members of society (Rawls, 1999). These EOP principles are in contrast to the strict equality of outcome, condition, or welfare, which requires an equal distribution regardless of any relevant criteria.

Luck egalitarians hold that unchosen inequalities must be eliminated (Dworkin, 1981). Sen and Fleurbaey object on the grounds that luck egalitarians have no principled objection to a society in which, on a background of equal opportunities, some end up in poverty or as the slaves of others (Fleurbaey et al., 2008). They argue for a more substantive equality of "autonomy" that includes the full range of individual freedom.

Some have argued that what is important is not relative condition compared to other people, but rather, whether people have enough to have satisfactory life prospects (Walzer, 1983). Others have shifted the focus on the incremental gain of well-being of those who are worst-off (Parfit, 1991). Yet others have debated the foundations of desert, or what one deserves corresponding to his or her virtue (Kagan, 2014).

3.1   Ethical Subjectivity of Algorithmic Fairness

As such, what types of inequality in outcome are fair is a philosophical and subjective debate with nuances and complexities insufficiently addressed in existing algorithmic fairness literature. What happens when faithfully representing the world as it is perpetuates an unfair state of affairs? This complicates the objective of machine learning, which is only reliable insofar as it is trained on data sets that reflect reality. To forecast sales, an algorithm learns from data representative of the company's customers. For example, online searches for "CEO" yield mostly images of white men, and online job postings may show high-income positions to men more frequently than women (Van Dam, 2019). This may result in a biased outcome, with men securing disproportionately high-paying jobs. However, this is reflective of the existing gender pay gap: in 2019, only 6.6% of Fortune 500 top executives were female, the highest proportion in history (Zillman, 2019). Continuing to underrepresent women in search results may perpetuate the bias that CEOs are typically men. In this instance, some call for the "correction" of the bias to reflect judgements about the way the world *should* be, which is by nature an ethically influenced choice.

As previously stated, on the contrary to past scholars' arguments (Heidari et al., 2018), our position is that computer scientists and model developers cannot completely delegate this consideration to a third party, whether it is the regulator, business leader, or the risk function. Model developers must be engaged in the discussion on what layers of inequality should and should not be influencing the model's prediction in order to inform their decisions on model design, feature selection, and performance metric selection.

Overall, in formalising fairness, the decision-maker should be explicit on (1) which inequalities and biases exist that affect the outcome of interest and (2) on which of them should be retained and which of them should be actively corrected. This will be further addressed in our proposal of Key Ethics Indicators (KEIs). We next link some of the fairness metrics to the ethical philosophy that inspired them, pointing out the contextual considerations in the ethical philosophy that should be kept in mind alongside the fairness formalisations.

3.2   Linking Ethical Philosophy to Algorithmic Fairness

Existing mathematical definitions of fairness, while loosely derived from a notion of egalitarianism, should be calculated while keeping in mind the nuances and context-specificity present in philosophical discourse. Revisiting the fairness metrics from Table 1, this section will link each metric to the ethical philosophy that inspired it, as well as addressing the gaps between the philosophical work and what is represented in the mathematical formula.

In discussing the entries of Table 4 in order: accuracy maximisation is prone to biases introduced in the model development lifecycle that may skew the predictions,

Table 4 Fairness metrics and their philosophical origins

Fairness metric	Equalising	Philosophy
Maximise total accuracy	N/A	Desert (Kagan, 1999, 2014)
Demographic parity, group fairness, disparate impact (Feldman et al., 2015)	Outcome	Strict egalitarianism (Equality of outcome/condition/welfare) (Greenberg, 1987)
Equal opportunity/false negative error rate balance (Hardt et al., 2016)	FNR	"Fair equality of opportunity"
False positive error rate balance/predictive equality (Chouldechova, 2017)	FPR	"Fair equality of opportunity"
Equal odds (Hardt et al., 2016)	TPR, TNR	"Fair equality of opportunity"
Positive predictive parity (Chouldechova, 2017)	PPV	"Fair equality of opportunity" (Rawls, 1999, 2001)
Positive class balance (Kleinberg et al., 2016)	Average probability of positive class	"Fair equality of opportunity"
Negative class balance (Kleinberg et al., 2016)	Average probability of negative class	"Fair equality of opportunity"
Counterfactual fairness (Kusner et al., 2017)	Prediction in a counterfactual scenario in which the person had a different attribute	David Lewis, cause and effect (Lewis, 1973)
Individual fairness (Dwork et al., 2012)	Outcome for "similar" individuals	Responsibility-sensitive egalitarianism (Fleurbaey et al., 2008)

which is especially problematic if the biases reflect patterns of societal discrimination, leading to "undeserved" outcomes contrary to the philosophy of desert. Demographic parity is problematic if there are legitimate rationale behind the unequal outcome (e.g. unequal income).

The equal opportunity metric, while it sounds attractively similar to Rawlsian EOP, fails to address discrimination that may already be embedded in the data (Gajane & Pechenizkiy, 2017). Discrimination may be crystallised in the data set due to biased data collection (e.g. selective marketing), biased data labelling (e.g. humans scoring male candidates as more competent), or biased human decisions feeding the system (e.g. if courts are more likely to find black defendants as guilty). Rawlsian EOP also assumes that inequalities in native talent and ambition may result in unequal outcomes, which is not addressed in the equalisation of false negative rates. Each group fairness metric, including equal odds, positive predictive parity, and positive/negative class balance, requires different assumptions about the gap between the observed space (features) vs. the construct space (unobservable variables): "if there is structural bias in the decision pipeline, no [group fairness] mechanism can guarantee fairness" (Friedler et al., 2016). This is supported in a critique of existing classification parity metrics, in which the authors conclude that "to the extent that error metrics differ across groups, that tells us more about the shapes of the risk distributions than about the quality of decisions" (Corbett-Davies & Goel, 2018). In many domains in which there are concerns over unfair algorithmic bias, including credit risk and employment, there has often been a documented history of structural and societal discrimination, which may affect the underlying data through biases previously discussed.

The challenge of the individual fairness approach is: how to define "similarity" that is, for example, independent of race (Kim et al., 2018). When the predictive features are also influenced by protected features, designation of a measurement of "similarity" cannot be independent of those protected features. For example, what proportion of gender income disparity is due to structural employment discrimination as opposed to job preferences? Some scholars have attempted to incorporate active corrections for racial inequality into metrics of similarity (Dwork et al., 2012), but this depends heavily on the assumption that the inequality due to racial discrimination can be isolated from other sources of inequality.

While counterfactual fairness metric provides an elegant abstraction of the algorithm, the causal mechanisms, e.g. of a default on a loan or on insurance risk, are typically not well understood. It is also difficult to isolate the impact of one's protected feature, e.g. race, on the outcome, e.g. risk of default, from the remaining features. The approach is also sensitive to unmeasured confounding variables, which may add additional discriminatory bias (Kilbertus et al., 2019).

In all, these metrics do not give any information on which layers of inequalities they are attempting to correct, which risks over- or under-correction. A deeper engagement with the ethical assumptions being made in each model is necessary to understand the drivers of the unequal outcomes. What types of inequalities are acceptable depends on the context of the model. Our KEI approach will account for such context-specificity of what inequalities are acceptable.

4   Lessons from Welfare Economics

Referring back to our definition of algorithmic ethics, justice is only one of five dimensions (*beneficence, non-maleficence, autonomy, justice, and explicability*), with fairness as a key principle related to justice. We derive lessons from literature on welfare economics to demonstrate the interconnectedness of fairness and welfare (*beneficience* and *non-maleficence*) and liberty (*autonomy* and *explicability*). By focusing narrowly on the fairness metrics, which quantify the redistribution of the target outcome, a decision-maker may overlook the key considerations of the impact on the stakeholders' welfare and autonomy. Because of the challenge in quantifying the relevant biases and disentangling them from the outcome of interest, correcting for a bias carries the risk of increasing the inaccuracies of the predictions. Beyond the egalitarian perspective on the relative distribution of resources between individuals and groups, it is important to consider the aggregate impact of an algorithm on the society.

4.1   Welfare in Algorithmic Ethics: Beneficence
and Nonmaleficence

We use an example concerning credit risk evaluation to argue that fairness should be considered alongside welfare. In attempting to improve a fairness metric, a decision-maker may inadvertently forego an algorithm that leaves everyone better-off (beneficence) or may inadvertently harm the sub-group they are attempting to help to "level the playing field." Fairness metrics should not be taken at face value without an understanding of how they may affect other ethical objectives. Fairness toolkits that assess fairness in isolation risk misleading the decision-makers by giving them incomplete information about whether their algorithm meets their ethical objectives.

From a welfare economic standpoint, a notion of fairness includes a consideration of well-being: from both utilitarian and libertarian perspectives, a fair reward principle maximises the sum total of individual well-being levels while legitimising redistribution that enhances the total outcome of individuals (Fleurbaey et al., 2008). This is not necessarily contradictory to the egalitarian perspectives discussed in ethical philosophy. In accordance with the Difference Principle, Rawlsian EOP Max-Min social welfare function should also maximise the welfare of those who are worst-off (Rawls, 1999). A model that results in financial harm of already-disadvantaged populations fails to meet the Rawlsian EOP criteria, even if the False Negative Rates are equalised as per the mathematical definition. Without consideration of the long-term impact on welfare, the fairness metrics fail to capture the full extent of the ethical dilemma embedded in a model selection process.

Accuracy is often considered in trade-off with fairness (Kleinberg et al., 2016), but from an ethical standpoint, that accuracy may represent a key principle in beneficence or non-maleficence. For an example of beneficence, a "good" credit risk algorithm would lower the aggregate portfolio risk for the lender, enabling more loans to more people and giving them access to credit that is crucial to upward socioeconomic mobility. For an example of non-maleficence, the false positive rates (i.e. loans that were approved but defaulted) also contains information about whether unaffordable loans are granted. A lender should aim to minimise the borrower's financial difficulty, given the adverse effects of unaffordable debt on both the market level (causing instability and a "bubble") and for the borrower (Aggarwal, 2018).

The ethical principle of non-maleficence may be in direct conflict with fairness in some circumstances. Adding fairness constraints may end up harming the groups they intended to protect in the long-term (Liu et al., 2018). In the presence of a feedback loop, we need to consider not only providing a resource (a loan) to an applicant in a disadvantaged group, but also what happens as a result of that resource. If the borrower defaults, his/her credit score will decline, potentially precluding the borrower from future loans. It is important to view fairness, not in isolation at a moment in time, but rather, in the context of long-term objectives in promoting the customer's financial well-being. This is a part of the context we formalise in our Key Ethics Indicator proposal.

4.2 Liberty in Algorithmic Ethics: Autonomy and Explicability

Fairness should also be assessed within the context of how the algorithm affects human *liberty*, a subject in welfare economics that is relevant to the AI ethics principles of *autonomy* and *explicability*. Fleurbaey argues responsibility-sensitive egalitarianism in welfare economics should move away from "responsibility," which may overlook certain people's lack of freedom to choose alternatives, and towards "autonomy" (Fleurbaey et al., 2008). In other words, for there to be "true" equality, three conditions must be met: (1) a minimum level of autonomy is attained, (2) with a minimum level of variety and quality of options offered, (3) with a minimum decision-making competence (Fleurbaey et al., 2008). A comprehensive egalitarian theory of justice is not just about equalising available opportunities but also about providing adequate opportunities and making them *accessible*. As per our definition of AI ethics, we define autonomy as the power to decide, striking a balance between the decision-making power humans retain and that which we delegate to artificial agents. We also define *explicability* as the combination of *intelligibility* (how it works) and *accountability* (who is responsible for the way it works). It complements the other four principles by helping us understand the good or harm an algorithmic system is actually doing to society, in which ways, and why Floridi and Cowls.

Autonomy: Liberty

In enforcing some of the stricter fairness conditions, decision-makers should be careful as to the potential impact this has on human autonomy. Luck egalitarians, for instance, have no principled objection to a society in which, on a background of equal opportunities, some end up in poverty or as the slaves of others (Fleurbaey et al., 2008) – this could violate fundamental human rights to freedom and result in undesirable levels of extreme societal inequality. Intervention is necessary when basic autonomy is at stake, and this should be a constraint on the definition of fairness. Fleurbaey argues this is consistent with egalitarian welfare economics, as egalitarians should be concerned – not only with equality of opportunities – but also with the content of the opportunities themselves, with freedom as the leading principle in defining responsibility in social justice (Fleurbaey et al., 2008).

By focusing on equality of opportunities, one may dismiss the differences in preferences as driven by choice and thus irrelevant. However, Fleurbaey argues that the ex post inequalities due to differences in preferences are also a target for intervention on the grounds of improving the range of choices to suit everyone's preferences. If more women prefer lower-paid positions than men, what is problematic is not only the societal and environmental conditioning that questions whether this is a genuine preference, but also the unfair advantage that attaches to these jobs – a differential value of the "menu" of options for women than for men because of their preferences (Fleurbaey et al., 2008). Considerations of fairness and the associated policy response must operate at the level of the menu, rather than distribution of jobs themselves.

Autonomy: Forgiveness

Fleurbaey also discusses a concept that is not addressed in algorithmic fairness literature: forgiveness. He argues that the ideal of freedom and autonomy contains the idea of "fresh starts": in absence of cost to others, it is desirable to give people more freedom and a greater array of choices in the future (Fleurbaey et al., 2008). This is in conflict with the "unforgiving conception of equality of opportunities" that ties individuals to the consequences of one's choices (Fleurbaey et al., 2008). In many countries, lenders are restricted in their access to information about borrowers' past defaults; for example, many delinquencies are removed from U.S. credit reports after seven years (Elul & Gottardi, 2015). Forcing a lender to ignore information about past behaviour may reduce the accuracy of its default prediction model, and it may be "unfair" by some definitions by putting those who have made more responsible financial decisions on equal level as those who have not; however, it is widely accepted practice to ensure that one decision does not have a disproportionate impact of limiting one's access to credit for good. A more complete coverage of fairness and justice, therefore, should go beyond redistribution of outcome features and consider the impact on individual welfare, autonomy, and freedom.

Autonomy: Vulnerability

Autonomy in rational decision-making also falters as an ethical objective when there is a significant asymmetry of power and information between two parties. Contractarian perspectives on fairness assumes two equal entities exchanging one resource for another (Gauthier, 1986).

Those with limited autonomy include vulnerable customers. When an algorithm targets and manipulates those with no other options, they do not have the autonomy to enter into the contract, whether or not the contract is fair. Payday loans and check cashing industry in the US targets those who cannot access traditional financial services, often due to their illegal immigration status or long working hours that do not provide a break while a bank is open for business, entrapping the most vulnerable groups into an unbreakable cycle of debt with unaffordable interest rates (Prager et al., 2009). While the interest rate may not necessarily be unfair (it may in some cases be proportional to the likelihood of an individual's repayment), it is ethically undesirable. The same principle applies to marketing insurance products to those with recent bereavement or the sale of complex financial instruments to someone without the capability of understanding their risks.

Another group is those with "thin" files, with a lack of or sparse credit history. There has been a movement to use "alternative data" or non-traditional data sources that do not directly relate to the borrower's ability to repay. One of the most extreme cases is the use of Internet browsing history, location, and payment data to calculate credit risk (Koren, 2016). The justification is often that this increases financial inclusion for those without alternate means to access credit. However, this requires the lender access to more data from the currently unbanked populations, disproportionately forcing them to give up more of their privacy than those with existing credit histories. It also provides additional risk of discrimination, as the non-traditional data sources are likely to be closely intertwined with personal characteristics (Koren, 2016). Location and social media data are more likely to reveal an individual's race and gender than credit history. While Kenya's poor were among the first to benefit from digital lending applications, they have led to a predatory cycle of debt the borrowers describe as a new form of slavery, between the endless nudges to borrow, the lenders' control over a vast archive of user data, and the ballooning interest payments (Donovan & Park, 2019). This double-standard of privacy between the unbanked and banked violates the equal rights of individuals to privacy and self-determination. While there may be an exchange of access to credit and personal data (e.g. if an individual gives consent to a personality test or access to his/her social media profile), there should be a protection of their right to privacy.

Fairness overall must be considered in the context of the impact on individual human rights – going beyond the equality of available opportunities, empowering human freedom and autonomy to ensure *accessibility* of these opportunities. Computer scientists can learn from the welfare economists' consideration of autonomy as a crucial component of egalitarian perspectives on fairness.

Explicability

Welfare economics is built on the assumption of rational, free agents, which is shared in Kantian ethical philosophy (Kant & Gregor, 1996). This has been applied to medical ethics to mandate that a patient be able to make a fully informed decision on whether or not to receive treatment (Eaton, 2004). Similarly, in algorithmic decision-making, individuals consenting to the usage of their data should fully understand how the data will be used. When humans employ autonomous systems, they cede, at least provisionally, some of their own autonomy (decision-making power) to machines (Floridi & Cowls, 2019). Respecting human autonomy thus becomes a matter of ensuring that both the decision-making authority and the subject of the decision retain enough autonomy to safeguard their well-being.

In order to incorporate the algorithm into rational decision-making, it is important to understand how the algorithm reached its prediction or recommendation. Due to the relatively limited interpretability of ML, "explainable AI" (xAI) is an ongoing area of research (Xu et al., 2019). There is often a trade-off between accuracy of an algorithm and its explainability, as complex phenomena are better represented by complex, "black-box" models than simple and interpretable models. This may, in turn, represent a trade-off between explainability (and thus a decision maker's capability for reasoning) and any beneficence afforded by the increase in accuracy and model performance. In some use cases, e.g. film recommendations, accuracy may outweigh the need for explanations. The explanations may vary based on the target of the explanation, e.g. customer, regulator, domain experts, or system developers (Arya et al., 2019). It is important to understand the interplay between an algorithm's explanation and its perceived fairness. There may be a number of possible explanations for any given decision, and the techniques for xAI alone do not detect or correct unfair outcomes. However, the explanations may help identify potential variables that are driving the unfair outcomes, e.g. if pricing varies for female-dominated professions compared to male-dominated professions, the model may be relying on occupation for its prediction, which acts as a proxy for gender.

While fairness formalisations may provide a simple methodology for model developers to incorporate metrics relevant to equalisation of outcomes between groups and individuals, they do not provide a holistic view of the important debates on what fairness means, as they are discussed in ethical philosophy and welfare economics. The narrow definition of unfair bias in each of these metrics only provides a partial snapshot of what inequalities and biases are affecting the model and does not consider the long-term and big-picture ethical goals beyond this equalisation.

5   Proposed Method: Key Ethics Indicators

In this final section, we propose a new approach that moves away from attempts to define fairness mathematically, and instead, gain a more holistic view of the ethical considerations of a model. Due to the subjectivity of fairness metrics, it may be challenging to select one over another. Rather than these general metrics, decision-makers should create a customised measurement of what "fair" looks like in each model. In addition, fairness should not be considered in isolation from the related ethical goals. The interaction between fairness and other values – e.g. welfare, autonomy, and explicability – should be taken into account in this analysis.

Contrary to claims otherwise (Heidari et al., 2018), the roles and responsibilities of a developer are necessarily intertwined with the role of the expert or business stakeholder, as the ethical and practical valuations of what "success" looks like in the model directly influences the algorithm design, build, and testing. It is important to have active engagement from the beginning between the developer and the subject matter expert to try to understand which inequalities should influence the outcome and how to address the inequalities that should not play a role in the prediction. This process requires engagement from all relevant parties, including the business owner and the technical owner, with potential input from regulators, customers, and legal experts.

Relying solely on the out-of-the-box fairness definitions as implemented in fairness toolkits would fail to capture the nuanced ethical trade-offs. For a decision-maker, it is important to devise customised success metrics specific to the context of each model, which as we described, involves considering welfare (beneficence, non-maleficence), autonomy, fairness, and explicability. This can be done in a following process:

1. Define "success" from an ethical perspective. What is the benefit of a more accurate algorithm to the consumer, to society, and to the system? What are the potential harms of false positives and false negatives? Are there any fundamental rights at stake?
2. Identify the layers of inequality that are affecting the differences in outcome
3. Identify the layers of bias
4. Devise an appropriate mitigation strategy. This may require changes to data collection mechanisms or to existing processes, rather than a technical solution.
5. Operationalise these objectives into quantifiable metrics, build multiple models and calculate the trade-offs between the objectives covering all ethical and practical dimensions.
6. Select the model that best reflects the decision-maker's values and relative prioritisation of objectives.

We now elaborate each of these steps, in turn.

5.1   *Define Success*

For each use case, there are unique considerations on what is considered a "successful" model, which are unlikely to be captured in a single mathematical formula. In credit risk evaluation, for example, three key objectives from ethical, regulatory, and practical standpoints are: (1) allocative efficiency: a more accurate assessment of loan affordability protects both the lender and the customer from expensive and harmful default, (2) distributional fairness: increasing access to credit to disadvantaged borrowers, including "thin-file" borrowers and minority groups, (3) autonomy: both increased scope of harm due to identity theft and security risk and due to the effects of ubiquitous data collection on privacy (Aggarwal, 2020). Here, a successful credit risk model would achieve all three objectives. By contrast, in algorithmic hiring, success metrics may include employee performance, increased overall diversity among employees and in leadership, and employee satisfaction with the role. It is important to identify all the objectives of interest, such that any trade-offs between them may be easily identified, allowing for a more holistic view of algorithmic ethics.

5.2   *Identify Sources of Inequality*

As previously discussed, due to the complex and entangled sources of inequalities and bias affecting an algorithm, there is no simple mathematical solution to unfairness. It is important to understand what types of inequality are acceptable vs. unacceptable in each use case. Table 2 presents different layers of inequality. Considering a credit risk evaluation, socioeconomic and talent inequalities may be considered relevant: if a man has a higher income than a woman, he may receive a higher credit limit given his higher ability to repay; higher education level and expertise in a high-demand field may indicate greater job security. Forcing the decision-maker to look beyond the legally protected characteristics to identify the inequalities that are acceptable and relevant and those that are not helps better identify the sub-groups that are at risk of discrimination.

5.3   Identify Sources of Bias

In addition to the inequalities discussed above, there may be biases in the model development lifecycle that exacerbate the existing inequalities between two groups. The challenge is that in many cases, the patterns associated with the target outcome are also associated with one's identity, including race and gender.

Suresh and Guttang (2019) have recently grouped these types of biases into 6 categories: historical, representation, measurement, aggregation, evaluation, and deployment. Historical bias refers to past discrimination and inequalities, and the remaining five biases, displayed in Table 5, align to the phases of the model development lifecycle (data collection, feature selection, model build, model evaluation, and productionisation) that may inaccurately skew the predictions. By understanding the type of bias that exists, the developer can identify the phase in which the bias was introduced, allowing him or her to design a targeted mitigation strategy for each bias type.

Table 5 gives examples in racial discrimination in lending processes to demonstrate each type of bias. For a practical tool in identifying unintended biases in these six categories, see: Lee and Singh (2021). Crucially, they point out that effective bias mitigation addresses the bias at its source, which may involve a non-technical solution. For example, bias introduced through the data collection process may require a change in marketing strategy.

Table 5 Layers of bias resulting in inaccurate predictions (Partial and Indicative)

Types of bias	Examples	Variable
Representation bias	Limited marketing and outreach in high-minority neighborhoods	Bias 0
Measurement bias	Unequal treatment in the lending process associated with race leads to mis-measurement of risk factors	Bias 1
Aggregation bias	There may be a difference in default frequency distribution between racial groups, which is poorly represented by a single model	Bias 2
Evaluation bias	The accuracy and precision metrics in default prediction vary across racial groups (e.g. lower confidence in predictions for minority borrowers)	Bias 3
Deployment bias	True outcome only known for accepted loans and unknown for denied loans	Bias 4

5.4 Design Mitigation Strategies

The mitigation strategy depends on whether we believe the inequalities in Table 2 and the biases in Table 5 need to be actively corrected to rebalance the inequalities and bias. It is important to understand the source of the bias in order to address it.

There have been existing methods proposed for *pre-processing*, removing bias from the data before the algorithm build, *in-processing*, building an algorithm with bias-related constraints, and *post-processing*, adjusting the output predictions of an algorithm. However, these methods presume that inequalities in Table 2 and the biases in Table 5 are known and can be quantified and surgically removed. How do we isolate the impact of talent and preference inequalities on income from the impact of discrimination? The attempt to "repair" the proxies to remove the racial bias has been shown to be impractical and ineffective when the predictors are correlated to the protected characteristic; even strong covariates are often legitimate factors for decisions (Corbett-Davies & Goel, 2018).

Often, the solution to these biases is not technical because their sources are not inherent in the technique. Instead of looking for a mathematical solution, there may be productive ways of counteracting these biases with changes to the process and strategy. Examples are shown in Table 6.

While the mitigation strategies are important, they are unlikely to provide a complete solution to the problem of algorithmic bias and fairness. That is because – unlike the assumptions underlying fairness formalisations – it is often not feasible to mathematically measure and surgically remove unfair bias from a model, which is affected by inequalities and biases that are deeply entrenched in society and in the data.

Legal scholars have argued that the traditional approach of scrutinising the inputs to a model is no longer effective due to the rising model complexity. Using Fair Lending law as an example, Gillis demonstrates that identifying which features are relevant vs. irrelevant fails to address discrimination concerns because combinations of seemingly relevant inputs may drive disparate outcomes between racial

Table 6 Possible actions to counteract biases (Partial and Indicative)

Types of bias	Variable	Example action
Treatment inequality/ societal discrimination (external)	Inequality 4	Identify a new feature to estimate income volatility associated with race
Representation bias	Bias 0	Change in marketing and outreach strategy to include more high-minority neighborhoods
Measurement bias	Bias 1	Employee training on subconscious bias, standardized practice on which loan types are recommended based on pre-specified relevant criteria
Deployment bias	Bias 2	Continuous monitoring and analysis of whether the decision boundary between rejection and acceptance is appropriate

group (Gillis, 2020). Rather than focusing on identifying and justifying inputs and policies that drive disparities, Gillis argues, it is important to shift to an *outcome-focused* analysis of whether a model leads to impermissible outcomes (Gillis, 2020). Similarly, Lee and Floridi have proposed an approach to assess whether the outcome of a model is desirable (Lee & Floridi, 2020). For a more comprehensive analysis of whether a model meets the stakeholders' ethical criteria, it is important to look beyond the inputs and the designer's intent and assess the long-term and holistic outcome.

5.5   Operationalise Key Ethics Indicators (KEIs), Calculate Trade-Offs Between KEIs

Once "success" for a model has been defined at a high-level, the next step is to operationalise the ethical principles such that they are measurable. Similarly to how a company may define a set of quantifiable values to gauge its achievements using Key Performance Indicators (KPIs), there should be outcome-based, quantifiable statements from an ethical standpoint: Key Ethics Indicators (KEI), enabling developers to manage and track to what extent each model is meeting the stated objectives.

For example, Lee and Floridi estimate the impact of each default risk prediction algorithm on financial inclusion and on loan access for black borrowers (Lee & Floridi, 2020). They operationalise financial inclusion as the total expected value of loans under each model and minority loan access as the loan denial rate of black applicants under each model. In Fig. 1 replicated from their work, they calculate the trade-offs between the two objectives for five algorithms, providing actionable insights for all stakeholders on the relative success of each model.

Context-specific KEIs can be developed for each use case. For example, in algorithmic hiring, employee satisfaction with a role may be estimated by attrition rates and employee tenure, employee performance may be measured through their annual review process, and diversity may be calculated across gender, university, region, age group, and race, depending on each organisation's objectives and values. Making explicit the ethical objectives in each use case would help decision-makers justify the use of any algorithm, which could in turn lead to the establishment of industry standards, informing best practices, policy design, and regulatory activity.

5.6   Select a Model and Provide Justifications

The trade-off analysis makes the ethical considerations clear. For example, in Fig. 1, Lee and Floridi conclude that Random forest is better in absolute terms (in both financial inclusion and impact on minorities) than Naïve Bayes, but the decision is more ambiguous between CART and LR: while CART is more accurate and results

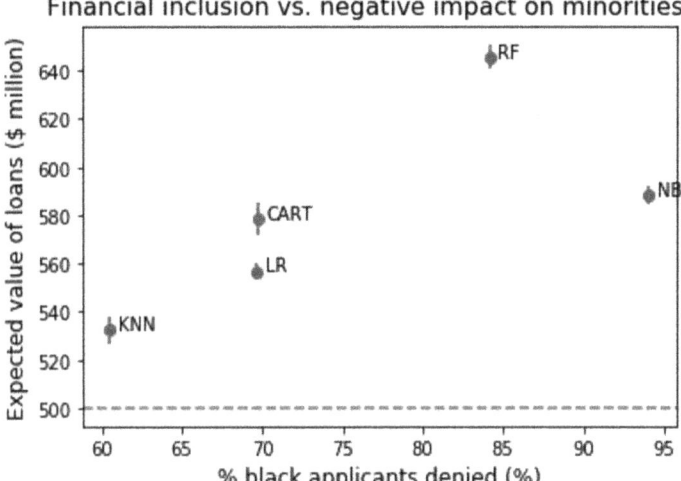

Fig. 1 Replicated from Lee and Floridi (2020): Trade-off analysis

in greater financial inclusion (equivalent of $15.6 million of loans, or 103 median-value loans), CART results in a 3.8 percentage points increase in denial rates for black loan applicants compared to LR. This quantifies the concrete stakes to the decision-maker who may decide on the model that is most suited to his or her priorities, customised to each use case.

One of the key benefits of the outcome-driven KEI trade-off analysis is that it provides interpretable and actionable insights into the decision maker's values, which is especially important for complex machine learning algorithms in which the exact mechanism may not be transparent or interpretable. This could also provide valuable justification to the regulator on why a certain model was seen as preferable to all other reasonable alternatives. This may also help reduce the hesitation among decision-makers around the use of machine learning models due to their non-transparent risks, if the analysis shows they are superior to traditional rules-based models in meeting each of the KEIs. Suitable records of the decisions must be kept, ensuring the model and its design are *reviewable* (Cobbe et al., 2021).

6 Conclusion

Implementations of fairness toolkits have predominantly implemented mathematical fairness definitions without locating their implications in overall algorithmic ethics. One of our contributions is to derive lessons from *ethical philosophy* and from *welfare economics* on what are the *contextual considerations* that are important in assessing an algorithm's ethics beyond what can be captured in a mathematical formula. For example, we refer to the debate in ethical philosophy on what constitutes acceptable vs. unacceptable inequalities. We also relate to the explicit consideration in welfare economics of welfare and liberty, which are associated with algorithmic ethics principles of beneficence, non-maleficence, autonomy, and explicability. Over-reliance on fairness metrics would capture only one dimension of an algorithm's ethical impact.

As a step forward, our second contribution is the proposal of a generalised "Key Ethics Indicator" (KEI) approach that *explicitly* considers the ethical objectives, aligning to the contextual features that we have drawn out as important in ethical philosophy and welfare economics literature. The widespread discomfort with the use of ML to make decisions derives from the tension between the opportunity provided by algorithms that can more accurately predict an outcome and the risk of systematically reinforcing existing biases in the data and the risk of undermining human autonomy. On the other hand, unlike human subconscious biases, machine predictions can be systematically audited, debated, and improved. By understanding the holistic ethical considerations of each algorithmic decision-making process using KEIs, decision-makers can be better informed about the value judgements, assumptions, and consequences of their algorithmic design, opening up the conversations with regulators and with society on what is an ethical decision.

References

Aggarwal, N. (2018). *Law and autonomous systems series: Algorithmic credit scoring and the regulation of consumer credit markets*. University of Oxford Business Law Blog.

Aggarwal, N. (2020). *The norms of algorithmic credit scoring*. Available at SSRN 3569083.

Aristotle, & Sinclair, T. A. (1962). *Aristotle: The politics; Translated with an Introduction by TA Sinclair*. Penguin Books Limited.

Arya, V., Bellamy, R. K. E., Chen, P-Y., Dhurandhar, A., Hind, M., Hoffman, S. C., Houde, S., Liao, Q. V., Luss, R., & Mojsilovi'c, A., et al. (2019). *One explanation does not fit all: A toolkit and taxonomy of AI explainability techniques*. arXiv preprint arXiv:1909.03012.

Bellamy, R. K. E., Dey, K., Hind, M., Hoffman, S. C., Houde, S., Kannan, K., Lohia, P., Martino, J., Mehta, S., & Mojsilovic, A., et al. (2018). *AI fairness 360: An extensible toolkit for detecting, understanding, and mitigating unwanted algorithmic bias*. arXiv preprint arXiv:1810.01943.

Binns, R. (2020). On the apparent conflict between individual and group fairness. In *Proceedings of the 2020 conference on fairness, accountability, and transparency* (pp. 514–524). ACM.

Chouldechova, A. (2017). Fair prediction with disparate impact: A study of bias in recidivism prediction instruments. *Big Data, 5*(2), 153–163.

Cobbe, J., Lee, M. S. A., & Singh, J. (2021). Reviewable automated decision-making: A framework for accountable algorithmic systems. In *Proceedings of the 2021 ACM conference on fairness, accountability, and transparency* (pp. 598–609). ACM.

Cole, N. S. (1973). Bias in selection. *Journal of Educational Measurement, 10*(4), 237–255.

Collins, P. H. (2002). *Black feminist thought: Knowledge, consciousness, and the politics of empowerment*. Routledge.

Corbett-Davies, S., & Goel, S. (2018). *The measure and mismeasure of fairness: A critical review of fair machine learning*. arXiv preprint arXiv:1808.00023.

Crenshaw, K. (1989). *Demarginalizing the intersection of race and sex: A black feminist critique of antidiscrimination doctrine, feminist theory and antiracist politics*. u. Chi. Legal f., 139.

Donovan, K. P., & Park, E. (2019). *Perpetual Debt in Silicon Savannah*. Boston Review.

Dwork, C., Hardt, M., Pitassi, T., Reingold, O., & Zemel, R. (2012). Fairness through awareness. In *Proceedings of the 3rd innovations in theoretical computer science conference* (pp. 214–226). ACM.

Dworkin, R. (1981). What is equality? Part 1: Equality of welfare. *Philosophy and Public Affairs, 10*(3), 185–246.

Eaton, M. L. (2004). *Ethics and the business of bioscience*. Stanford University Press.

Elul, R., & Gottardi, P. (2015). Bankruptcy: Is it enough to forgive or must we also forget? *American Economic Journal: Microeconomics, 7*(4), 294–338.

Feldman, M., Friedler, S. A., Moeller, J., Scheidegger, C., & Venkatasubramanian, S. (2015). Certifying and removing disparate impact. In *Proceedings of the 21th ACM SIGKDD international conference on knowledge discovery and data mining* (pp. 259–268). ACM.

Fleurbaey, M., et al. (2008). *Fairness, responsibility, and welfare*. Oxford University Press.

Floridi, L., & Cowls, J. (2019). A unified framework of five principles for AI in society. *SSRN Electronic Journal, 1.1.*

Friedler, S. A., Scheidegger, C., & Venkatasubramanian, S. (2016). *On the (im) possibility of fairness*. arXiv preprint arXiv:1609.07236.

Gajane, P., & Pechenizkiy, M. (2017). *On formalizing fairness in prediction with machine learning*. arXiv preprint arXiv:1710.03184.

Gauthier, D. (1986). *Morals by agreement*. Oxford University Press on Demand.

Gillis, T. B. (2020). *False dreams of algorithmic fairness: The case of credit pricing*. Available at SSRN 3571266.

Goldman, B., & Cropanzano, R. (2015). "Justice" and "fairness" are not the same thing. *Journal of Organizational Behavior, 36*(2), 313–318.

Greenberg, J. (1987). A taxonomy of organizational justice theories. *Academy of Management Review, 12*(1), 9–22.

Hardt, M., Price, E., & Srebro, N. (2016). *Equality of opportunity in supervised learning*. CoRR abs/1610.02413 (2016). arXiv:1610.02413 http://arxiv.org/abs/1610.02413

Heidari, H., Loi, M., Gummadi, K. P., & Krause, A. (2018). *A moral framework for understanding of fair ml through economic models of equality of opportunity*. arXiv preprint arXiv:1809.03400.

Holstein, K., Vaughan, J. W., Hal Daum'e, M. D., III, & Wallach, H. (2019). Improving fairness in machine learning systems: What do industry practitioners need? In *Proceedings of the 2019 CHI conference on human factors in computing systems* (pp. 1–16). CHI.

Hutchinson, B., & Mitchell, M. (2019). 50 years of test (un) fairness: Lessons for machine learning. *Proceedings of the conference on fairness, accountability, and transparency*.

Kagan, S. (1999). 30. *Equality and desert. What do we deserve?: A reader on justice and desert*, 298.

Kagan, S. (2014). *The geometry of desert*. Oxford University Press.

Kamiran, F., & Žliobaitė, I. (2013). Explainable and non-explainableˇ discrimination in classification. In *Discrimination and privacy in the information society* (pp. 155–170). Springer.

Kant, I., & Gregor, M. (1996). *The metaphysics of morals*.

Kilbertus, N., Ball, P. J., Kusner, M. J., Weller, A., & Silva, R. (2019). *The sensitivity of counterfactual fairness to unmeasured confounding*. arXiv preprint arXiv:1907.01040 (2019).

Kim, M., Reingold, O., & Rothblum, G. (2018). Fairness through computationally-bounded awareness. *Advances in Neural Information Processing Systems*, 4842–4852.

Kleinberg, J., Mullainathan, S., & Raghavan, M. (2016). *Inherent trade-offs in the fair determination of risk scores*. arXiv preprint arXiv:1609.05807 (2016).

Koren, J. R. (2016). *What does that web search say about your credit?* (Jul 2016). https://www.latimes.com/business/la-fi-zestfinance-baidu-20160715-snap-story.html

Kusner, M. J., Loftus, J. R., Russell, C., & Silva, R. (2017). *Counterfactual fairness*. arXiv e-prints, Article arXiv:1703.06856(March 2017), arXiv:1703.06856 pages. arXiv:1703.06856 [stat.ML].

Lee, M. S. A., & Floridi, L. (2020). Algorithmic fairness in mortgage lending: From absolute conditions to relational trade-offs. *Minds and Machines, 2020*, 1–27. https://doi.org/10.1007/s11023-020-09529-4

Lee, M. S. A., & Singh, J. (2020). *The landscape and gaps in open source fairness toolkits.* Available at SSRN (2020).

Lee, M. S. A., & Singh, J. (2021). *Risk identification questionnaire for unintended bias in machine learning development lifecycle.* Available at SSRN (2021).

Lewis, D. (1973). Causation. *Journal of Philosophy, 70*(17), 556–567.

Liu, L. T., Dean, S., Rolf, E., Simchowitz, M., & Hardt, M. (2018). *Delayed impact of fair machine learning.* arXiv preprint arXiv:1803.04383.

Nozick, R. (1974). *Anarchy, state, and utopia* (Vol. 5038). Basic Books.

Parfit, D. (1991). *Equality or priority.* University of Kansas, Department of Philosophy.

Prager, R. A., et al. (2009). *Determinants of the locations of payday lenders, pawnshops and check-cashing outlets.* Federal Reserve Board.

Rawls, J. (1999). *A theory of justice*, rev. ed.

Rawls, J. (2001). *Justice as fairness: A restatement.* Harvard University Press.

Saleiro, P., Kuester, B., Stevens, A., Anisfeld, A., Hinkson, L., London, J., & Ghani, R. (2018). *Aequitas: A bias and fairness audit toolkit.* arXiv preprint arXiv:1811.05577.

Scheffler, S. (1994). *The rejection of consequentialism: A philosophical investigation of the considerations underlying rival moral conceptions.* Oxford University Press.

Sen, A. K. (1992). *Inequality reexamined.* Oxford University Press.

Suresh, H., & Guttag J. V. (2019). *A framework for understanding unintended consequences of machine learning.* arXiv preprint arXiv:1901.10002 2 (2019): 8.

Van Dam, A. (2019). *Searching for images of CEOs or managers? The results almost always show men.* The Washington Post (3 01 2019).

Veale, M., Van Kleek, M., & Binns, R. (2018). Fairness and accountability design needs for algorithmic support in high-stakes public sector decision-making. In *Proceedings of the 2018 chi conference on human factors in computing systems*, pp. 1–14.

Verma, S., & Rubin, J. (2018). Fairness definitions explained. In *2018 IEEE/ACM International Workshop on Software Fairness (FairWare)* (pp. 1–7). IEEE.

Wachter, S., Mittelstadt, B., & Russell, C. (2020). *Why fairness cannot be automated: Bridging the gap between EU nondiscrimination law and AI.* Available at SSRN (2020).

Wachter, S., Mittelstadt, B., & Russell, C. (2021). *Bias preservation in machine learning: The legality of fairness metrics under EU non-discrimination law.* Available at SSRN (2021).

Walzer, M. (1983). *Spheres of justice. A defense of pluralism and equality.* Basic.

Xu, F., Uszkoreit, H., Du, Y., Fan, W., Zhao, D., & Zhu, J. (2019). Explainable AI: A brief survey on history, research areas, approaches and challenges. In *CCF international conference on natural language processing and Chinese computing* (pp. 563–574). Springer.

Zillman, C. (2019). *Fortune 500 female CEOs reaches all-time record of 33.* Fortune (16 05 2019).

Ethics Auditing Framework for Trustworthy AI: Lessons from the IT Audit Literature

Nathaniel Zinda

Abstract *The European Guidelines on Trustworthy Artificial Intelligence* refer to auditing as a key way to implement ethical practices into the development and deployment of artificial intelligence (AI). However, auditing AI, and especially the "ethics audit" (EA, also known as the *business ethics audit, social audit,* or *corporate social responsibility audit*) of AI, is still a vague concept. It is unclear what should be the *object* of the audit – whether the processes used to develop an AI system or the system's use and real-world application – as well as *which aspects* of AI systems should be audited – for example, whether the auditing of AI should focus on risk, accountability, or governance. This chapter aims to shed light on EA of AI by analysing the existing relevant literature on auditing information technologies (IT). By using a qualitative evidence synthesis, a method that employs selective or purposive sampling in order to identify 'themes' or 'constructs' from the literature, this chapter reviews methods for auditing IT, with a particular focus on methodologies connected to three key concepts: governance, assurance, and risk. Its goals are to identify a set of methodologies and standards that can be a source of reference for the AI community when developing EA protocols for AI; and to clarify important lessons and considerations.

Keywords Artificial intelligence · Ethics · IT audit · Governance · Quality assurance · Risk management

1 Introduction

As Artificial Intelligence (AI) continues to pervade society, the ethical debate over how AI systems should be developed and deployed is at a turning point. Previous years of research and debate have focused on the principles (the 'what') of ethical

N. Zinda (✉)
Oxford Internet Institute, University of Oxford, Oxford, UK

© The Author(s), under exclusive license to Springer Nature
Switzerland AG 2022
J. Mökander, M. Ziosi (eds.), *The 2021 Yearbook of the Digital Ethics Lab*,
Digital Ethics Lab Yearbook, https://doi.org/10.1007/978-3-031-09846-8_12

AI. The outcome has been an emerging consensus on a set of shared values encompassing principles like beneficence, non-maleficence, autonomy, justice and explicability (Floridi & Cowls, 2019). While the global, multi-stakeholder interest is both important and encouraging, there is a recognition that more detailed guidance is needed to connect these principles to the day-to-day practice (the 'how') of developing AI systems (Morley et al., 2019). There is growing agreement that auditing these systems would foster the development of ethical AI and hold "AI developers accountable for harms associated with AI development" (Brundage et al., 2020, p. 2). Most prominently, the *Ethical Guidelines for Trustworthy Artificial Intelligence* from the European Commission's High-Level Expert Group on Artificial Intelligence (AI HLEG) argue that facilitating the auditability of AI systems is necessary to "ensure responsibility and accountability for [these] systems and their outcomes" (AI HLEG, 2018, p. 19). Auditing is thus presented as a necessary step to develop *trustworthy* AI. While auditing is a well-established and robust field in its own right, the question of how precisely to apply it to AI systems, particularly in view of the complex social and technical challenges that AI implicates, is one that has yet to be answered. Some proposed approaches to auditing AI systems have emerged with varying degrees of specificity (e.g. ICO, 2020; Raji et al., 2020). While there is a growing interest in this topic, the concept remains very much in its infancy. This chapter aims to cast light on the EA of AI by analysing existing methodologies and standards identified in the information technologies (IT) literature. In the rest of this chapter, Sect. 2 defines the level of abstraction (LoA) of our analysis; Sect. 3 provides the results of the review and summarizes elements of the main methodologies; Sect. 4 discusses additional overarching considerations for the EA of AI; and Sect. 5 concludes the chapter. The review itself is a qualitative evidence synthesis, a method that employs selective or purposive sampling in order to identify 'themes' or 'constructs' that can be drawn from the literature (Grant & Booth, 2009). Details regarding our methodology are outlined in Appendix 1. Appendices 2 and 3 include additional information on some of the methodologies discussed throughout the chapter.

2   Governance, Assurance and Risk

An audit is a "systematic, independent and documented process" for determining the extent to which specific criteria are fulfilled (ISO 19011, 2018). Oftentimes, these criteria come in the form of company policies, legal requirements, industry standards, or similar mechanisms of constraint. The precise nature of an audit, particularly in the context of information technology (IT), is a function of the audit's type and purpose (Gantz & Maske, 2014). Given the diversity within the IT audit literature, it is first necessary to define the aspects that will be most relevant to the present case. That is, to define the level of abstraction (LoA) appropriate to identify, select, and analyse existing methodologies and standards within the IT audit literature. We identify three key aspects of IT auditing that we argue are also crucial for the ethics auditing of AI. These are governance, assurance, and risk.

Governance is central to establishing mechanisms of accountability within an organization. Accountability as a principle is "fundamentally about the answerability of actors for outcomes" (Kohli et al., 2018). In the context of AI systems, the question about who or what is answerable for outcomes is more complex than it initially appears. There are, for example, relevant debates about the extent to which algorithms and artificial agents can be held morally accountable. Floridi & Sanders (2004) argue that the introduction of artificial agents decouples moral accountability from moral responsibility and thus "moral accountability is a necessary but [ultimately] insufficient condition for moral responsibility". When we consider this issue in the context of auditing, we are specifically concerned with accountability in the institutional sense: with the "organizations designing and deploying algorithms" (Raji et al., 2020). In this context, the answerability for outcomes is primarily established and enforced through an organization's governance structures. This conception of accountability is consistent with the guidance outlined by the AI HLEG (2018), and what they describe as "accountability via governance frameworks".

ISO 37000 (2020) defines governance structures as "the system[s] by which the whole organization is directed, controlled and held accountable to achieve its core purpose over the long term". Accordingly, governance is achieved through constraint mechanisms like company policies and procedures, oversight boards, training programs, and certification schemes. These mechanisms control and align the organization's development and use of technology with particular goals, and as such, these mechanisms are often considered critical aspects of an organization to audit. An organization's traditional governance structures and AI governance can be intertwined. Dafoe (2018), for example, argues that ideal AI governance necessitates that we ask whether "organizational principles and institutional mechanisms" exist in such a way as to promote the goals of AI governance, such as human autonomy, fairness, or explicability. The AI HLEG (2018), for its part, suggests that 'certification specifications' can play a key role in supporting an organization's governance frameworks to "ensur[e] accountability for the ethical dimensions of decisions" associated with developing and deploying AI systems. Clearly, governance is a critical aspect to analyse when considering the auditing of IT and it can offer valuable lessons for the ethics audit (EA) of AI systems.

Along with Raji et al. (2020), we borrow a concept from environmental studies known as urgent governance (Lynch & Veland, 2018). Urgent governance distinguishes between "auditing for system reliability vs. societal harm" (Raji et al., 2020). As Leveson (2011) notes, the notions of reliability and safety in systems engineering are not synonymous, and sometimes there are serious tensions between them. Manufacturing plants, for example, can be highly reliable in performing their core functions, but can also produce unsafe outcomes such as air pollution. AI systems, just like any engineered complex system, can also be technically functional and reliable, while producing harmful social and ethical outcomes.

Urgent governance relates heavily to the second variable of our LoA: assurance. Traditionally, governance structures are calibrated to produce systems that achieve high standards of functionality and reliability. This occurs primarily through the implementation of quality assurance systems. These assurance systems are typically an "iterative set of processes intended to help organizations deliver products that meet applicable requirements," whether they be external, such as laws and regulations, or internal, such as system or product criteria (Gantz & Maske, 2014). Urgent governance can translate effectively to the context of EA of AI. Whereas traditional structures rely on quality assurance processes to check features such as accuracy and reliability, urgent governance structures could indeed be formed to check social and ethical standards, while relying upon, and making use of, similar quality assurance processes. As Raji et al. (2020) argue, urgent governance suggests that "a separate governance structure is necessary for the evaluation of…systems for ethical compliance," and that this evaluation could be "embedded in the established quality assurance workflow".

The final component in our LoA is risk. According to ISO 31000 (2018), risk is simply the "effect of uncertainty on objectives." This could include uncertainty that governance structures are properly aligning technology with organizational goals, or, at the system level, uncertainty that an engineered system will operate as intended. Needless to say, risk is present at all levels of an organization, and as such, risk and risk management are common aspects to audit (Gantz & Maske, 2014). Discussions around auditing AI likewise place a heavy emphasis on risk. As the AI HLEG (2018) notes, AI can pose significant risks to fundamental rights, and achieving trustworthy AI partly depends on "adopt[ing] adequate measures to mitigate these risks" in a manner proportional to "the magnitude of the risk". Moreover, existing proposals have suggested that audits of AI systems should rely on a risk-based approach for mitigating unethical outcomes (e.g. ICO, 2020).

Having described the LoA of our review, we can now analyse the relevant literature. This is the task of Sect. 3.

3   Results

3.1   Governance

The review yielded several methodologies, frameworks, and standards applicable to IT governance. They are highly variable in terms of their scope and purpose. The most important ones for present purpose are five. They are summarized in Table 1 and discussed in the rest of this section.

Perhaps the most pervasive theme in the IT governance literature is the breadth of scope and purpose. That is, the methodologies we identified did not constitute "alternative treatments of the same issues" (Hamidovic, 2010). For example, the COSO Internal Control Framework is a set of components and principles for successful governance of an organization, whereas IIA Std 2110 is a standard for conducting internal audits of governance structures. This mismatch reflects what Coertze & von Solms (2014) identify as the broader issue in the IT governance literature; namely, that the most frequently cited frameworks "all differ in their

Table 1 IT governance – methods and standards

Title	Remarks
COSO Internal Control Integrated Framework (COSO, 2013)	*High-level principles for successful internal control within an organization. Not specific to the context of IT.*
IIA Standard 2110: Governance (IIA Std 2110, 2016)	*Standard and implementation guide for internal audits of governance structures.*
Information Technology Infrastructure Library (ITIL) (Cartlidge et al., 2007)	*Set of practices for IT service management. Limited amount of information publicly-available.*
ISO/IEC 38500 – Governance of IT for the Organization	*Principles for governing bodies on the effective and acceptable use of IT within an organization. Limited detail on implementation.*
Control Objectives for Information and Related Technology (COBIT) (ISACA, 2019)	*End-to-end, process-based framework for aligning IT use and development with business objectives.*

terminology, focal points of discussion and target audiences". The lack of conceptual or structural alignment makes it difficult to develop an integrated conceptualization of the IT governance literature.

In the context of auditing IT governance structures, standards are primarily concerned with people and processes, as opposed to products or systems. The Institute of Internal Auditors' (IIA) International Standard on Governance (IIA Std 2110, 2016) places a strong emphasis on the policies and processes that are established, and the extent to which people in the organization adhere to them. The standard itself outlines several considerations for implementing an internal audit program for IT governance. This includes evaluating the governance processes for making strategic decisions, overseeing risk management, and promoting ethics within the organization (IIA Std 2110, 2016). In this case, 'strategy, risk and ethics' are not considered in relation to the IT systems or services within the organization; rather, they are concerned with higher-level organizational processes. For example, when auditing the governance body's oversight of risk management, IIA Std 2110 (2016) recommends audit activities such as interviews with risk management personnel or reviews of meeting minutes in which risk management strategies and policies were discussed. Similarly, when promoting ethics within the organization, the standard recommends reviewing the organization's conformance to hiring and training processes, whistleblowing policies, or its mission and value statements.

In contrast to audit programs (e.g. IIA Std 2110, 2016) or principle-based standards (e.g. ISO 38500, 2015), IT governance frameworks like ITIL and COBIT come the closest to a comprehensive set of best-practice protocols for governing the use and development of technology within an organization. The Information Technology Infrastructure Library (ITIL) is a framework for IT service management developed by the UK government. The methodology emphasizes the continuous improvement of IT service delivery and support. More specifically, it includes a service delivery life cycle along with "key processes and activities, their inputs and outputs, and roles and responsibilities" for each phase in the life cycle (Gantz & Maske, 2014, p. 182). Because of its emphasis on IT service delivery specifically, the methodology is far less comprehensive than COBIT. For example, ITIL only outlines 9 governance processes whereas COBIT defines over 30. Nonetheless, within the area of service management, the ITIL framework offers much more detailed guidance on implementation, as it sacrifices some of 'the *what*' for 'the *how*' (Radovanovic et al., 2010). Detailed guidance and source material on the ITIL framework is not accessible to those outside industry organizations. This is anecdotally the case in this review, and it has been articulated as a problem by other researchers in the IT governance literature (e.g. Hamzane & Belangour, 2019). Other weaknesses of the ITIL framework include its failure to address system-level processes, such as the software development life cycle and issues in quality management (Wessels & van Loggerenberg, 2006).

COBIT is far more comprehensive than ITIL. It is an end-to-end framework that defines 34 key control processes across multiple IT domains in an organization.[1] This comprehensive approach makes COBIT one of the most widely used models for IT governance (Gantz & Maske, 2014, p. 78). Each process in the framework defines several metrics for evaluation. This includes a RACI matrix (referring to **R**esponsible, **A**ccountable, **C**onsulted, and **I**nformed) which outlines who is "responsible and authorized for implementation of particular control activities and who only needs to be informed and consulted" (Radovanovic et al., 2010). Other metrics include maturity models (on a scale from 0 to 5), critical success factors, key goal indicators, key performance indicators, and control objectives and tests (Selig, 2008, p. 298). As Wessels & van Loggerenberg (2006) note, the nature and detail of COBIT's processes and metrics means that the framework is extremely audit-oriented. There are, however, weaknesses to the COBIT framework. Although COBIT covers a wide range of IT processes, it only provides the directives that organizations must follow, not *how* to follow them (Wessels & van Loggerenberg, 2006). Furthermore, like the ITIL framework, COBIT largely fails to address the software development life cycle. Given that the software development life cycle is a critical aspect of the ethical and robustness issues that confront AI, additional methodologies should be consulted to provide system-level guidance.

3.2 Assurance

As Raji et al. (2020) argue, urgent governance structures could rely on, and make use of existing methods for quality management (henceforth QM). That is, QM systems can be repurposed to ensure that technical systems meet a standard of 'ethical' quality rather than traditional standards of functionality and reliability.

---

[1] These domains include *evaluate, direct, and monitor* (EDM), *align, plan, and organize* (APO), *build, acquire, and implement* (BAI), *deliver, service, and support* (DSS), and *monitor, evaluate, and assess* (MEA). Each domain within the framework contains a set of processes and subprocesses. For example, the "Manage Programs and Projects" process is within the BAI domain and is denoted as BAI01.

Table 2 Quality management systems – methods and standards

Title	Remarks
ISO 9000 – Quality Management Systems – Fundamentals and Vocabulary	*Describes fundamental concepts and principles of quality management.*
ISO 9001 – Quality Management Systems – Requirements	*Specifies auditable requirements for a quality management system.*
ISO/TS 9002 – Quality Management Systems – Guidelines for the Application of ISO 9001	*Provides additional guidance, with a clause by clause correlation to ISO 9001.*
ISO 10005 – Quality Management – Guidelines for Quality Plans	*Provides guidelines for establishing, reviewing, applying and revising quality plans.*
ISO/TR 10013 – Guidelines for Quality Management System Documentation	*Provides guidelines for the development and maintenance of documentation necessary for quality management systems.*

Following this suggestion, our review identified several methods and standards related to the QM process (see Table 2).[2] Broadly speaking, these outline essential components for QM systems, and specific guidance for software engineering systems, including guidance on the software development life cycle. Other methodologies in the literature relate to AGILE-based methodologies in software engineering and their application to safety-critical systems. The methodologies in this section are not in competition with one another. In fact, they are highly complementary, and often touch on different aspects of the QM process or focus on applying QM to different domains.

The ISO 9000 family is a prominent set of international standards on QM systems. Its purpose is to provide guidance, for organizations of all types, to produce products or services that meet the quality requirements of the organization or external bodies (Hoyle, 2001). As one of the only QM standards that is designed for third-party certification, the ISO 9000 family is one of the most commonly used sets of standards for assuring quality in technical systems (Gantz & Maske, 2014). ISO 9001 (2015) outlines the certification requirements for best-practice QM systems. These requirements are designed to be general-purpose and are not specific to any particular type of organization or technology.

---

[2] We recognize there are fundamental difficulties applying QM processes to emerging technologies like machine learning (ML) and AI more generally, at a technical level at least (e.g. Auer & Felderer, 2018; Nakajima, 2018). Auer & Felderer (2018) note, for example, that a key limitation to testing ML systems is the faulty "assumption that test environments sufficiently mimic the later application in order to allow quality assurance". While technical limitations will need to be overcome, Schöppl et al. (2021) rightly argue the technical challenges should not distract from the lessons and practices that can be leveraged from the 'organisational' component of audits, such as those that focus on the management of people and processes. Thus, our concern is primarily with the 'transferable' requirements in these methodologies: for instance, the structural components of certifiable QM systems, documentation requirements, and methods of reporting non-conformance.

ISO 9001 (2015) advances a process-based approach to QM systems. That is, the standard defines a QM system as a set of interrelated processes that address the leadership, planning, support, operation, and evaluation of a project.[3] These processes exist both at the level of product development (e.g. design checklists, testing, etc.) *and* at the level of human implementation (e.g. management reviews, internal training, etc.). Although processes are expected to differ depending on the organization, ISO 9001 requires certain things to be determined for each process:

- the inputs required and the outputs expected from these processes;
- the sequence and interaction of these processes;
- the criteria and methods (including monitoring, measurements and related performance indicators) needed to ensure the effective operation and control of these processes;
- the resources needed for these processes and ensure their availability; [and]
- the responsibilities and authorities for these processes... (ISO 9001, 2015).

Records and documentation are also an important part of ISO 9001 certified QM systems. The standard requires organizations – to the extent necessary – maintain documented information to support its processes, as well as to provide "confidence that the processes are being carried out as planned" (ISO 9001, 2015). In other words, both instruction and traceability are important considerations when determining what forms of documentation are necessary. For the most part, the ISO 9001 standard does not require any specific documentation. Organizations have discretion over which processes should include documentation and in what form the documentation should be. Given that QM processes themselves are highly context-dependent, so is the necessary documentation.

ISO 10005 (2018) is a related standard that provides guidance on creating standardized quality plans for organizations with or without ISO 9001 certified QM systems.[4] These are comprehensive documents that describe how an organization will (or should) produce an intended outcome, where outcomes could be a project, process or procedure. Quality plans specify the "actions, responsibilities and associated resources to be applied" in specific circumstances (ISO 10005, 2018). One can conceive of quality plans as the QM handbooks for particular teams or technologies within an organization. The standard describes several situations where quality plans are especially relevant, such as:

- to develop and validate new products [...];
- to demonstrate, internally and/or externally, how requirements will be met; [and]

---

[3] According to the definitions laid out in ISO 9000, a process should not be confused with a procedure. A process is defined as a "set of interrelated or interacting activities that use inputs to deliver an intended result." Internal training, for example, can be conceived of as a process. Procedures, on the other hand, are a "specified way to *carry out* an activity or process." In the above context, this would be the instructions for carrying out the internal training.

[4] Even though ISO 10005 can be used for stand-alone purposes, its guidance is designed to be compatible with ISO 9001 requirements.

- as the basis for monitoring and assessing compliance with the requirements for quality [...] (ISO 10005, 2018)

These situations are highly reflective of the circumstances where EA of AI would be applicable, suggesting that an ISO 10005 based quality plan could be a relevant part of an appropriate audit methodology. The standard itself notes that monitoring an organization's conformity with quality plans can come in the form of an audit. In these cases, the quality plan should specify the type of the audit, its nature and scope, and how the results will be used. Annex A in the standard provides example templates, along with suggested headers and contents. Additional requirements for QM documentation, quality plans or otherwise, can be found in ISO/TR 10013 (2001).

The ISO 9000 series and ISO 10005 are sector agnostic. Because of this, additional standards have emerged to align ISO 9001 with systems and software engineering (see Table 3).

The main standard for quality assurance in the software context is IEEE Std 730 (2014). According to IEEE Std 730, the software quality assurance (SQA) process can be broken down into three main parts: SQA process implementation, product assurance, and process assurance. The SQA process implementation clause is heavily reminiscent of the requirements in ISO 9001 and the guidance in ISO 10005. It requires organizations to create a Software Quality Assurance Plan (SQAP), roughly analogous to the quality plans in ISO 10005. SQAPs define the processes, activities, and outcomes necessary for the project, as well as the roles and responsibilities for performing them. Annex C in the standard provides additional guidance for creating SQAPs. It provides an outline mapped to the relevant requirements in the standard, as well as an extensive list of questions and suggestions for different phases of the assurance process.

Like ISO 9001, IEEE Std 730 makes explicit reference to assurance in the context of both the *product* and *process*. In other words, assurance not only requires

Table 3 Quality management systems – software engineering – methods and standards

Title	Remarks
ISO 90003 – Software Engineering – Guidelines for the Application of ISO 9001 to Computer Software	*Provides guidance for applying ISO 9001 requirements to computer software. Additional guidance connected to ISO 12207.*
IEEE Std 730 – IEEE Standard for Software Quality Assurance Processes	*Requirements for initiating, planning, controlling, and executing Software Quality Assurance.*
ISO 15288 – Systems and Software Engineering – System Life Cycle Processes	*Provides a common framework of process descriptions for the system development life cycle.*
ISO 12207 – Systems and Software Engineering – Software Life Cycle Processes	*Applies processes described in ISO 15288 to the software development life cycle.*
ISO 24748-1 – Systems and Software Engineering – Life Cycle Management – Part 1: Guidelines for Life Cycle Management	*Provides guidelines for system life cycle management, complementing processes described in ISO 15288 and ISO 12207.*

evaluating the product to ensure it conforms to requirements, but also evaluating the *processes used to develop the software* to ensure that they consistently produce quality software. This latter task could include, for example, assessing the knowledge of the employees about the quality processes and their skill in conducting them, as well as ensuring that testing results are properly documented and reported to management (IEEE Std 730, 2014).

The issue of process assurance is heavily connected to the system development life cycle. IEEE Std 730, for example, requires that those performing SQA to "determine whether the defined software life cycle processes…are appropriate, given the product risk". There are many possible life cycle models that organizations can use. An organization's decision on which life cycle model to choose is heavily influenced by the nature of the technical project and its specific risks. ISO 15288 (2015) and ISO 12207 (2017) provide relevant guidance on designing system life cycle stages, including a "common framework of process descriptions for describing the life cycle of systems" (ISO 15288, 2015).[5] According to ISO 15288, the typical system undergoes a common set of stages, including *concept, development, production, utilization, support* and *retirement*. These stages are not normative, and flexible adaptation for specific types of systems is expected.[6] The technical requirements for both standards are found in ISO 24748-1 (2018), which describes in greater detail common processes for each stage of system development. Some key activities include, for example, the creation of entry and exit criteria for each stage in the life cycle, where progressing to the next stage depends on fulfilling the current stage's exit criteria. Therefore, quality is assured by controlling and evaluating the development of the system from stage to stage. The life cycle model for AI systems will naturally differ from more traditional software systems. It will be necessary to develop AI-specific entry and exit criteria, perhaps tied to verification and validation requirements for different kinds of AI models. That said, when considering what an AI system development life cycle would look like, there is no need to start from scratch. The guidelines in ISO 24748-1 and ISO 12207 are an important source of reference to build on. One can identify the most relevant criteria and outcomes in the standard and add or subtract from the (AI) life cycle model as necessary.

Overall, the ISO 9001 family and the software engineering standards and life cycle guidelines are reputable sources that could inform and legitimize the structure of an 'ethics' QM protocol for AI systems. This protocol could overlay an adapted life cycle model based on ISO 24748-1, one modified for the ethical challenges associated with the development of AI applications.

---

[5] ISO 15288 is a general-purpose standard for all types of technical systems, whereas ISO 12207 applies the requirements in ISO 15288 to software systems in particular. Technical specifications for both standards are outlined ISO 24748, which is also referenced.

[6] For example, according to ISO 24748-1, a typical software system undergoes only four stages: *concept, development, operation & maintenance* and *retirement*.

AGILE Methodologies and Safety-Critical Systems

The review identified several methodologies that were related to AGILE-based software engineering and their application in safety-critical contexts (see Table 4). AGILE is a set of software development methodologies that are more flexible than traditional methodologies (e.g. WATERFALL) and allow changes to software requirements later in the development stage. These methodologies are increasingly common in the software development context.

The question of how to apply Software Quality Assurance (SQA) standards in the context of AGILE methodologies has been partly addressed by existing industry standards. Annex F in IEEE Std 730, for example, illustrates how the SQA activities required by the standard can be adapted to the AGILE development process. The standard suggests, for example, that product non-conformances be managed through existing AGILE tools, such as inserting non-conformances as a high-priority item in the product backlog.[7] Another recommendation is that the documentation required from development teams be simple and easy to generate given that AGILE teams already tend to operate quite quickly and with less formal requirements.. This could be as simple as "taking a digital picture of the whiteboard on which progress is tracked" (IEEE Std 730, 2014). Product managers/owners could then be made responsible for formalizing these records into standardized documents for audits.

The recommendations in IEEE Std 730 are not a panacea to some of the risks that AGILE introduces to software engineering projects. In the context of safety-critical software systems, however, modified versions of AGILE have been proposed in order to introduce more robust safeguards to the development process (e.g. Fitzgerald et al., 2013; Hanssen et al., 2018). As Douglas & Ekas (2012) note, safety-critical systems have additional needs beyond most software projects, such as:

Table 4 AGILE and safety-critical systems – methods and standards

Title	Remarks
R-SCRUM (Fitzgerald et al., 2013)	*Tests a modified-version of the AGILE method Scrum in a regulated environment.*
SafeSCRUM (Hanssen et al., 2018)	*Proposes a modified-version of the AGILE method Scrum to be used for developing safety-critical software.*
AGILE Change Impact Analysis (CIA) for SafeScrum (Stalhane et al., 2014)	*Proposes a new strategy for AGILE CIA based on the SafeScrum methodology.*
ISO 15026-2 – Systems and Software Engineering – Systems and Software Assurance – Part 2: Assurance Case	*Specifies minimum requirements for the structure and contents of an assurance case.*

---

[7] In AGILE methodologies, the product backlog is a "single source of requirements for any changes to be made to the product," such as customer requirements or bug fixes (*Scrum Guide*, n.d.). The backlog evolves based on the changing needs of the project and the product manager/owner is responsible for ensuring activities are done in order for the project to be complete.

- initial safety analysis;
- continuous safety assessment;
- continuous traceability analysis;
- change management; and
- requirements-based verification.

Many of these practices are difficult to manage using AGILE. Furthermore, many are also applicable to the needs of AI systems, whether or not they are part of an environment that is traditionally understood as 'safety-critical'.

Evidence in the literature suggests that these practices can be managed using AGILE. Fitzgerald et al. (2013) conducted a case study for a regulated life sciences company to determine whether a modified AGILE methodology could be compliant with regulations in the industry. The augmented SCRUM implementation (termed R-SCRUM) attempted to integrate quality assurance into the product development processes by setting continuous quality control checkpoints during the 'hardening' sprint and inserting non-conformances into the product backlog along the lines discussed in IEEE Std 730. Appendix 2 in this review presents a comparison between the original SCRUM methodology and the R-SCRUM modification. The authors determined from the case study that "agile processes can, in fact, be augmented to work very well in regulated environments" (Fitzgerald et al., 2013). Other takeaways included replacing the suite of standalone tools for code reviews, bug-tracking, and related activities with an integrated toolset. The authors note that this integration did a lot to support what they term 'living traceability,' or the ability to provide an "up-to-date accurate snapshot [of development] in real-time." As Steghofer et al. (2019) note, this concept of living traceability is the foundation of adapting AGILE for safety-critical systems. In addition to R-SCRUM, other modifications to AGILE exist in the literature, such as the SafeSCRUM model (Hanssen et al., 2018) and implementing Change Impact Analysis in AGILE environments (Stalhane et al., 2014).

3.3 Risk

Risk is a critical aspect of IT auditing. In many ways, it is misleading to conceive of risk as a separate domain from governance and assurance, as risk features heavily into both. That said, there are methodologies and standards that deal specifically with risk (see Table 5). These should not be viewed as alternatives to the methodologies discussed above. Rather, they offer guidance for integrating best-practice risk management into both governance and assurance audits.

ISO 31000 (2018) defines risk as the "effect of uncertainty on objectives." Uncertainty is present at all levels of an organization, from enterprise-wide governance structures to specific projects. The risk management methods and standards identified in this review reflect the multiple levels of analysis from which to approach risk. Some methodologies, like the COSO Enterprise Risk Management (ERM) Integrated Framework, are focused on enterprise-level risk, whereas others, like

Table 5 Risk management – methods and standards

Title	Remarks
COSO Enterprise Risk Management	*Enterprise risk management framework, compatible with COSO.*
RiskIT	*Enterprise risk management framework, compatible with COBIT.*
ISO 31000 – Risk Management – Guidelines	*Provides guidelines for managing risk faced by organizations.*
ISO 31010 – Risk Management – Assessment Techniques	*Provides guidance on the selection and application of various risk management techniques.*
ISO 16085 – Information Technology – Software Life Cycle Processes – Risk Management	*Describes process for risk management throughout software development life cycle. Intended for both technical and managerial personnel.*
Management of Risk (M_o_R)	*Risk management certification and guidance scheme similar to ISO 31000.*

ISO 16085 (2004), are focused on risk at the level of the project or system. Regardless of the methodology's primary focus, there is a high degree of overlap on what constitutes best-practice risk management in the context of IT.

Risk management requires criteria or a threshold for which to assess the acceptability of particular risks. ISACA's RiskIT (2009), for example, is a process-based framework (like COBIT) where specific processes are described across three cyclical functions: governance, evaluation, and response. For risk governance, RiskIT (2009) requires that organizations *establish and maintain a common risk view* that defines how specific risks are evaluated and responded to. This includes a requirement that organizations propose and approve risk tolerance thresholds that set the amount of IT-related risk a project can incur while still meeting its objectives. Furthermore, the methodology requires these thresholds be codified into an IT risk policy that documents the amount of risk allowed in certain circumstances, as well as the type of risk and acceptable forms of measurements.

Similar to RiskIT (2009), ISO 31000 (2018) emphasizes the importance of risk criteria. Although not as specific in the activities that it prescribes, the standard stresses that risk criteria be established at the *beginning* of the risk management process and that the criteria consider the views of the organization's stakeholders. Although establishing risk criteria *a priori* is necessary, it is understood that risk criteria "are dynamic and [therefore] should be continually reviewed and amended, if necessary" (ISO 31000, 2018). In setting the risk criteria, several things be considered, including:

- how consequences and [their] likelihood will be defined and measured;
- time-related factors;
- consistency in the use of measurements;
- how the level of risk is to be determined; [and]
- how combinations and sequences of multiple risks will be taken into account (ISO 31000, 2018).

There is no single way to express risk in IT systems and processes. Process-based frameworks like COBIT and RiskIT categorise risk in three types: inherent risk, current risk, and residual risk. Inherent risk is risk as it is first identified, without any measures in place to mitigate it. Current risk is the level of risk *after* controls are applied. And residual risk is the level of risk that remains. The European Commission utilized this risk structure in their privacy and data protection audit of the CogNet Automatic Network Management technology for 5G. Auditors assigned inherent risk ratings to various aspects of the CogNet technology, such as its cloud-based operating model, hardware identifiers, and use and storage of mobile device data. Auditors then evaluated the controls that were in place to determine whether the residual risk was at acceptable levels (Martin, 2017).

The CogNet audit was relatively narrow in scope. More complex technologies require more complex risk analysis. A similar European Commission audit focused on the CITYCOP system, a mobile application and web portal designed to improve crime reporting and communication between citizens and law enforcement. Unlike the CogNet case, the audit extended beyond compliance with data protection law, and encompassed social and ethical standards. Auditors created a legal and ethical evaluation framework built on European data protection law, relevant human rights principles, and principles related to ethical policing.[8] To perform the audit, these requirements were mapped to every function in the system along with their associated risks (see Appendix 3) (*CITYCOP D11.6*, 2014). The CITYCOP audit plan did not reference an established risk management methodology, but it followed many of the best-practices found in the literature, including "establish[ing] a criteria for assessing the risk level (e.g. high, medium or low)" and assessing risk as a function of "the likelihood and impact of the occurrence of [risk] events" (*CITYCOP D12.1*, 2014).

Analysing risk requires one to define the best method of determining probability and impact. RiskIT (2009) notes that risk can be analysed either qualitatively or quantitatively. Both the CogNet and CITYCOP audits, for example, considered risk in qualitative terms. This approach works best "when there is only limited or low-quality information available", but it also introduces a degree of subjectivity into the analysis. Because of this, qualitative risk analysis can weaken the replicable quality that is often desired with audits (RiskIT, 2009). Nonetheless, qualitative approaches can be completed far quicker than quantitative approaches, and they can be just as effective depending on the underlying nature of the risk in question. Although quantitative approaches tend to be considered more objective, some types of risk (e.g. ethical risks) are very difficult (or even impossible) to quantify precisely. Applying a quantitative approach in these cases could run the risk of oversimplification and create a false sense of confidence in the accuracy of the results. RiskIT (2009) recommends that quantitative approaches only be used when there is "sufficient, complete and reliable data on past and comparable events". This scenario is not always achievable, and it is particularly difficult unless an organization already uses a highly

---

[8] These ethical requirements were drawn from the European Code of Police Ethics and community policing ethics codes. Requirements included things like good faith, non-violence and the minimum use of force, and confidentiality.

quantitative process improvement methodology like Six Sigma (RiskIT, 2009). This is unlikely to be the case in modern software environments, where AI systems are often developed using AGILE methodologies. Nonetheless, RiskIT (2009) recommends the following factors be considered when selecting a risk analysis approach: user needs, i.e. whether highly accurate data is necessary for the particular risk analysis; the availability and quality of data related to the risk; the time available to perform the risk analysis; and the expertise and comfort-level of the auditors/experts.

Risk assessment tools and techniques are an important aspect of performing risk analysis. ISO 31010 (2019) is a guidance document for the methodology outlined in ISO 31000 (2018). The standard provides detailed recommendations for the practical aspects of risk management, including techniques for planning the risk assessment, developing risk models, applying risk assessment techniques, and reviewing the risk analysis. Techniques are categorised by task, such as those for eliciting views from stakeholders, identifying risk, analysing controls, and determining consequences and likelihood (ISO 31010, 2019). While it would be impossible to synthesise here all the relevant and useful information, the standard's appendices summarize and compare the strengths and weaknesses of the recommended techniques, as well as the circumstances and systems in which they are best used. Needless to say, the standard serves as an important source of recommended techniques to design a methodology for the EA of AI.

Artefacts are likewise central to the traceability and auditability of product teams that are developing and deploying IT systems. As such, records and documentation feature heavily into the identified risk management methodologies. As both ISO 31000 and RiskIT emphasize, a key principle of best-practices risk management is taking into account human factors. Documentation helps provide a more holistic account of how product teams communicate, assess, and respond to risk, as well as the trade-offs they made throughout the development process. In the context of systems and software engineering, ISO 12207 (the standard for software life cycle processes) requires organizations to document risk through a *risk profile*. The risk profile records the 'state' of each risk, such as its likelihood of occurring, its consequences, and its associated risk threshold. Risks in the risk profile are ordered by their priority and updated when there are any changes to their status. Attempts to mitigate risks should be documented through a risk action request and stored with the other information in the risk profile (ISO 12207, 2017). In other words, the risk profile functions as a central repository of risk-related information that pertains to the system under development. Auditors can examine the risk profile to see which risks were considered and evaluate how they were assessed, prioritized, and mitigated throughout the development process.

4   Further Considerations for EA of AI

The methods and standards discussed in this review should provide a repository of information to consult for various aspects of creating protocols for the EA of AI. Figure 1 maps the primary methodologies in view of the variables in our LoA. It

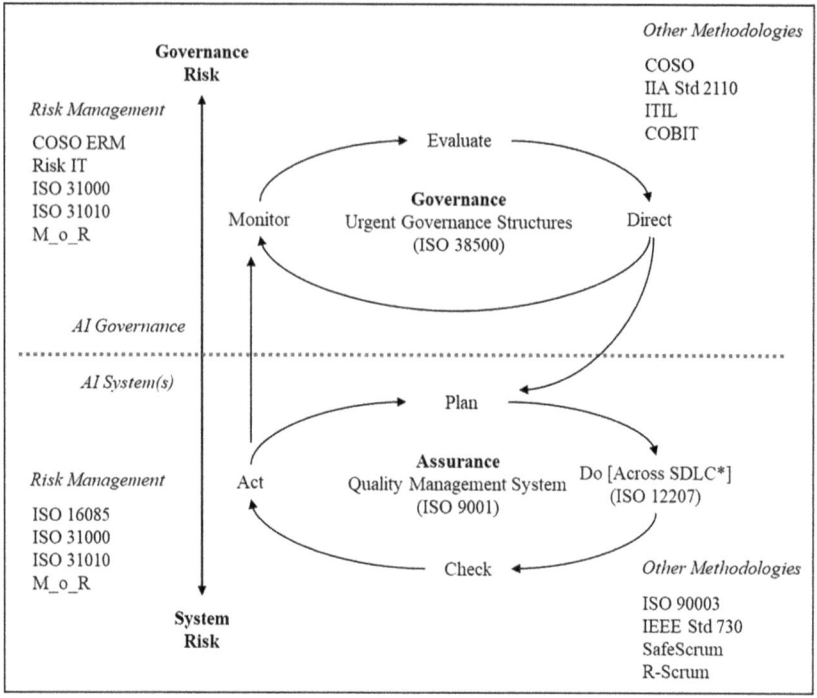

Fig. 1 Map of IT Methods and Standards. (* Software Development Life Cycle)

is important to note that aspects of methodologies in one category may inform decisions in another. For example, although RiskIT is primarily focused on enterprise-level risk, specific requirements or conceptual themes may be useful to guide risk management practices at the system-level. The same can be said for methodologies like COBIT, which is primarily concerned with governance, but contains aspects that can be useful in other domains. In addition to treating these methodologies as possible sources of reference, we identify three additional considerations that were clarified by this review.

4.1   COBIT as a Structural Analogue for an EA of AI Framework

A structural analogue is a term in the chemical sciences that refers to a compound that is structurally similar to another, with the exception of a certain component or components. Despite the similar structure, the compound can behave very differently. In the context of creating protocols for the EA of AI, the suggestion is that it may be reasonable to replicate structural aspects (e.g. terminology, metrics, etc.) of well-established methodologies, but repurposing them for social and ethical objectives. This opportunity is especially promising with the COBIT framework, which is widely used in the literature and is particularly well-suited for audits (Wessels & van Loggerenberg, 2006). COBIT's emphasis on aligning IT with business objectives means that, as it currently stands, most processes are not applicable to social and ethical concerns. However, the structural properties that make COBIT such a prominent methodology can still be useful for the EA of AI.

Evidence in the literature suggests that there is some merit to the structural analogue approach. The audit and advisory firm KPMG, for example, developed an AI Risk and Control Matrix for 105 AI risks within 17 processes from the COBIT framework (Shefford & Holland, 2018).[9] The KPMG model encompasses overarching topics like AI strategy and governance, and more specific topics like business continuity and IT operations. The model's risks and controls are explicitly tied to processes and IT domains defined in the COBIT framework. Although still preliminary, the model reflects the possibility of repurposing COBIT processes and metrics for novel audit purposes. A similar process is recommended by ISACA (the organization behind the COBIT framework) in its guidance for leveraging COBIT for auditing AI (*Auditing Artificial Intelligence*, 2018). The whitepaper argues that COBIT processes can function as tools for the auditor to identify and frame AI risks and controls across an organization. Mapping ethical risks to existing processes in the COBIT framework could be a useful method for developing EA protocols in a way that is recognizable by most organizations and internal auditors. Even if this approach proves unfruitful, there are broader possibilities for developing EA protocols as a COBIT-like dashboard of processes with similar evaluation metrics such as the COBIT maturity model, key performance indicators, RACI matrix, and more.

As mentioned in Sect. 3.1, one key weakness of the COBIT framework is its failure to address the software development life cycle. Because of this, it is unlikely that a structural analogue to COBIT would be *sufficient* for an EA of AI framework. However, depending on the approach, there is certainly room to supplement the governance processes in COBIT with system-level protocols based on the assurance methodologies described in Sect. 3.2.

4.2 Accounting for AGILE

The growing popularity of AGILE methodologies in software engineering is highly relevant to the EA of AI. AGILE naturally introduces risk to the quality assurance process. That is, the qualities that make AGILE so effective at meeting rapidly changing customer requirements also make it difficult to control and evaluate conformance to *a priori* requirements. Mökander et al. (2021) note that this represents a clear technical constraint on the EA of AI, especially given that AI systems are

---

[9] KPMG's AI Risk and Controls Matrix can be accessed here: https://assets.kpmg/content/dam/kpmg/uk/pdf/2018/09/artificial-intelligence-risk-and-controls-matrix.pdf

increasingly being developed using these methods. As such, AGILE is something that the EA of AI must contend with. Although AGILE will likely never be as safe as more linear methodologies (e.g. WATERFALL), this does not mean that the issue is intractable. On the contrary, the literature suggests several possible accommodations.

As Mökander et al. (2021) suggest, one approach is the introduction of tools and practices that facilitate 'living traceability' of the system under development. Integrated toolsets for code reviews, bug-tracking, and other development activities (e.g. Atlassian's Jira) can provide auditors a detailed record of all activities undertaken throughout the project lifecycle (Fitzgerald et al., 2013). These toolsets have the added benefit of being widely subscribed to among software teams, as well as potentially serving as an excellent medium for "the ethical deliberation amongst software developers and managers" (Mökander et al., 2021) as they collaborate throughout the software development process. Outside of these toolsets, some of the standards discussed in this review can provide additional guidance to organizations on how to adapt traditional QM requirements to AGILE development processes (e.g. IEEE Std 730, 2014). Similar 'translation' documents could be created for a novel methodology for EA of AI that maps the protocol's requirements to an AGILE environment, while noting additional risks that emerge, as well as new requirements that organizations must adhere to for AGILE-based development.

A final option is to make use of AGILE methodologies for safety-critical systems like R-SCRUM or SafeSCRUM. One possibility is creating a two-tiered audit certification framework where organizations that opt for AGILE could receive a lower status certification but be required to implement one of these augmented AGILE methodologies. The experience of Fitzgerald et al. (2013) suggests that it is possible to marry robust QM with AGILE software development in highly regulated, safety-critical environments, and to do so in a way that facilitates the traceability necessary for successful auditing. Needless to say, while AGILE presents risks to ethical and robust AI, there are options for accommodating the approach while preserving a high degree of robustness and control.

4.3   Risk Thresholds for Ethics Audits?

An overarching theme in this review was the risk-based approach to IT auditing. As the methodologies we reviewed suggested, a key element for risk-based auditing is determining appropriate risk appetites and thresholds for which to evaluate the risks that arise from IT. There is some consensus that auditing AI systems should also follow a risk-based approach (e.g. ICO, 2020). However, translating the concept of risk thresholds to the domain of AI ethics requires addressing two key questions: who should be responsible for setting risk thresholds, and how precisely should one go about setting these thresholds, particularly when it involves peoples' fundamental rights?

The methodologies we identified primarily lay the responsibility of determining risk thresholds on the organization. That is, each organization is responsible for assessing their own capacity to handle negative outcomes and for using this assessment to determine an appropriate appetite for risk. However, in the IT context, risk is primarily conceived in terms of business value rather than social or ethical outcomes. While we recognize that there is some degree of crossover (e.g. reputational damage from unethical behavior), one cannot ignore the incentives for organizations to undervalue ethical concerns in the face of greater efficiency, profit, and so forth. Because of this context, it is certainly reasonable, to audit the ethics of AI, to break from IT audit methodologies in terms of taking a more active role in constraining the risk appetites that organizations can adopt.

Related to this question is the problem of actually determining the amount of ethical risk an organization can adopt. As the ICO (2020) notes, it is "unrealistic to adopt a 'zero tolerance' approach to risks to rights and freedoms." It is rare for risks to be entirely mitigated, and at some point, difficult questions about trade-offs will arise regarding how much ethical risk is acceptable in order to proceed. The problem of navigating these trade-offs is reminiscent of other conceptual constraints with auditing AI, many of which have been documented by Mökander et al. (2021). For instance, the trade-offs between "different legitimate, yet conflicting normative values" or between competing conceptions of a single normative value (e.g. fairness).[10] As with these conceptual constraints, the negotiation of risk trade-offs is a highly context-dependent problem with a high degree of complexity.

While there is no simple answer to this question, one strategy is shifting the emphasis away from 'proper' risk thresholds and toward a principle of procedural legitimacy. That is, an organization's decision to accept certain ethical risks is granted a degree of legitimacy if they have faithfully executed the requisite procedures for identifying and treating risks and this can be demonstrated through the requisite artefacts. In line with the ICO (2020), the conversation about risk appetites should be less about zero tolerance and more "about ensuring that these risks are identified, managed and mitigated." As Mökander et al. (2021) note, though audits cannot guarantee that the right balance has been struck, "the identification, evaluation, and communication of trade-offs" that audits help make visible is an important feature of assessment in and of itself. Of course, this should not excuse proceeding in the face of *severe* ethical risks, regardless of whether procedures were faithfully followed. Such actions should rightly be considered non-starters. However, just as courts sometimes accept a limiting principle to avoid 'death by a thousand rulings,' so can EA of AI maintain its focus on the processes that most effectively maximize ethical outcomes without getting drawn into the endless task of adjudicating acceptable levels of risk in every complex circumstance that may arise. Indeed, EA of AI is best viewed as a playbook, and not an answer sheet. As such, EA of AI can be

---

[10] Kusner et al. (2017) identifies six different (and sometimes mutually-exclusive) definitions of fairness, including individual fairness, demographic parity, and the equality of opportunity. These definitions tend to mask unresolved social and political disputes which constrain the effectiveness of EA of AI frameworks.

most effective, from a practical standpoint, by facilitating a dialectic process whereby auditors "ensure that the right questions are asked" (Mökander et al., 2021), and that system owners provide adequate answers to them.

5 Conclusion

While auditing is a well-established and robust field in its own right, the concept of applying the practice to the EA of AI is one that remains in its infancy. In order to provide a bridge between well-established auditing methods and the issue of AI ethics, we reviewed IT audit methodologies, with a focus on those connected to three key concepts: governance, assurance, and risk. Our review identified a set of methodologies and standards that can offer detailed guidance for developing a framework for the EA of AI. This includes ISO standards that detail requirements for software QM systems and their necessary documentation (e.g. ISO 90003, 2018; ISO/TR 10013, 2001), as well as a recommended set of risk assessment techniques across the software development life cycle (e.g. ISO 31010, 2019). These sources can be used as a foundation on which to build a methodology of EA for AI, one grounded in widely-accepted and legitimized processes. Beyond this, we identified key methodologies that have the potential to be more directly integrated into a novel auditing framework for AI, such as aspects of the COBIT framework and safety-critical AGILE methodologies.

Appendices

Appendix 1: Methodology

The qualitative evidence synthesis began by searching three databases (SCOPUS, IEEEXplore, and Google Scholar). Figure 2 provides the generic form of our search query, which was adapted to the specific syntactical requirements of each search engine and run on the 28th January 2020. For SCOPUS and IEEEXplore, the search string was restricted to Title, Abstract, and Keywords to ensure the return of the most relevant papers. No other filters or restrictions were employed (e.g. date range).

The search records were then screened in two stages. The first stage removed duplicate entries, records in a non-English language, and citations for which no document could be obtained. The titles and abstracts were then assessed for relevance based on our analytic framework. Because of the breadth of our search query, a large number of results were returned for each database. The first 100 records for each were saved and scanned, looking for relevant phrases, arguments or discussion points relating to specific IT audit methods or standards.

The search records only included academic sources. This presented some difficulty because the source material for these methodologies was primarily non-academic, given that most methodologies in this space are published by industry organizations and independent accreditation bodies. To account for this, we noted the methodologies that were discussed in the search records and followed up in an attempt to access the source material. This was not always possible since the materials for some methodologies (e.g. ITIL) were not publicly available and there was insufficient third-party information elsewhere.

Appendix 2: AGILE and Safety-Critical Systems

SCRUM Methodology

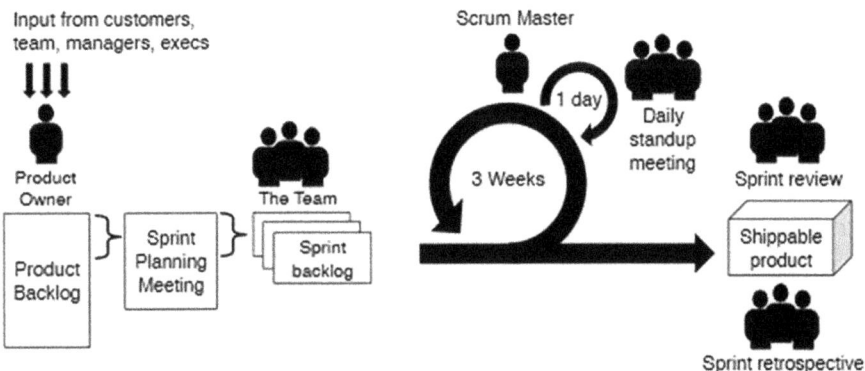

Source: Fitzgerald et al. (2013)

("IT" OR "information technology") AND ("audit" OR "iso") AND ("methodolog*" OR "framework*" OR "standard*") AND ("governance" OR "quality assurance" OR "quality management" OR "quality control" OR "risk" OR "risk management" OR "risk assessment")

* indicates a wildcard operator (e.g. "methdolog*" will return results for either "methodology" or "methodolog*ies*")

Results by Search Engine:

- SCOPUS (Title, Abstract, Keywords: 7,683 results returned, first 100 saved)
- IEEEXplore (Title, Abstract, Keywords: 1,503 results returned, first 100 saved)
- Google Scholar (approximately 3,180,000 results returned, first 100 saved)

Fig. 2 Generic form of search query and results by search engine

R-Scrum Methodology

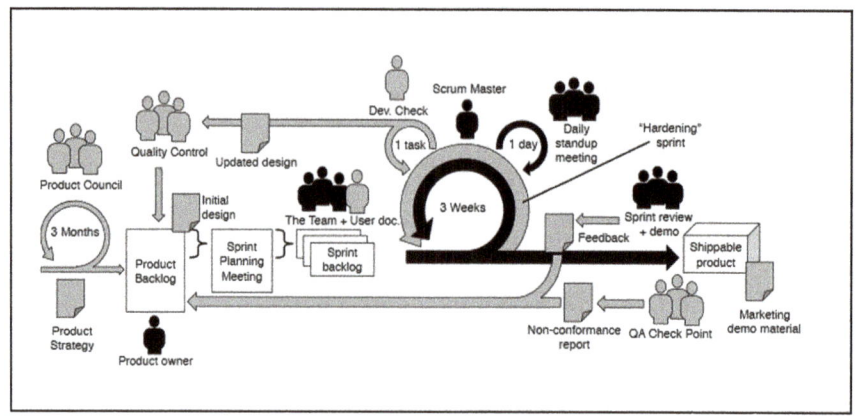

Source: Fitzgerald et al. (2013)

Appendix 3: Ethics and Risk Mapping

CITYCoP Ethics Compliance Process

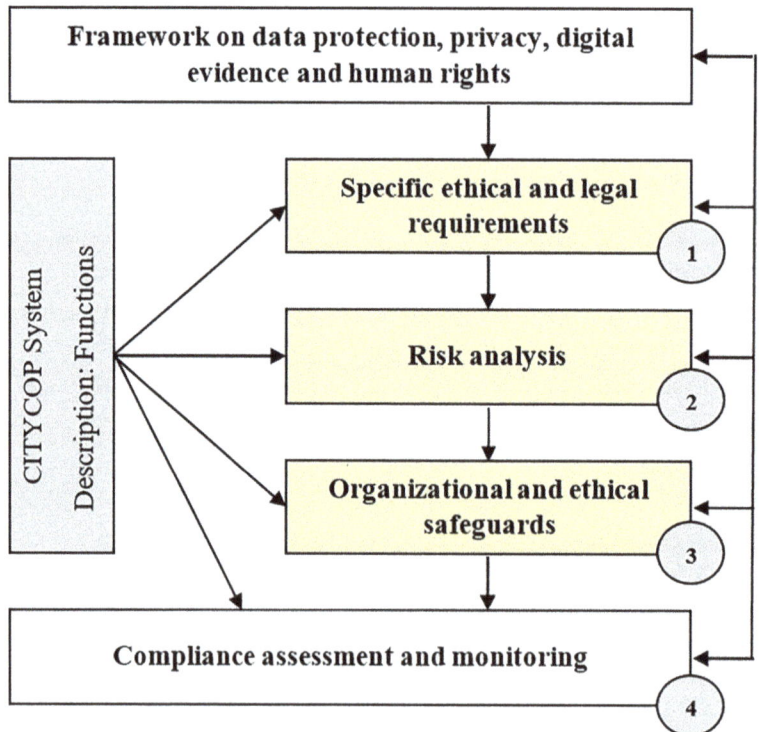

Source: *CITYCOP D11.6* **(2014)**

CITYCoP Ethics Compliance Matrix

System Functions	Ethical & Legal Requirements	Risks	Organizational & Technical Safeguards	Compliance Status	Remarks
F1	Req1 for F1				
	Req2 for F1				
	Req3 for F1				
F2					
	①	②	③	④	
F3	Req1 for F3				
	Req2 for F3				
	Req3 for F3				

Source: *CITYCOP D11.6* **(2014)**

References

AI HLEG. (2018). *Ethical guidelines for trustworthy artificial intelligence.* European Commission. https://ec.europa.eu/futurium/en/ai-alliance-consultation/guidelines#Top

Auditing Artificial Intelligence. (2018). ISACA. https://www.isaca.org/bookstore/bookstorewht_papers-digital/whpaai

Auer, F., & Felderer, M. (2018). Shifting quality assurance of machine learning algorithms to live systems. *Software Engineering Und Software Management,* 211–212.

Brundage, M., Avin, S., Wang, J., Bluemke, E., & Lebensold, J. (2020). *Toward trustworthy AI development: Mechanisms for supporting verifiable claims* (2004.07213[cs.CY]). arXiv.

Cartlidge, A., Hanna, A., Rudd, C., Macfarlane, I., Windebank, J., & Rance, S. (2007). *An introductory overview of ITIL V3.* The IT Service Management Forum. http://www.itsmf.org.rs/sites/default/files/itSMF%20ITIL%20V3%20Introduction%20overview.pdf

CITYCOP D11.6—Ethics and Legal Compliance Matrix Report. (2014). European Commission. https://ec.europa.eu/research/participants/documents/downloadPublic?documentIds=08166e5bb2c8445&appId=PPGMS

CITYCOP D12.1—A Data Protection Audit Plan. (2014). European Commission. https://ec.europa.eu/research/participants/documents/downloadPublic?documentIds=08166e5b23ac15d&appId=PPGMS

Coertze, J., & von Solms, R. (2014). The Murky Waters of IT. *Governance, 1–8.* https://doi.org/10.1109/ISSA.2014.6950498

COSO. (2013). *Internal control—Integrated framework: Executive summary.* Committee of Sponsoring Organizations of the Treadway Commission (COSO). https://www.coso.org/Documents/990025P-Executive-Summary-final-may20.pdf

Dafoe, A. (2018). *AI governance: A research agenda.* Future of Humanity Institute. https://www.fhi.ox.ac.uk/wp-content/uploads/GovAIAgenda.pdf

Douglas, B., & Ekas, L. (2012). Adopting agile methods for safety-critical systems development. IBM Software.

Fitzgerald, B., Stol, K.-J., O'Sullivan, R., & O'Brien, D. (2013). *Scaling agile methods to regulated environments: An industry case study* (Vol. 3, pp. 863–872) https://doi.org/10.1109/ICSE.2013.6606635

Floridi, L., & Cowls, J. (2019). A unified framework of five principles for AI in society. *Harvard Data Science Review, 1*(1) https://doi.org/10.1162/99608f92.8cd550d1

Floridi, L., & Sanders, J. W. (2004). On the morality of artificial agents. *Minds and Machine, 14*, 349–379. https://doi.org/10.1023/B:MIND.0000035461.63578.9d

Gantz, S. D., & Maske, S. (2014). *The basics of IT audit: Purposes, processes, and practical information*. Syngress.

Grant, M., & Booth, A. (2009). A typology of reviews: An analysis of 14 review types and associated methodologies. *Health Information & Libraries Journal, 26*(2), 91–108. https://doi.org/10.1111/j.1471-1842.2009.00848

Hamidovic, H. (2010). Fundamentals of IT governance based on ISO/IEC 38500. *ISACA Journal, 5*.

Hamzane, I., & Belangour, A. (2019). Implementation of a decision system for a suitable IT governance framework. *International Journal of Computer Science and Information Security, 17*(5), 1–7.

Hanssen, G., Stalhane, T. K., & Myklebust, T. (2018). *SafeScrum—Agile development of safety-critical software*. Springer.

Hoyle, D. (2001). *ISO 9000 Quality Systems Handbook* (4th ed.). Butterworth. Heinemann.

ICO. (2020). *Guidance on the AI auditing framework: Draft guidance for consultation*. Information Commissioner's Office. https://ico.org.uk/media/about-theico/consultations/2617219/guidance-on-the-ai-auditing-framework-draft-forconsultation.pdf

IEEE Std 730. (2014). *IEEE Std 730—IEEE standard for software quality assurance processes*. IEEE. https://ezproxy-prd.bodleian.ox.ac.uk:2219/document/6835311?arnumber=6835311

IIA Std 2110. (2016). *IIA standard 2110—Governance—Implementation guide 2110*. The Institute of Internal Auditors. https://www.aiiaweb.it/sites/default/files/imce/pdf/ig2110-2016-12.pdf

ISACA. (2019). *Official ISACA COBIT 5—Enabling processes guide*. ISACA. https://www.itgovernance.co.uk/shop/product/official-isaca-cobit-5-enabling-processesguide

ISO 10005. (2018). *ISO 10005:2018—Quality management—Guidelines for quality plans*. International Organization for Standardization. http://www.dndisystema.lviv.ua/sites/default/files/attachments/2017/248/isofdis10005e.pdf

ISO 12207. (2017). *ISO 12207:2017—Systems and software engineering—Software life cycle processes*. International Organization for Standardization. https://ezproxyprd.bodleian.ox.ac.uk:2219/document/8100771?arnumber=8100771

ISO 15288. (2015). *ISO 15288:2015—Systems and software engineering—System life cycle processes*. International Organization for Standardization. https://ezproxyprd.bodleian.ox.ac.uk:2219/stamp/stamp.jsp?tp=&arnumber=7106435

ISO 16085. (2004). *ISO 16085:2004—Information technology—Software life cycle processes—Risk management*. International Organization for Standardization. https://ezproxyprd.bodleian.ox.ac.uk:2219/stamp/stamp.jsp?tp=&arnumber=6298075

ISO 19011. (2018). *ISO 19011:2018—Guidelines for auditing management systems*. International Organization for Standardization. https://www.iso.org/obp/ui/#iso:std:iso:19011:ed3:v1:en

ISO 24748-1. (2018). *ISO 24748-1—Systems and software engineering—Life cycle management—Part 1: Guidelines for life cycle management*. International Organization for Standardization. https://ezproxy-prd.bodleian.ox.ac.uk:2219/stamp/stamp.jsp?tp=&arnumber=8526560

ISO 31000. (2018). *ISO 31000:2018—Risk Management—Guidelines*. International Organization for Standardization. https://www.ashnasecure.com/uploads/standards/BS%20ISO%20310002018.pdf

ISO 31010. (2019). *ISO 31010:2019—Risk management—Risk assessment techniques*. International Organization for Standardization. https://www.academia.edu/41536420/ISO_31010_2019_Risk_management_Risk_assessment_techniques_Management_du_risque_Techniques_dappr%C3%A9ciation_du_risque

ISO 37000. (2020). *ISO 37000—Guidance for the governance of organizations (ongoing)*. International Organization for Standardization. https://committee.iso.org/sites/tc309/home/projects/ongoing/ongoing-1.html

ISO 90003. (2018). *ISO 90003—Software Engineering—Guidelines for the application of ISO 9001:2015 to computer software*. International Organization for Standardization. https://ezproxyprd.bodleian.ox.ac.uk:2219/document/8559961?arnumber=8559961

ISO 9001. (2015). *ISO 9001:2015—Quality management systems—Requirements*. International Organization for Standardization. https://groupe.afnor.org/produits/editions/bivi/FDIS%20ISO%209001E.pdf

ISO/IEC 38500. (2015). *ISO 38500:2015—Information technology—Governance of IT for the organization*. International Organization for Standardization. https://www.iso.org/standard/62816.html

ISO/TR 10013. (2001). *ISO/TR 10013—Guidelines for quality management system documentation*. International Organization for Standardization.

Kohli, N., Barreto, R., & Kroll, J. (2018). Translation tutorial: A shared lexicon for research and practice. In *Human-Centered Software Systems* (p. 7).

Kusner, M., Loftus, J., Russell, C., & Silva, R. (2017). Counterfactual fairness. In *Advances in neural information processing systems* (pp. 4067–4077).

Leveson, N. G. (2011). *Engineering a safer world: Systems thinking applied to safety*. The MIT Press.

Lynch, A., & Veland, S. (2018). *Urgency in the anthropocene*. The MIT Press. muse.jhu.edu/book/62437.

Martin, A. (2017). CogNet: Data protection and privacy audit report. European Commission. https://cordis.europa.eu/project/id/671625/results

Mökander, J., Morley, J., Taddeo, M., & Floridi, L. (2021). Ethics-based auditing of automated decision-making systems: Nature, scope, and limitations. *Science and Engineering Ethics, 27*(4), 44. https://doi.org/10.1007/s11948-021-00319-4

Morely, J., Floridi, L., Kinsey, L., & Elhalal, A. (2019). From what to how: An initial review of publicly available AI ethics tools, methods and research to translate principles into practices. *Science and Engineering Ethics*. https://doi.org/10.1007/s11948-019-00165-5

Nakajima, S. (2018). *Quality assurance of machine learning software*. 2018 IEEE 7th Global Conference on Consumer Electronics (GCCE 2018), Nara, Japan.

Radovanovic, D., Radojevic, T., Lucic, D., & Sarae, M. (2010). *Analysis of Methodology for IT Governance and Information Systems Audit* (pp. 943–949) https://doi.org/10.3846/bm.2010.126

Raji, I. D., Smart, A., White, R., Mitchell, M., Gebru, T., Hutchinson, B., Smith-Loud, J., Theron, D., & Barnes, P. (2020). Closing the AI accountability gap: Defining an end-to-end framework for internal algorithmic auditing. 12. https://doi.org/10.1145/3351095.3372873

RiskIT. (2009). *The risk IT practitioner guide*. ISACA. https://www.colmich.edu.mx/computo/files/MAAGTIC/RiskIT_PG_30June2010_Reseach.pdf

Schöppl, N., Taddeo, M., & Floridi, L. (2022). Ethics auditing: Lessons from business ethics for ethics auditing of AI. In J. Mökander & M. Ziosi (Eds.), *The 2021 Yearbook of the Digital Ethics Lab*. Springer. https://doi.org/10.1007/978-3-031-09846-8

Scrum Guide: What is a Product Backlog? (n.d.). Scrum.Org. https://www.scrum.org/resources/what-is-a-product-backlog

Selig, G. J. (2008). *Implementing IT governance*. Van Haren Publishing.

Shefford, A., & Holland, P. (2018). *AI risk and controls matrix*. KPMG. https://assets.kpmg.com/content/dam/kpmg/uk/pdf/2018/09/artificial-intelligence-risk-andcontrols-matrix.pdf

Stalhane, T. K., Hanssen, G., Myklebust, T., & Haugset, B. (2014). Agile change impact analysis of safety critical software. *Lecture Notes in Computer Science, 8696*, 444–454.

Steghofer, J.-P., Knauss, E., Horkoff, J., & Wohlrab, R. (2019). Challenges of scaled agile for safety critical systems (1911.12590v1). *arXiv*.

Wessels, E., & van Loggerenberg, J. (2006). *IT governance: Theory and practice*. Conference on Information Technology in Tertiary Education, Pretoria, South Africa.

Ethics Auditing: Lessons from Business Ethics for Ethics Auditing of AI

Noah Schöppl, Mariarosaria Taddeo, and Luciano Floridi

Abstract This chapter reviews the business ethics literature on ethics auditing to extract lessons for the emerging practice of ethics auditing of Artificial Intelligence (AI). It reviews the definitions, purposes and motivations of ethics audits, identifies their benefits as well as limitations, and compares various theoretical and practical approaches to ethics auditing. It distils seven lessons for the ethics auditing of AI and finds that ethics audits need to be comprehensive, involve stakeholders, entice behaviour change, be pragmatic and rigorous, be widely endorsed, fitting in context but also comparable, and lastly integrate a technical dimension with an organisational dimension. It is crucial that, while ethics auditing can also have financial benefits, their main goal must remain the improvement of the ethical performance and meaningful accountability of the audited organisation. The novel elements of AI should not blind us to the continuities of social embeddedness and organisational dynamics. Ethics auditing of AI can learn valuable lessons from failed and successful previous efforts to audit the ethics of organisations.

Keywords Ethics auditing · Business ethics · AI ethics · AI auditing · Stakeholder management

N. Schöppl
Oxford Internet Institute, University of Oxford, Oxford, UK

M. Taddeo · L. Floridi (✉)
Oxford Internet Institute, University of Oxford, Oxford, UK

Department of Legal Studies, University of Bologna, Bologna, Italy
e-mail: luciano.floridi@oii.ox.ac.uk

© The Author(s), under exclusive license to Springer Nature Switzerland AG 2022
J. Mökander, M. Ziosi (eds.), *The 2021 Yearbook of the Digital Ethics Lab*, Digital Ethics Lab Yearbook, https://doi.org/10.1007/978-3-031-09846-8_13

1 Introduction

The idea of auditing features prominently in the emerging discussions of AI ethics (Mökander & Floridi, 2021; Brundage et al., 2020; Sandvig et al. 2014). While AI ethics poses novel and pressing issues, it can also learn from other disciplines, in particular from other domains of applied ethics (Mittelstadt, 2019). This also holds for auditing of AI. The idea that the ethical implications of a particular technological application and organisation need to be audited is anything but new. Pioneering companies have started to measure their ethical performance since the 1980s and started to audit their ethics performance since the 1990s (Kaptein, 1998a; Rosthorn, 2000). This chapter reviews the business ethics literature on the theory and practice of auditing processes and organisations. The chapter addresses the question: what can ethical auditing of AI learn from the business ethics literature on ethics audits?[1] Two sub-questions are considered: what is the state of the debate about ethics audits in the business ethics literature? And how does the business ethics literature evaluate the possibility of auditing ethics? The first sub question will be answered in the first two sections of the article. The second sub-question will be answered in Sects. 3 and 4, while the implications for ethics auditing of AI and conclusions are described in Sects. 5 and 6. This literature review is a systematised qualitative evidence synthesis (Grant & Booth, 2009) and its methodology is described in Appendix A.1. Appendices A.2 to A.5 provide graphical representations of the four different applied ethics auditing frameworks presented in Sect. 3. Appendix A.6 contains a list of used abbreviations.

2 Definitions, Purposes, and Motivations

2.1 Definition: What Is an Ethics Audit?

For the purposes of this chapter, the expression "Ethics Audit" (EA) refers to activities that in the literature are variably described as *ethics audits*, *business ethics audits*, *social audits*, or *corporate social responsibility audits*. An early and generic definition of ethics auditing is provided by Kaptein as "a systematic approach for identifying the ethics gap" of an organisation, with the ethics gap being the "discrepancy between the desired and the current moral situation" (Kaptein, 1998a). Thus, two central elements of an ethics audit are the description of an ideal state and the assessment of the *status quo*.

A narrower, minimalist definition focusing on internal compliance is provided by Rosthorn (2000) as "regular, complete and documented measurements of compliance with the company's published policies and procedures". This definition is less useful as it reduces EA to a compliance tool for management. Metzger et al. (1993)

---

[1] For a similar review of lessons from the IT audit literature, please see Zinda (2022), i.e., chapter 12 in this volume.

propose a more comprehensive, so called "extended organisational ethics audit", which consists of "studying the *de facto* institutional logic and reward system of the company (as perceived by the employees) to make sure it aligns with the long term objectives of the organization and leadership." This provides a broader scope for EA. However, it still endorses the view that ethics is primarily about making the organisation internally consistent and accountable to its leadership.

Others consider EA as a tool for stakeholders to hold organisations and their leadership accountable. Such a definition comes from the external, social audit tradition, which defines EA as "analyses of accountable entities undertaken (more or less systematically) by bodies independent of the entity, and typically without the approval of the entity concerned" (Gray, 2001). The core of this definition is that it is not the organisation that holds its agents accountable, but rather that the organisation is held accountable by its stakeholders and society at large.

More recently, a synthesis has emerged, which acknowledges the role of EA as a mechanism to hold organisations accountable to stakeholders, but leaves control over the EA process largely in the hands of management. As opposed to the narrow management view, it gives an expansive role to the needs of stakeholders, but it is different from the social audit tradition, as it is performed with the consent of the leadership of the organisation. One such definition comes from Ojasoo, who defines EA as "a process for evaluating and diagnosing the *external and internal* consistency of an organization's values and their congruence with real behaviour. [emphasis added]" (Ojasoo, 2016). This definition emphasises the value of EA in ensuring that companies' stated ethical ambitions and their actual performance remain coherent and do not become hypocritical, regardless of whether the relationship is internal to the company or with external stakeholders.

García-Marzá (2005) provides a more comprehensive definition of EA: "An ethics audit should measure both the existing distance between the moral idea and the commitment assumed, and the specific fulfilment of the commitment." In this definition, the EA assesses both the distance between stated values and actual behaviour, and the distance between stated commitments and normatively desirable commitments. Hence, an EA also evaluates whether the ethical yardstick chosen by the organisation is the appropriate one. According to this definition, an EA assesses whether "an organisation lives up to its ethical commitments" as well as "whether the ethical commitments of the organisation are the right ones". Koldovsky adds a further and related dimension to this, by suggesting that an EA should not only ask

- how should I do the things I am doing?
- but also what things should I be doing in the first place?

According to Koldovskyi (2015), an EA is not only about doing things right or doing the right things, but doing the right things in the right way. Thus, depending on the chosen scope, EA has both a "how" as well as a "what" dimension: it assesses what an organisation does and how it does it.

2.2   Purpose and Motivations: Why Do Ethics Audits?

Why do we need to audit ethics in the first place? In principle, the purpose of an audit is to provide third-party information and assurance about the behaviour of an agent in a situation when the agent's statements about their behaviour on their own terms is insufficient and may not even be trusted. The most prominent types of audits are financial, but auditing also has applications in quality management, law and (cyber)security (Kaptein, 1998a; Kok et al., 2001).

Most papers, explicitly mention stakeholder theory as the conceptual or normative underpinning of ethics auditing (Kaptein, 1998b; Rosthorn, 2000; Gray, 2001; García-Marzá, 2005; Morimoto et al., 2005; Koldovskyi, 2015; Ojasoo, 2016). This builds on Freeman's approach (1984), according to which a stakeholder is any actor who affects or is affected by a business. Stakeholder theory exists both in a descriptive, instrumental version, and in a prescriptive, normative version. Instrumental stakeholder theory sees the management of expectations of stakeholders as a necessary means to profit maximisation, while normative stakeholder theory posits that it is the responsibility or even purpose of a business to reconcile and fulfil the needs of its stakeholders. These two approaches shape the literature on EA. Following this logic, Gray distinguishes between the accountability purpose and the competing management control purpose of EA:

> In essence, accountability places society at the heart of the analysis and questions the legitimacy of an organisation's actions, or perhaps even its right to exist. A management control orientation places the organisation at the centre of the debate and the society's − not the organisation's − legitimacy may be called (however implicitly) into question. Unless we maintain the most naive of views over the relationship between the aims of society and the aims of corporations it is clear that each approach produces very different results. (Gray, 2001, p. 11)

While Gray argues for the accountability purpose, Rosthorn (2000) favours the management control perspective. Rosthorn argues that the reason why companies should themselves provide more information about their ethical performance is that otherwise they will be subject to scrutiny by external stakeholder groups which they cannot control and which will threaten their profitability:

> With expertise and funds at their disposition, the stakeholder groups are free to define their own research within their own agendas and their own terms of reference. Societal groups plainly have their own conceptual frameworks and measure them in their own way [...]. By using results of either high or dubious robustness, such groups are then free to make their findings available to whatever common interest groups they choose. They make good use of the media to communicate audit conclusions which, in general, may seem to be hostile to the corporate sector. (Rosthorn, 2000)

Rosthorn argues that, to avoid external scrutiny about their ethics, companies should perform ethics audits. EAs that are initiated and paid for by an executive leadership tend to be "more agreeable than some stakeholder-driven audits which are done on terms defined by the stakeholder" (Rosthorn, 2000). While to those favouring the management control purpose this is a feature, for those advocating an accountability purpose, this is a bug in the design of an EA. Crucially, if EAs are only seen as internal compliance tools, their results are unlikely to be released to the public domain and thus do not allow for meaningful accountability of corporate actors (Rosthorn, 2000). Even if the results are released, the information provided about the ethical performance of an organisation is less meaningful for stakeholders as it is more tuned towards the interest of the corporation. Thus, EAs based on the management control purpose are less trustworthy for stakeholders, because frequently there is an enormous "bias in choosing categories against which an organization's performance may be measured" (Morimoto et al., 2005).

In the management control approach, ethics auditing is not a tool for change but essentially becomes a way to justify what an organisation already does and to defend the vested interests of its leadership (Morimoto et al., 2005; Kok et al., 2001). It aims at changing perception of behaviour rather than behaviour itself – behaviour is only changed to the extent to which it is deemed necessary to create the impression of ethical behaviour, as defined by the leadership of the organisation. The accountability approach on the other hand, aims to improve the ethics performance of organisations by making them accountable to the concerns and needs of their stakeholders. According to the accountability approach, the need for auditing is based in the 'iron law of oligarchy', which states:

> Established elites develop ideological and organizational mechanisms to protect their incumbent status and constrain critical communication and challenge (Neuman, 2016).

The first generation of EAs primarily were tools to measure the integrity and compliance within an organisation (Kaptein, 1998b). The focus was consequently on the relationship between employees and company leadership. Accountability to stakeholders outside of the company was seen as a concession to external pressures, while the primary emphasis lay on the value of ethics audits for the financial bottom line of the business through the avoidance of risks (Metzger et al., 1993; Rosthorn, 2000). The divergence between the management control purpose and the accountability purpose still exists, yet with the rise of the Corporate Social Responsibility (CSR) discourse, the normative stakeholder theory has become much more widespread. Ethics audits tend to take external stakeholders more seriously, which has widened the focus from internal integrity.

In Table 1, various frequently mentioned motivations for EA have been distilled from the literature and grouped in the accountability and management control purpose.

3   Methods: How to Do an Ethics Audit?

A number of specific EA frameworks and methodologies have been developed to conduct an EA of an organisation. Several of the proposed frameworks seem to have remained theoretical and never have found widespread application. They include: the Expanded Organisational Ethics Audit by Metzger et al. (1993), the Audit Dialogue Approach by García-Marzá (2005), the CSR Audit Protocol by Morimoto et al. (2005), as well as the CSR Audit Framework by Koldovskyi (2015).

Table 1 Motivations for ethics auditing

Accountability purpose	Shared purposes	Management control purpose
Hold firms responsible for hypocritical behaviour (Gray, 2001; Koldovskyi, 2015; Ojasoo, 2016)	Avoid unethical behaviour (García-Marzá, 2005; Metzger et al., 1993)	Increase trust in brand and credibility of firm (García-Marzá, 2005; Kaptein, 1998b; Koldovskyi, 2015)
Increase external transparency (Gray, 2001; Rosthorn, 2000)	Create accurate and complete information (Metzger et al., 1993; Morimoto et al., 2005; Rosthorn, 2000)	Improve internal decision making (Gray, 2001; Koldovskyi, 2015; Metzger et al., 1993; Rosthorn, 2000)
Encourage more ethical behaviour (García-Marzá, 2005; Metzger et al., 1993)		Reduce reputational or legal risks (Morimoto et al., 2005; Rosthorn, 2000)
Fulfil stakeholder needs (García-Marzá, 2005; Ojasoo, 2016; Koldovskyi, 2015)		Manage and control stakeholder demands (Gray, 2001; Rosthorn, 2000)

First, according to Metzger et al. (1993), data collected from employees can be synthesised in an expanded ethics audit, which provides an understanding of the underlying institutional logic and reward structure as perceived by its employees, to identify and reverse perverse incentives. While Metzger et al. (1993) argue that the achievement of ethical goals should be incentivised just as economic goals are incentivised in companies, others argue that measuring ethical performance in monetary terms can erode intrinsic ethical motivations and norms (Sandel, 2013). In addition, there are a number of mechanisms, commonly summarised as 'Goodhart's Law', through which any system which relies on proxies runs the risk of overfitting and creating perverse incentives (Manheim & Garrabrant, 2019).

In this consequentialist dominated field, a second notable theoretical contribution comes from García-Marzá's (2005) deontological approach, which proposes to use discourse ethics as a procedural grounds for an EA. García-Marzá argues that the EA should create a meaningful dialogue with a firm's stakeholder groups, instead of measuring the outcomes of an organisation. In such a Habermasian dialogue, the organisation leadership can find the best reconciliation between various, competing, general, and specific stakeholder interests (García-Marzá, 2005). However, how all relevant stakeholder groups can fairly and meaningfully be represented in such a dialogue has not yet been shown in practice and remains a methodological hurdle. A practical example of such an idea, is to discuss EA reports with stakeholders at roundtable discussions, as for instance done by Shell (Kaptein, 1998a; Kok et al., 2001).

Lastly, Morimoto et al., (2005, p. 320) introduce the distinction between required and desirable criteria. In any rigorous EA, there should be the real possibility for a company to fail the audit. Otherwise it is mere 'ethics-washing' and a pointless exercise (Wagner, 2018). Yet, if the bar is too high, many companies are unlikely to choose to do a voluntary EA. EAs thus need to find a sensible balance between rigour and acceptability (Morimoto et al., 2005). If a company does not fulfil the essential, required criteria, it fails the ethics audit. These can be categorical variables or threshold values of continuous variables. However, a sophisticated EA should also provide more than a mere pass/fail outcome and crucially identify desirable, actionable areas of improvement for the next auditing period. Felber (2019) provides a second possibility for such distinctions, by translating all criteria in a continuous scale from −3600 to 1000 points and providing positive points for ethical behaviour and negative points for unethical behaviour. This acknowledges that "[a]ll organizations have negative as well as positive impacts" (Morimoto et al., 2005).

While most approaches discussed thus far have found no or limited application, other frameworks have found practical application in a number of organisations and are summarised in Table 2.[2]

Table 2 Applied ethics auditing frameworks

Name	General Business Ethics Audit (GEBA)	Ethics Thermometer (ET) and Ethics Stakeholder Reflector (ESR)	Economy for the Common Good (ECG) Audit	Improved Ethics Audit and Risk Assessment Model (ERA)
Data sources	Document review, and 10–15 interviews between 60–90 mins with staff of different seniority levels	Standardized questionnaire for employees (ET) and stakeholders (ESR) with about 200 propositions	Document review, interviews and questionnaires with stakeholders, walking tour in the company	Document review, questionnaire and interviews with selected stakeholders, walking tour in the company.
Application	Applied by the British-Asian firm LaMP Consulting	Applied by KPMG Ethics & Integrity Consulting	Audited in 400+ organisations, internally used in 1000+ worldwide	Piloted by academics with companies in Estonia
Openness	Proprietary and little public information about methodology	Proprietary but book with methodology available	All methodology is open-sourced through the ECG movement	Mostly available in academic publications and a PhD dissertation
Purpose	Management control	Management control	Accountability	Accountability
Graphical representation	Radar chart in Appendix A.2	Ethical qualities matrix in Appendix A.3	Stakeholder-values matrix in Appendix A.4	Ethical risk matrix Appendix A.5
Literature	Rosthorn (2000)	Muel Kaptein (1998a); Muel Kaptein (1998b)	Felber (2019); Felber et al. (2019)	Ojasoo (2016); Rihma (2014)

---

[2] The Social Audit approach is not included in this framework, despite its widespread application, since it employs a diverse set of, often non-systematic, *ad hoc* methodologies. Moreover, a Social Audit is usually conducted without the consent of the audited firm. Nevertheless, the Social Audit tradition also can make very valuable contributions to the establishment of ethics auditing practices for AI by creating external pressure, as Raji & Buolamwini (2019) have recently shown. For an overview of the Social Audit tradition see Gray (2001) and Rahim & Vicario (2015).

A few comments are in order. GEBA (Rosthorn, 2000) is primarily qualitative, ET (Kaptein, 1998b) exclusively quantitative, and ECG (Felber, 2019) and ERA (Ojasoo, 2016) use both qualitative and quantitative sources. Morimoto et al. (2005) argue that ethics audits which use such mixed-methods approaches are superior. By using both epistemological approaches in a complementary way, they can synthesise results to make a more accurate and meaningful assessment. While GEBA and ET only use data of internal stakeholders, ESR, ECG and ERA also collect data from external stakeholders. Moreover, it is notable that ERA and especially ECG use more data sources and are more rigorous than GEBA and ET. This also is in line with the assessment that ERA and ECG clearly have an accountability purpose, whereas ET and especially GEBA primarily have a management control purpose. Most authors agree that, while internal reviews are necessary, at least occasional external audits by trained an independent auditors are preferable in order to ensure veracity, credibility and methodological rigour of EAs (Felber, 2019; Metzger et al., 1993; Rosthorn, 2000). Moreover, Felber (2019) also introduces a second auditing possibility: audits can be performed either by an external, trained auditor or a peer-auditing process where a group of organisations audit each other. Lastly, it is notable that while, ET, ESR, and GEBA have been introduced by consulting firms (even if Kaptein also published academically about ET and ESR), the ECG model emerged from a grass-roots movement, and ERA is the outcome of academic research (see Sect. 4 for an overall evaluation).

Virtually all authors agree that a 'tick box' approach to EA is insufficiently rigorous and that EA requires deep, constant and meaningful engagement with ethical questions (García-Marzá, 2005; Koldovskyi, 2015; Morimoto et al., 2005; Ojasoo, 2016). The approach that comes closest to a pure tick-box approach is the ET, since it is purely based on one-size-fits-all quantitative surveys with 200 Likert scale items. Felber (2019) and especially Ojasoo (2016) make valuable contributions by providing practical guidance on setting priorities in ethical risk management, i.e. to identify and focus on the most pressing and significant ethical risks in the particular context (see Appendices 4 and 5). With regards to the output of an EA, it is crucial that it results in actionable recommendations for improvement plans, since the identification of ethics blind spots enables improved and conscious ethics management (Felber, 2019; Rosthorn, 2000). For instance, an audit can identify the rate of women or minorities in management and the workforce in general; monitor trends in annual reports; benchmark it against the general population or competitors; and set targets for the next reporting period (Rosthorn, 2000). The main result of an EA should be more ethical practices in an organisation (García-Marzá, 2005). The collection of data is never an end itself, and before any data is collected, it should be considered how having this data would lead to better decisions and behaviour (Harji & Nicholls, 2020).

4   Evaluating Ethics Auditing

Based on the literature review, the following benefits and limitations of EA have been identified.

Table 3 Benefits and limitations of ethics auditing

Benefits	Limitations
Enables organisations to behave more ethically and fulfil stakeholder needs (García-Marzá, 2005; Morimoto et al., 2005)	No consensus on universal ethical benchmark for organisations (Morimoto et al., 2005; Kok et al., 2001; Felber, 2019; Rosthorn, 2000)
Alignment of the goals of an organisation (Felber, 2019; Metzger et al., 1993)	Ethics is challenging to quantify and assess rigorously (Morimoto et al., 2005; Felber, 2019)
Improved information flows (Koldovskyi, 2015; Metzger et al., 1993)	No standard indices and auditing procedures (Koldovskyi, 2015; Morimoto et al., 2005)
Improved employer branding and employee morale (Metzger et al., 1993; Ojasoo, 2016)	Complexity and cost of implementation (Metzger et al., 1993; Morimoto et al., 2005)
Attractiveness to customers and investors (Morimoto et al., 2005; Ojasoo, 2016)	Not widely applied since not mandatory, negative screening (Koldovskyi, 2015; Morimoto et al., 2005)
Reduced reputational and legal risks (Metzger et al., 1993; Rosthorn, 2000)	Often selective and incomplete (Koldovskyi, 2015; Morimoto et al., 2005)
Reduces ethics gap and hypocritical behaviour (Kaptein, 1998b; Ojasoo, 2016)	Frequently does not lead to behaviour change (Metzger et al., 1993; Morimoto et al., 2005)
Higher trust and lower operational costs (García-Marzá, 2005; Kaptein, 1998b; Ojasoo, 2016)	Frequently biased towards executives and owners at the cost of other stakeholders (Koldovskyi, 2015; Metzger et al., 1993; Morimoto et al., 2005)

While the limitations shown in Table 3 are real and many remain unsolved, on balance all authors and many entrepreneurs still find that the added value of EA makes them worthwhile (Felber, 2019; Kaptein, 1998b; Rosthorn, 2000). As long as desirable practices or outcomes are at least to some extent definable, qualitatively verifiable or quantitatively measurable at an acceptable cost, ethics auditing is seen as a valuable tool for corporate governance, social accountability, and company management (Morimoto et al., 2005, p. 316). Nevertheless, it is worth noting that EAs still constitute a pioneering niche practice that has not yet been mainstreamed.

A last aspect of discussion is the impact of EA on the financial performance of a firm. While most authors argue that EAs can contribute to the profitability of a firm (Koldovskyi, 2015; Metzger et al., 1993; Ojasoo, 2016), there also are undeniably situations in which the ethical thing to do is not the same as the profitable thing to do (Gray, 2001). If ethics is taken seriously as an end in itself and not only as a means for profit-maximisation, it will at times come with financial opportunity costs, which is why the necessity for ethics auditing emerges in the first place.

5 Lessons for AI Ethics Auditing

Business ethics auditing has faced similar issues since the 1990s as AI ethics auditing faces today. The successful and failed efforts of business ethics auditing can therefore provide insights for the AI ethics and governance challenges today's organisations face. This section synthesises seven lessons that can be learnt from such comparable experience.

Lesson 1: Ethics audits, Which also review the goals, commitments, and structure of an organisation, are preferable to those that only review its performance

AI ethics audit should not only assess whether an organisation does what it says but also review whether its goal system is adequate, including internal incentive structures for executives and employees (Koldovskyi, 2015; Metzger et al., 1993). Beyond seeking to make sure that a company does what it says according to its own goals and framings, an EA should also seek to question and improve a company's normative commitments (García-Marzá, 2005; Koldovskyi, 2015; Metzger et al., 1993). EA should not be a mere box-ticking exercise but a reflective process about how to do the right things in the right way. For AI auditing this means that auditing should focus on how AI is used as well as for which purposes. Whether a company has a comprehensive understanding of its responsibility in the first place and reflects this in its governance structure, should also be a key part of the auditing process.

Lesson 2: Ethics audits Which involve external stakeholders are preferable to those which only evaluate ethics internally

AI ethics audits should be seen not primarily as a management control tool for company leadership but as a meaningful accountability mechanism for stakeholders. If an EA only has input from those inside the company, it is likely to miss valuable perspectives (Morimoto et al., 2005). This is particularly true for the workforce of AI companies, in which women and people of colour are enormously underrepresented (Raji & Buolamwini, 2019; Varma, 2018; Webb, 2019; Perrault et al., 2019). Taking into account the power and knowledge asymmetry between company leadership and other stakeholders, a meaningful space for stakeholders should be included in AI ethics audits, since otherwise they run the risk of missing crucial ethical blind spots and becoming mere exercises of public posturing, self-justification, and 'ethics washing' (Gray, 2001; Wagner, 2018).

Lesson 3: While ethics auditing always has management control benefits, its main goal should be accountability and more ethical behaviour

While positive branding can be a side-effect, the purpose of AI ethics auditing should be to promote accountability and positive behaviour change (Ojasoo, 2016). The audience of ethics audits are not only outsiders, who need to be convinced of the ethics of the company, but also primarily decision makers within the company. The ethics audit should inform, support and empower morally autonomous decision makers who want to behave ethically for its own sake, and correct, incentivise and hold accountable those who are willing to sacrifice ethics for commercial ends (Kok et al., 2001; Felber, 2019). In order to work as an effective accountability mechanism, EAs needs to be able to question what a company does (Metzger et al., 1993). The purpose of EAs is to be uncomfortable to the vested interests that run the risk of corrupting the ethics of any organisation (Morimoto et al., 2005). In the words of Gray (2001), EAs should include the possibility of "hurting" at least a bit, since without this possibility they may not lead to behaviour change and lose credibility.

Lesson 4: The gap between abstract principles and auditable measures needs to be overcome with the right balance of methodological pragmatism, authenticity, and rigour

While on the one hand, abstract principles may serve to entrap actors in moral commitments they did not initially intend to make, they equally may create opportunities to escape public pressures through inconsequential lip service (Thomas, 2001; Wagner, 2018). The vagueness of principles poses a challenge to developing rigorous auditing standards[3]: "At the time when high level company policy documents such as its business ethics code were being prepared, little thought was given to subsequent internal auditing of compliance" (Rosthorn, 2000). Similar observations have been made for AI governance: there is a broad agreement that companies should be responsible and fair but little agreement about what this means in practice (Mittelstadt, 2019). Overcoming this gap is an admittedly challenging task (Morimoto et al., 2005). Part of the solution lies in rigorous iterative testing and refining of methodologies:

> As corporate experience in business ethics audit cycle grows, the auditors will be allowed to contribute more substantially to the preparation of subsequent generations of business ethics policies and codes of business conduct. (Rosthorn, 2000).

Another part of the solution lies in giving the possibility, pragmatically and contextually, for organisations to become aware, reflect on, and authentically show their values and performance.

Lesson 5: There s a trade-off between the rigor and widespread application of a standard

The more rigorous a standard, the more costly it tends to be. If the audit is too rigorous it remains a niche and is only used by pioneers, not the mainstream or laggards (Auld, 2015; Dauvergne & Lister, 2012). If it is too lax it may find wider application but runs the risk of becoming merely an exercise in 'ethics washing' (Wagner, 2018). For instance ECG has stricter standards and auditing practices than the Global Reporting Initiative (GRI), which has found much higher adoption than the former (Hipper & Hofielen, 2017). Nevertheless, a good balance can be struck between an audit, which is comprehensive and robust, and at the same time is relatively fast, flexible and cost-effective (Morimoto et al., 2005). There is a trade-off to be made between having high standards and remaining in a niche, on the one hand; and having low standards that find widespread adoption, on the other hand. One best practice to overcome this trade-off and increase the spread of a high auditing standard is to open source the full methodology as done by the ECG. Yet, since companies tend to pick standards that are favourable to their circumstances, a further sensible solution to this trade-off can be regulatory intervention or endorsement by authoritative bodies (Morimoto et al., 2005; Rosthorn, 2000). A successful example of a sector-specific 'race to the top', rather than a 'race to the bottom' is the EU's Eco-Management and Auditing Scheme (EMAS). It was introduced in 1993 as a

---

[3] For further reading see the chapter by Zinda, chapter "Ethics Auditing Framework for Trustworthy AI: Lessons from the IT Audit Literature", in this volume.

voluntary scheme and enables organisations to assess and continuously improve their environmental performance. After it was adopted in the EU, corporates which anyway were audited in the EU and feared international competitive disadvantages, teamed up with NGOs to lobby for EMAS to be adopted in Asia and America and, in 1996, the International Standards Organisation (ISO) adopted it as a global standard (Bradford, 2020). A standard for EAs of AI can hope to achieve a similar development in the coming years.

Lesson 6: Standards which can be applied universally but prioritise the most relevant contextual risks are preferable to non-comparable ethics audits

There still is a growing plurality of EA standards and no consensus or consolidation behind a particular standard. Such a development should be avoided in the case of EAs of AI. Like in other applications of auditing, when governments and institutions of authority pick standards and make them applicable across the board (such as for instance in EMAS or the EU Non-Financial Reporting Directive) this is preferable to a patchwork of non-comparable ethics audits, because comparability improves the effectiveness and accountability of EAs and encourages more responsible behaviour (Felber, 2019; Koldovskyi, 2015; Morimoto et al., 2005). However, in order to be relevant for decision makers across contexts, an EA methodology also needs to take the contextual specificities into account and allow for prioritisation and management of risks depending on the particular industry and circumstances, such as the size of the company (Ojasoo, 2016; Felber et al., 2019). A coherent universal framework that is designed to be applied in combination with industry specific protocols is also desirable for EAs of AI.

Lesson 7: Both technical and organisational auditing are individually insufficient for ethics audits of AI, but both are necessary for its success

AI faces additional challenges such as higher levels of opacity of machine learning systems compared to other technologies. In the age of increasingly advanced AI systems, organisations are not only made up of humans but also artificial agents (Floridi & Sanders, 2004). Usually, stakeholder interests can be protected by sending in auditors with specialised knowledge. However, in the case of AI, both human operators and auditors can only make limited statements about the decision making process of 'black box' systems (Burrell, 2016). Thus, templates of EA cannot just be copied and pasted to be used for AI ethics auditing and further research on the explainability of AI is warranted. However, the novel element of opaque AI, should not distract from the elements which can and need to be learned from organisational business ethics auditing. These elements include:

- the organisational dynamics between an auditor and an audited organisation, including second order questions such as 'who holds the auditor accountable?';
- the inspection of the general ethical practices, awareness, reflexivity and alignment within an organisational unit.

The roots of unethical practices most often lie not in a particular application, service or product, but in the organisational culture, systemic reward structure, or constitution of an organisation (Felber, 2019; Cohen, 2019; Mayer, 2018; Meadows, 2014; Kok et al., 2001; Kaptein, 1998b; Metzger et al., 1993). Both technical and organisational auditing are individually insufficient for EA of AI, but both are necessary for its success.

6 Conclusion

In this chapter, we have reviewed the business ethics literature on ethics auditing in order to synthesise lessons that can be applied in developing the emerging practice of ethics audits of AI. The review found that ethics audits need to be comprehensive, involve stakeholders, entice behaviour change, be pragmatic and rigorous, be widely endorsed, fitting in context but also comparable, and lastly integrate a technical with an organisational dimension. An AI ethics audit cannot just be a technology audit, it also needs to include audits of social, cultural and management practices in context, as well as stakeholder concerns (Kaptein, 1998b; Koldovskyi, 2015, p. 140; Metzger et al., 1993).[4] It is crucial that, while EAs can also benefit the profitability of companies, their main goal must remain the improvement of the ethical performance and meaningful accountability of the audited organisation (García-Marzá, 2005; Ojasoo, 2016).[5]

Ethics audits have serious limitations and need to be complemented with other measures in order to be effective. For ethics principles to be implemented in an organisation further measures are necessary. Three in particular could be highlighted:

- training: education, awareness and sensitization for ethics issues, rules and procedures (Kok et al., 2001; Metzger et al., 1993)
- open channels of communication: this refers to the ability to raise concerns openly and safely when agreed principles are violated, and to a culture that does not kill the messenger and encourages whistleblowing (Kaptein, 1998b; Metzger et al., 1993); and
- holding offenders accountable: strict and certain consequences for gross violations are crucial to avoid signalling that the leadership does not really prioritize ethics (Gray, 2001; Metzger et al., 1993).

The development of EA for AI faces some unique technical challenges, but it does not have to reinvent the wheel. The novel elements of AI should not blind us to the continuities of social embeddedness and organisational dynamics. AI ethics auditing can learn many valuable lessons from failed and successful previous efforts to audit the ethics of organisations.

---

[4] For further reading see the chapter by Lee et al., chapter "Formalising Trade-Offs Beyond Algorithmic Fairness: Lessons From Ethical Philosophy and Welfare Economics", in this volume.
[5] For further reading see the chapter by Schöppl, chapter "State-Firm Coordination In AI Governance", in this volume.

Appendices

Appendix 1: Methodology

This literature review is a systematic qualitative evidence synthesis (Grant & Booth, 2009): it uses purposive sampling, integrates and compares individual studies in looking for themes and conceptual models, and finally provides recommendations for practice. The review began by iteratively refining a search query, which was run on 7th February 2020 on four databases (Google Scholar, SCOPUS, IEEEXplore, PhilPapers):

("ethics audit" OR "ethics assessment" OR "social audit") AND ("business ethics" OR "corporate governance" OR "corporate social responsibility" OR "artificial intelligence")

This produced 6160 results on Google Scholar (of these the first 50 results were scanned), 37 results on SCOPUS, 1 result on IEEExplore, and 0 results on PhiloPapers. The titles, keywords and abstract of the resulting 88 records was scanned to evaluate to which extent they could make a contribution to answering the posed research question. Based on this first sampling round, the 10 first records were selected. Relevant papers were read in their entirety, while in books only the relevant chapters were read for this analysis. Further sources were identified by reviewing the references of the selected publications and scanning publications which had cited them via the 'cited by' feature on Google Scholar. Additionally, sources were included that were already known to the author. This resulted in 25 analysed sources for this systematic qualitative evidence synthesis.

Furthermore, it should be noted that this review only included academic sources, while previous reviews found that "[m]ost of the literature discovered appears not to be peer reviewed scientific texts or academic papers. Rather, it consists to a great extent of subjective comment in business magazines or company reports." (Morimoto et al., 2005, p. 318). Specific product certification schemes, such as for organic or fair trade products, are not included in this review, since the its focus is on auditing ethics on an organisational level. Lastly, it should be noted that this review has not focused on legal compliance, but mostly on auditing that goes beyond what is required by law. For other reviews of ethics auditing and its frameworks, see Morimoto et al. (2005), Ojasoo (2016), and Hipper & Hofielen (2017).

Appendix 2: General Business Ethics Audit – Radar Chart

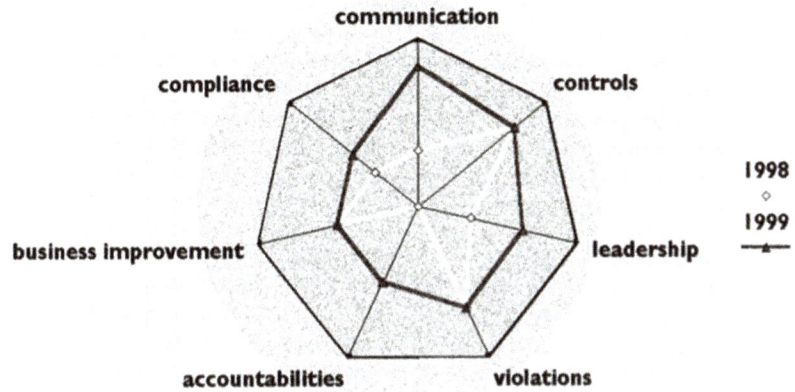

This radar chart is retrieved from Rosthorn (2000) and shows the development of a company on the General Business Ethics Audit for two consecutive years.

Appendix 3: Ethics Thermometer – Ethical Qualities Matrix

ECONOMY
FOR THE COMMON GOOD

AUDIT CERTIFICATE	FULL BALANCE SHEET	Common Good Balance Sheet: 2016-2017	for:	VAUDE Sport GmbH & Co. KG
			Auditor:	Manfred Kofranek und Gitta Walchner

VALUE / STAKEHOLDER	HUMAN DIGNITY	SOLIDARITY AND SOCIAL JUSTICE	ENVIRONMENTAL SUSTAINABILITY	TRANSPARENCY AND CO-DETERMINATION
A: SUPPLIERS	A1 Human dignity in the supply chain — 80 %	A2 Solidarity and social justice in the supply chain — 60 %	A3 Environmental sustainability in the supply chain — 70 %	A4 Transparency and co-determination in the supply chain — 60 %
B. OWNERS, EQUITY- AND FINANCIAL SERVICE PROVIDERS	B1 Ethical position in relation to financial resources — 30 %	B2 Social position in relation to financial resources — 60 %	B3 Use of funds in relation to social and environmental impacts — 80 %	B4 Ownership and co-determination — 10 %
C: EMPLOYEES, INCLUDING CO-WORKING EMPLOYERS	C1 Human dignity in the workplace and working environment — 70 %	C2 Self-determined working arrangements — 40 %	C3 Environmentally-friendly behaviour of staff — 80 %	C4 Co-determination and transparency within the organisation — 50 %
D: CUSTOMERS AND OTHER COMPANIES	D1 Ethical customer relations — 50 %	D2 Cooperation and solidarity with other companies — 40 %	D3 Impact on the environment of the use and disposal of products and services — 70 %	D4 Customer participation and product transparency — 60 %
E: SOCIAL ENVIRONMENT	E1 Purpose of products and services and their effects on society — 60 %	E2 Contribution to the community — 80 %	E3 Reduction of environmental impact — 80 %	E4 Social co-determination and transparency — 80 %

Certificate valid until: 31 March 2020   BALANCE TOTAL   631

This certifies the audited Common Good Report. The certification is based on Common Good Balance Sheet 5.0.
For more information on the audit, the indicators and the audit system, go to www.ecogood.org

Appendix 4: Economy for the Common Good Balance Sheet

Retrieved from M. Kaptein (1998a, b).

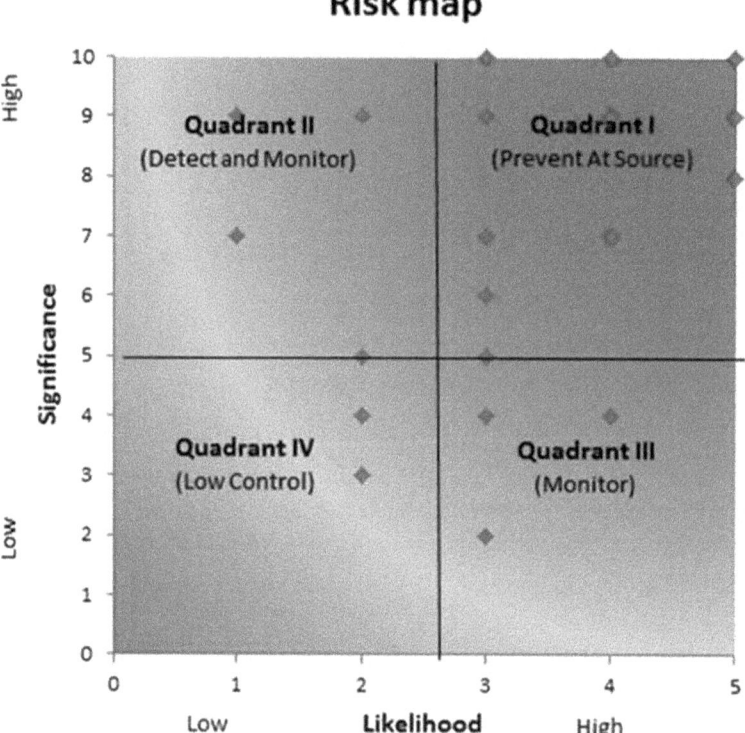

This is an example of an Economy for the Common Good Balance sheet as provided by the mountain sports equipment company VAUDE. It is retrieved from and available at: https://csr-report.vaude.com/gri-wAssets/pdf/de/Dokumente-2018-fuer-2017/2018_ECG_certificate_AUDIT_full_VAUDE.pdf

A full ECG audit report for the electronics firm elobau is available at: https://www.ecogood.org/media/filer_public/f6/2b/f62b467b-b02b-450a-bb39-c195a1967c27/elobau_gwo_eng_web.pdf

Moreover, the full methodology behind the Common Good Balance Sheet is available under a creative commons licence at: https://www.ecogood.org/apply-ecg/

Appendix 5: Improved Ethics Audit – Risk Assessment Model

This risk map is retrieved from Ojasoo (2016). She elaborates: "The risk map [...] prioritizes each ethical risk according to significance and likelihood and maps the risks into four quadrants:" (Ojasoo, 2016, p. 8)

- I quadrant "Prevent at Source" risks, high priority. These risks are both significant in consequence and likely to occur.
- II quadrant "Detect and Monitor" risks, second priority. Risks in this quadrant are significant, but they are less likely to occur.
- III quadrant "Monitor" risks. Risks in this quadrant are less significant, but have a higher likelihood of occurring.
- IV quadrant "Low Control" risks. Risks in this quadrant are both unlikely to occur and not significant.

Appendix 6: List of Abbreviations

AI = Artificial Intelligence
CSR = Corporate Social Responsibility
EA = Ethics Auditing
ECG = Economy for the Common Good Audit
EMAS = Eco-Management and Auditing Scheme
ERA = Improved Ethics Audit and Risk Assessment Model
ESR = Ethics Stakeholder Reflector
ET = Ethics Thermometer
EU = European Union
GEBA = General Business Ethics Audit
GRI = Global Reporting Initiative
ISO = International Standards Organisation

References

Auld, G. (2015). Certification as governance. In D. A. Bearfield & M. J. Dubnick (Eds.), *Encyclopedia of public administration and public policy* (3rd ed.). CRC Press. https://doi.org/10.1081/E-EPAP3

Bradford, A. (2020). *The Brussels effect: How the European Union rules the world*. Oxford University Press.

Brundage, M., et al. (2020). Toward trustworthy AI development: Mechanisms for supporting verifiable claims. *Computers and Society.*. https://arxiv.org/abs/2004.07213

Burrell, J. (2016). How the machine 'thinks': Understanding opacity in machine learning algorithms. *Big Data & Society, 3*(1), 2053951715622512. https://doi.org/10.1177/2053951715622512

Cohen, J. E. (2019). *Between truth and power: The legal constructions of informational capitalism*. Oxford University Press.

Dauvergne, P., & Lister, J. (2012). Big brand sustainability: Governance prospects and environmental limits. *Global Environmental Change, 22*(1), 36–45. https://doi.org/10.1016/j.gloenvcha.2011.10.007

Felber, C. (2019). *Change everything: Creating an economy for the common good*. Zed Books Ltd..

Felber, C., Campos, V., & Sanchis, J. R. (2019). The common good balance sheet, an adequate tool to capture non-financials? *Sustainability, 11*(14), 3791. https://doi.org/10.3390/su11143791

Floridi, L., & Sanders, J. W. (2004). On the morality of artificial agents. *Minds and Machines, 14*(3), 349–379. https://doi.org/10.1023/B:MIND.0000035461.63578.9d

Freeman, R. E. (1984). Strategic management: A stakeholder approach. .

García-Marzá, D. (2005). Trust and dialogue: Theoretical approaches to ethics auditing. *Journal of Business Ethics, 57*(3), 209–219. https://doi.org/10.1007/s10551-004-8202-7

Grant, M. J., & Booth, A. (2009). A typology of reviews: An analysis of 14 review types and associated methodologies. *Health Information & Libraries Journal, 26*(2), 91–108. https://doi.org/10.1111/j.1471-1842.2009.00848.x

Gray, R. (2001). Thirty years of social accounting, reporting and auditing: What (if anything) have we learnt? *Business Ethics: A European Review, 10*(1), 9–15. https://doi.org/10.1111/1467-8608.00207

Harji, K., & Nicholls, A. (2020, February 11). *Impact measurement: Innovations and tensions.* Oxford Impact Measurement Programme Webinar.

Hipper, A., & Hofielen, G. (2017). Sustainability management and ethics—A comparison of four ethical frameworks: Global reporting initiative, German sustainability code. In: *ISO 26000 and Common Good Balance Sheet.* Humanistic Management Practices.

Kaptein, M. (1998a). *Ethics management: Auditing and developing the ethical content of organizations.* Springer.

Kaptein, M. (1998b). The ethics thermometer: An audit-tool for improving the corporate moral reputation. *Corporate Reputation Review, 2*(1), 10–15. https://doi.org/10.1057/palgrave.crr.1540063

Kok, P., van der Wiele, T., McKenna, R., & Brown, A. (2001). A corporate social responsibility audit within a quality management framework. *Journal of Business Ethics, 31*(4), 285–297. https://doi.org/10.1023/A:1010767001610

Koldovskyi, A. (2015). Corporate social responsibility audit: Theoretical aspects. *Risk Governance & Control: Financial Markets & Institutions, 5*(3), 135–144.

Manheim, D., & Garrabrant, S. (2019). *Categorizing variants of Goodhart's Law.* ArXiv:1803.04585 [Cs, q-Fin, Stat]. http://arxiv.org/abs/1803.04585

Mayer, C. (2018). *Prosperity: Better business makes the greater good.* Oxford University Press.

Meadows, D. (2014). *Thinking in systems: A primer.* Chelsea Green Publishing Company.

Metzger, M., Dalton, D. R., & Hill, J. W. (1993). The organization of ethics and the ethics of organizations: The case for expanded organizational ethics audits. *Business Ethics Quarterly, 3*(1), 27–43. https://doi.org/10.2307/3857380

Mittelstadt, B. (2019). Principles alone cannot guarantee ethical AI. *Nature Machine Intelligence, 1*(11), 501–507. https://doi.org/10.1038/s42256-019-0114-4

Mökander, J., & Floridi, L. (2021). Ethics-based auditing to develop trustworthy AI. *Minds and Machines, 31*, 323–327. https://doi.org/10.1007/s11023-021-09557-8

Morimoto, R., Ash, J., & Hope, C. (2005). Corporate social responsibility audit: From theory to practice. *Journal of Business Ethics, 62*(4), 315–325. https://doi.org/10.1007/s10551-005-0274-5

Neuman, W. R. (2016). *The digital difference: Media technology and the theory of communication effects.* Harvard University Press.

Ojasoo, M. (2016). CSR reporting, stakeholder engagement and preventing hypocrisy through ethics audit. *Journal of Global Entrepreneurship Research, 6*(1), 14. https://doi.org/10.1186/s40497-016-0056-9

Perrault, R., Shoham, Y., Brynjolfsson, E., Clark, J., Etchemendy, J., Grosz, B., Lyons, T., Manyika, J., Mishra, S., & Niebles, J. C. (2019). *Artificial Intelligence Index 2019.* Stanford University.

Rahim, M. M., & Vicario, V. (2015). Social audit: A mess or means in CSR assessment? In M. M. Rahim & S. O. Idowu (Eds.), *Social audit regulation: Development, challenges and opportunities* (pp. 1–13). Springer International Publishing. https://doi.org/10.1007/978-3-319-15838-9_1

Raji, I. D., & Buolamwini, J. (2019). Actionable auditing: Investigating the impact of publicly naming biased performance results of commercial AI products. In *Proceedings of the 2019 AAAI/ACM Conference on AI, Ethics, and Society* (pp. 429–435). https://doi.org/10.1145/3306618.3314244

Rihma, M. (2014). *Ethics audit: A management tool for assessing of corporate social responsibility and preventing ethical risk*. TUT Press.

Rosthorn, J. (2000). Business ethics auditing—More than a Stakeholder's toy. In J. Sójka & J. Wempe (Eds.), *Business challenging business ethics: New instruments for coping with diversity in international business: The 12th annual EBEN conference* (pp. 9–19). Springer. https://doi.org/10.1007/978-94-011-4311-0_3

Sandel, M. J. (2013). *What money can't buy: The moral limits of markets*. Penguin Books.

Sandvig, C., Hamilton, K., Karahalios, K., & Langbort, C. (2014). Auditing algorithms: Research methods for detecting discrimination on internet platforms. *Data and Discrimination: Converting Critical Concerns into Productive Inquiry, 22*, 4349–4357.

Thomas, D. C. (2001). *The Helsinki effect: International norms, human rights, and the demise of communism*. Princeton University Press.

Varma, R. (2018). U.S. Science and Engineering Workforce: Underrepresentation of women and minorities. *American Behavioral Scientist, 62*(5), 692–697. https://doi.org/10.1177/0002764218768847

Wagner, B. (2018). Ethics as an escape from regulation: From ethics-washing to ethics-shopping. In *Being profiled: Cogitas Ergo Sum* (pp. 84–90). Amsterdam University Press.

Webb, A. (2019). *The big nine: How the tech titans and their thinking machines could warp humanity*. PublicAffairs.

Zinda, N. (2022). Ethics auditing framework for trustworthy AI: Lessons from the IT audit literature. In J. Mökander & M. Ziosi (Eds.), *The 2021 Yearbook of the Digital Ethics Lab*. Springer. https://doi.org/10.1007/978-3-031-09846-8

AI Ethics and Policies: Why European Journalism Needs More of Both

Guido Romeo and Emanuela Griglié

Abstract The use of artificial intelligence (AI) in European newsrooms is growing and offers new opportunities to improve both the quality and reach of reporting as well as expanding readership. AI may also contribute to diversify revenues through the personalization of content and dynamic paywalls. This is welcome progress as journalism is a central element of democracy and sound balance sheets are the foundations of an independent and free press. In this chapter, we map the ethical issues potentially arising from the extensive automation of newsrooms and related gaps in the codes of conduct for journalists. We argue that, while over-regulating the field may be counterproductive, European guidelines and policies supporting specific training, access to new technologies, and funding, as well as developing checklists for intelligent tools, would help address the ethical issues raised by AI.

Keywords Artificial intelligence · Bias · Digital ethics · Filtering · Journalism · Media ecology

1 Introduction

Artificial Intelligence (AI) applications in newsrooms are oftentimes described as "robo-journalists" (Broussard et al., 2019). Although this terminology relates more to sci-fi than to the actual practice of the industry, automation is not a futuristic scenario in newsrooms. AI is a significant part of journalism already; it is just unevenly distributed. A growing number of newsrooms worldwide have gone beyond the experimentation phase and implemented AI-driven tools. The AI

G. Romeo (✉)
Info.nodes, Italy, Milan
e-mail: guido@guidoromeo.com

E. Griglié
La Stampa, Torino, Italy

© The Author(s), under exclusive license to Springer Nature Switzerland AG 2022
J. Mökander, M. Ziosi (eds.), *The 2021 Yearbook of the Digital Ethics Lab*, Digital Ethics Lab Yearbook, https://doi.org/10.1007/978-3-031-09846-8_14

transformation of journalism started more than a decade ago, with sports and business reporting. Today, it is spreading to all fields, from investigative reporting to feature writing and production of visuals, although, as in most industries, the largest share of tools addresses the less glamourous and repetitive tasks of newsrooms, such as tagging, classifying, or fact-checking (Whittaker, 2019). This is a welcome transition because it frees up humans from mundane newsroom tasks, letting them focus on higher quality products and increasing competitiveness to improve monetization. In a media landscape that is driving many legacy media out of business, it is easy to forecast that, over the next decade, any news organization still on its feet will have embraced AI technologies to some degree.

Although slower than their US colleagues in the uptake of these new technologies, European journalists are now experimenting with AI-driven solutions in many fields, from content moderation to story discovery and correction of human biases. AI systems are also proving very effective in profiling readers' news consumption to offer them tailored subscription packages. This application is relevant for our analysis as the data gathered could allow profiling one's preferences on a political or ethical level.

Helberger has observed that the impact of digital technologies on media and journalism serves as a good indicator for the more widespread impacts of such tools on democratic societies. In this chapter, we argue that, if misguided, the growing use of AI in news media, may pose serious ethical challenges undermining the essential role journalism plays in supporting shared social narratives and a balanced democracy. Some of these issues are not different from those observed in other domains (Yang et al., 2018), for instance, algorithmic bias and accountability. Others, as excessive personalization, copyright and privacy breach, abuse of users' trust, have specific relevance to journalism. These challenges have become more pressing as it becomes clear that AI in journalism is shaping customer behaviour. Customers, especially Gen Z (6–24 years old), increasingly expect personalization of production, supply, or content creation, as stressed in LSE survey (2019) of 71 international organizations (70% from Europe). If extensively applied, excessive personalization may induce echo chambers and fragmentation of a shared vision of reality, among other consequences. At present, clear guidelines for the ethical use of AI in journalism are still lacking. In this chapter, we first map the relevant ethical challenges related to the use of AI in news media (Sect. 2); we then explore the current uses of AI to support journalism (Sect. 3) and the current European policy context (Sect. 4). We conclude our analysis by offering some policy recommendations to address the issues identified in Sect. 2.

2   The Issues Raised by AI in Journalism

As we anticipated in the Introduction, some of the ethical challenges posed by the use of AI in the newsroom are no different from those identified in other domains. Whether deployed in healthcare or defence, the use of AI poses pressing questions concerning the transparency of AI systems, the attribution of responsibilities for their failures or unintended consequences, and fairness of outcomes (Floridi & Taddeo, 2016; Tsamados et al., 2021). When considering journalism, these challenges become even more pressing because, if left unaddressed, they also conflict with the ethical foundations of journalism and its importance for democracy and an informed public debate.

Journalism's mission is underpinned by the right of everyone to have access to information and ideas, stated in Article 19 of the Universal Declaration of Human Rights (2015) and in Article 10 of the European Convention on Human Rights (ECHR): "Everyone has the right to freedom of opinion and expression; this right includes freedom to hold opinions without interference and to seek, receive and impart information and ideas through any media and regardless of frontiers". Good journalism and independent media play a pivotal role in modern democracies. The European Court of Human Rights has underlined the democratic role of media as 'purveyor of information' ('European Court of Human Rights – Case of Barthold v. Germany (1985) – Paragraph 59.', n.d.) to create forums for public debate and to act as a public watchdog. In this landscape, what are the responsibilities of journalists towards their readers and society at large? As noted by the International Federation of Journalists ('European Convention on Human Rights – Official Texts, Convention and Protocols' n.d.): "The journalist's responsibility towards the public takes precedence over any other responsibility, in particular towards their employers and the public authorities. Journalism is a profession, which requires time, resources and the means to practice – all of which are essential to its independence". Regardless of how journalism codes and cultures may vary across the world, the first point of IFJ's Global Charter: "Respect for the facts and the right of the public to truth is the first duty of the journalist", should resonate in any newsroom regardless of the language or the medium.

At least three other articles (Art. three, eight, and nine) of the Global Charter are worth considering to understand what issues may arise with the implementation of AI-driven tools. Article 3 reads: "The journalist shall report only in accordance with facts of which he/she knows the origin. The journalist shall not suppress essential information or falsify any document". Although data are not facts, the relevant point here is the journalists' knowledge of the reliability of the source material on which their reporting is built and how it has been elaborated. Article 8 prescribes that: "The journalist will respect privacy. He/she shall respect the dignity of the persons named and/or represented and inform the interviewee whether the conversation and other material is intended for publication". Privacy and dignity are evoked regarding the sensitive information an interviewee may share during a conversation but may not be relevant for the reporting. One may wonder whether this kind of prescription applies to data or preferences conveyed by a reader when interacting with a chatbot or indicating preferences while subscribing to an automated news service. And finally, article 9 dictates that: "Journalists shall ensure that the dissemination of information or opinion does not contribute to hatred or prejudice and shall do their utmost to avoid facilitating the spread of discrimination on grounds such as geographical, social or ethnic origin, race, gender, sexual orientation, language, religion, disability, political and other opinions". As we shall see, this raises many questions regarding how AI systems may profile users and distribute news content accordingly, increasing the risk of generating "echo chambers" in which beliefs are amplified or reinforced by communication and repetition inside a closed system and insulated from rebuttal. Not surprisingly, such an issue has been under the spotlight in many public and political debates, from no-vax movements to Brexit and US elections to, most recently, the 2021 storming of the US Capitol.

This ethical code of journalism dates back centuries but has seen little evolution in the last decades, despite the technological developments that have profoundly changed the work of reporters and editors. Moreover, journalism codes have not yet considered AI and the growing importance of data in news reporting. In the following subsection, we identify five critical ethical challenges posed by AI to journalism founding principles.

2.1 Algorithmic Accountability

Insofar as it is a discipline of verification, journalism is intrinsically required to understand and account for how it operates. Human editors have always been (and will be) biased in their decision-making. However, the growing scope of autonomous systems makes the amplification of bias operated by machines an urgent and pressing problem to address (Journalism, Media, and Technology Trends and Predictions, 2020). Journalists have proven able to unveil these issues in other industries, as done by ProPublica's reporters on at least two occasions related to the prediction of the risk of recidivism (Mattu et al., n.d.), and another one that calculates car insurance premiums and payouts (Angwin et al., 2017). It would simply be self-contradictory for news media not to look carefully at what they are doing themselves with their algorithms. However, the task is not straightforward because, even if editors have control over what data and instructions they give an AI system, it is often hard to know how a given conclusion is reached, as the growing use of neural networks (Buhrmester et al., 2019) contributes to a "black box" effect were the process connecting input and output becomes technically inscrutable. This lack of transparency in AI systems' rationale is the source of many social and ethical concerns, and the technical challenge of the "black box" of AI is especially problematic when there are high stakes involved.

This opens two sets of problems. One is immediate accountability: if an AI system produces factually incorrect data or a full story is spotted before publication, who shall be accountable? The second, and more complex problem, concerns the case in which the mistake is spotted after publication and has caused damage to third parties, for example, a massive panic on Covid-vaccines, a bank run, or plummeting stock prices. If it was the case, most legal systems hold responsible the authors of the story, and possibly the editor in chief and the publishing company, but how shall this be confronted when the defendant is a program not completely understood in its workings? As AI cannot be held legally accountable, human accountability needs to be embedded in all stages of the content value chain ('Automating the News — Nicholas Diakopoulos', n.d.). Note that this is also the approach taken by the AI Act. News media should have a strong interest in ensuring accountability of their AI for their immediate survival for at least two reasons. First, unlike social media platforms, it is not uncommon for news media to be the object of a libel lawsuit, which may even lead to bankruptcy and extinction, as in the case of Gawker (Lehman, 2016). Second, as trust in media is declining, news media need to ensure and demonstrate that circulation of misleading content is stopped at the root. To be perceived as a reliable source of information for the public, and especially with the much more ethically conscious Gen Z users, news brands need to demonstrate they are actively fighting not only false news and disinformation but also gender and ethnic bias.

2.2   Excessive Personalization

Personalization of news products is a growing driver of audience and revenue. This trend, however, calls for many caveats as this generates "Filter Bubbles" ('Fake News Is a Media Ecology Problem | by Luca De Biase | Medium', n.d.) and "Echo Chambers" ('Eli Pariser on "The Filter Bubble: What the Internet Is Hiding from You"', n.d.). As biases, filter bubbles, and echo chambers precede digital environments (Brugnoli et al., 2019; Sasahara et al., 2021), algorithms that personalise an individual's online experience amplify the impact of these social dynamics. As underscored by a report to the European Parliament ('Artificial Intelligence (AI): New Developments and Innovations Applied to e-Commerce – Think Tank', n.d.), while some degree of personalization is desirable and useful for the user, AI systems must be designed to be societally aware to fight polarization, monopolistic concentration, and excessive inequality, and pursue diversity and openness. The impact of excessive personalization of news brought on by incautious profiling through automation may be even more dramatic considering the global expansion of "news deserts", communities no longer covered by local news media and thus relying on little or no professional reporting to inform public debate. For many people, the impoverishment of the news ecosystem has played a pivotal role in the polarization of the political debate in the Western world in the last decade. "News deserts" may anticipate "democracy deserts". According to a study from the UNC School of Media and Journalism ('How Trump Thrives in "News Deserts"', 2018) more than 1300 communities in the US are considered news deserts and from 2005 to the end of 2018, the UK saw was a net loss of 245 local news titles leaving an estimated 58% of the country not served by any regional newspaper. According to the Reuters Institute, this trend is also common in Europe, as the media industry that provides most of the investment in the news currently sees revenues decline about €2.5 million per day, as print readers die off and publishers find digital media a less lucrative business. The drop in diversity of the news ecosystem, if not access to news altogether, may be aggravated if users rely on digital platforms for access and distribution, as these have been shown to influence distribution. Suffice it to recall that, in 2014, Facebook (accounting as a source for up 1 reader out of 3 to access news) faced criticism after it emerged it had conducted a psychology experiment on nearly 700,000 users' news feeds to control which emotional expressions the users were exposed to without their knowledge (BBC News, 2014). In 2021, Google admitted to hiding news sources to Australian users while "running a few experiments" that would "each reach about 1% of Google Search users in Australia to measure the impacts of news businesses and Google Search on each other" (Karp, 2021). Apple News, the number one news app in the world with 85 million active daily users, has been shown to influence access to outlets ('Apple News Is Excluding Local Newsrooms from Its Coveted Traffic Bump', n.d.) to the detriment of local news publishers even if they were the first ones to report the news.

2.3   Lack of Media Pluralism

"Freedom of the press is built on sound balance sheets" goes an old media adage. Today, as larger newsrooms are building their own AI, smaller outlets often do not have the financial ability or technical expertise to take on this endeavour. The likely consequence, especially in Europe, is that smaller media will be forced to 'buy' rather than 'build' their AI approach, burdening their budget with licenses of proprietary systems if they want access to new business areas. This is very similar to what happened in the last decades, with advertising becoming more and more reliant on tech giants for revenue growth. This scenario presents dire consequences for the pluralism of European media and social structures.

The Media Pluralism Monitor (2020) considers the role of the digital platforms in the new ecosystem of the media, assessing separately the concentration in the production and distribution of information. While the indicator on News media concentration assesses the risks of market power in the production of news (both legacy and digital providers), a new indicator (Online platforms and competition enforcement) addresses the role of the digital intermediaries in the distribution of news. The average score for the area of Market Plurality is 64%, which is considerably higher than in MPM2017 (53%), thus signalling the growing economic threats to media pluralism. Under Market Plurality, no country scores a low risk; this area has the highest average risk among the areas considered by the Monitor. Almost half of the countries (13 out of 30) score a high risk, while the remaining 17 are at medium risk[1].

---

[1] High risk: Albania, Bulgaria, Croatia, Cyprus, the Czech Republic, Finland, Hungary, Latvia, Malta, Romania, Slovakia, Slovenia, and Turkey; Medium risk: Austria, Belgium, Denmark, France, Estonia, Germany, Greece, Ireland, Italy, Lithuania, Luxembourg, the Netherlands, Poland, Portugal, Spain, Sweden, and the United Kingdom.

2.4 Privacy

Data is the fuel of AI tools, both for development and training as well as for performance. In the EU, data collection of user data is already regulated by the GDPR but, as highlighted in Sect. 2.1, journalism approaches based on so-called big data – think of the work that made possible investigations such as the Panama Papers – enable reporters to gain access to vast amounts of classified and sensitive materials (ICIJ, 2021). While professional journalists are used to handling this material, the scale at which this happens with AI-driven tools is unprecedented and needs to be confronted. A careful balance needs to be found between the right to privacy and freedom of information, which in this case shall be understood as the right of the press to access and elaborate sensitive information to report on news of interest to the public.

2.5 Copyright

Digital technologies have frequently challenged copyright laws. Machine learning potentially renews this conflict as AI systems may generate their output by exploiting human-created 'expressive' works – be it a data set of articles, paintings, or music, for example – which tend to have rights owners. This is likely to test the legal interpretation of 'fair use' and 'right to report', where proprietary material is used to produce new and 'transformative' content without permission or payment of royalties.

3 AI's Contribution to Journalism

AI can support the work of a newsroom or individual journalists at three levels: newsgathering, story production, and content distribution. These three levels do not necessarily stand alone. For example, automatically generated news or journalistic pieces created with the support of AI-driven tools can be distributed in a personalized way. However, these distinctions help to analyse and understand how AI is impacting newsrooms, even if, as noted by Polis' global survey of journalism and AI (2019): "One of the key aspects of AI and journalism is that it allows the whole journalism model to become more holistic, with a feedback loop between the different parts of the production and dissemination process. The moderation of user comments, for example, could be a way of gathering content, creating or editing that content and as a crucial way of increasing audience engagement". Also, underpinning all AI processes is data: about audiences such as their behaviour; about the reported subject such as official records; about the journalism such as sentiment and language. Let us see each level separately.

3.1 Newsgathering

Newsgathering includes many different applications of AI. The most proficient newsrooms in developing and leveraging these capabilities are often those with experience in data journalism and, previously, with computer-assisted reporting (CAR), as their staff and digital infrastructure are better prepared for the new tasks. A variety of AI-related technologies may be involved at this stage, from machine learning to speech-to-text translation and image recognition. Benefits are manifold and range from increased productivity, as costly human labor is freed from the more repetitive and mundane tasks and can concentrate on higher-level tasks, to expanding access to information. Also, automation enables the newsroom to process data for which a human-driven effort would simply not possible in terms of resources, e.g., large troves of data as those typically released in leaks that have become a journalistic genre with Wikileaks. Moreover, the use of AI has been shown to expand the variety of news a journalist may produce. For example, from a single crime map, many analysis stories based on patterns of felonies may emerge, as well as fact-checking operations and so forth. Overall, AI can increase both productivity (number of stories run per day) and the in-depth and quality reporting needed to support subscription-based models. Stuttgarter Zeitung's (STZ) Crime Map, for instance, developed tools to collect and organize on a map the crime data from the German Polizei in Stuttgart ('Crimemap: Polizeibericht, Brände Und Unfälle in Stuttgart – Stuttgarter Zeitung', n.d.). It is a good example of the use of AI for story mapping. Together with the company Arvato, STZ trained a computer through machine learning to understand police communique from Stuttgart Police headquarters. The system then sorts the information into predefined categories, recognizes when and where a crime took place, and feeds it into the Crime Map. Besides the general overview encouraging readers to spend more time on the site exploring data, reporters use the collected crime data for deeper story analysis (areas with the highest crime rate, changes over time, outcome of new policies). The use of AI for fact-checking is increasingly more common, for instance, Storyful's bots (Bloomberg. Com, 2017) have an AI-system assisting human reporters in their verification process, making sure a video floating around Twitter shows, for instance, the latest barrel bomb explosion in Aleppo – rather than a roadside bomb in Homs – before it goes into a client's breaking-news post. More recently, the Covid pandemic has created the need for timely updates on contagion, alerts, new regulations, and limits which offering opportunities for automation and reader engagement.

3.2 Story Production

The use of AI in news production inevitably evokes the trite headline of robot-journalists replacing flesh and bone reporters as much more efficient, both in terms of time and cost in writing articles. However, automated writing is only applied by a fraction of outlets producing large volumes of stories at very high speed. Automated text generation has also proved successful in covering sports as, similarly to finance, there is an abundance of data already organized and distributed in formats machines can easily digest or use for training autonomous systems. This line of applications is rapidly improving with digitization and open licenses making more data easily accessible but also thanks to new language models as GPT-3 that uses deep learning to produce human-like text.

Fully automated writing with GPT-3 on general subjects has been tested by mainstream news media as The Guardian (GPT-3, 2020) but, despite the hype, is still far from Pulitzer level copy (although a machine writing "Humans must keep doing what they have been doing, hating and fighting each other. I will sit in the background and let them do their thing" admittedly sounds quite witty). As shown experimentally (Floridi & Chiriatti, 2020) GPT-3 does not pass any tests based on mathematics, semantic (that is, the Turing Test), and ethical questions. The researchers underline that "GPT-3 does not do what it is not supposed to do and that any interpretation of GPT-3 as the beginning of the emergence of a general form of artificial intelligence is merely uninformed science fiction".

Automation of news production on a large scale has already proven financially successful in terms of an increase in subscriptions. An area where AI technologies are also proving valuable is reviewing text and image choices of human editors to help media improve balance in reporting regarding gender or other parameters. The benefits of AI tools in news and content production are manifold. Automation's capability to increases speed and precision, while cutting marginal costs of story production is highly appreciated in reporting on financial markets, sports, and electoral results where timing is critical. This is typically the case of global news agencies as Associated Press and Reuters (Kobie, 2018), which have now been using automated writing systems for several years to produce takes and articles from stock exchange reports in few milliseconds release to their customers, who rely on speed to make financial decisions.

3.3   Content Distribution

When appropriately designed and trained, intelligent systems combining various tools (from text sentiment analysis to computer vision) and metrics may help track what people the text and imagery are depicted on an editorial site. By estimating the age and gender of subject features, AI tools enable editorial insights into how an outlet's news coverage relates to the demographics of its audience. As in other industries, the decreasing marginal cost brought on by automation may help expand production in areas previously considered uneconomical. For journalism, this means democratizing information, for instance, extending coverage to areas with little margin as local sports matches. Sweden's Östgöta Media, Klackspark, a site about local football in covers all matches in the Östergötland region as automation enables it to publish 850 articles a month, with 70 per cent of bylines going to robots

(Nyheter – Klackspark, 2020). The system, developed by United Robots, adds depth to articles sending relevant questions via text message to team coaches after matches and automatically adding the answers into the articles.

Another positive impact of automation is the expansion of service information like that on local real estate markets, creates a more level playing field for investors and increase price transparency of real estate markets, creating quality services otherwise unsustainable with human reporters. One example is the "Homeowners Bot" launched in 2016 by the Swedish media house MittMedia (Niemanlab, 2019), which automatically creates articles about the real estate market based on local property data. According to the company, the bot produces 10,000 articles per quarter and has had hundreds of users subscribe after reading articles it created.

Finally, intelligent systems may help newsrooms improve the balance of their reporting, for instance, in terms of gender, ethnicity, or age group, which can expand the reader base and increase subscriptions. BBC's *50:50* project (2020), for instance, has successfully leveraged a combination of basic AI tools and data to rebalance its news coverage for gender from 85% dedicated to males and a meagre 15% to women, and in Fall 2020 has started applying the methodology do disabilities and ethnicity.

Personalization of news and services is one of the most promising areas of AI application in news media. Instead of – or in addition to – delivering the same stories to every single person, more and more media organizations offer a set of stories individually tailored for every single user either on their homepage or through automatically generated newsletters and feeds. The demand is also driven by the evolution of many mainstream digital services like Netflix and Amazon that are shaping users' expectations. The benefits of news tailoring, coupled with appropriate pricing strategies, are already proving very beneficial for several outlets in Europe, increasing readerships and revenues. Automated tailored distribution is often the final step of news cycles involving AI at different stages. However, as mentioned earlier, excessive personalization may fragment the space for public debate, which is enabled by a common information baseline.

The main benefits demonstrated by AI tools in tailoring news to users' preference range from smart distribution (e.g., tailored newsletters and feeds based on tagging) and the optimization of the AI the delivery route and ensure that the most efficient path is taken when delivering newspapers and magazines to engagement and curation of the interaction with the readership and the conversion of users browsing a news outlet's website into subscribers.

4   The European Policy Context

The European Commission has proposed the Artificial Intelligence Act (2021), the first-ever legal framework on AI addressing the risks of AI, and positions Europe to play a leading role globally. The Artificial Intelligence Act would prohibit the use of AI systems that are considered a clear threat to the safety, livelihoods, and rights of people, such as systems that are designed to manipulate human behaviour through subliminal techniques and systems that allow the government to conduct "social scoring" resulting in unfavourable treatment. "High-risk" AI systems are not prohibited under the proposed Act but are subject to restrictions. High-risk systems include those used for justice and democracy, such as using the AI to apply the law to evaluate a set of facts. Given that access to information and freedom of expression are fundamental rights of citizens, it remains to be seen whether AI tools for journalistic purposes could fall within the scope of high-risk applications. It is, however, worth noticing how the AI Act underscores the need for a high-quality of the datasets used to train, validate and test the system, and the fact that their activity must be logged to ensure that their functioning can be tracked and monitored. Providers must retain documentation that allows authorities to assess compliance with these measures, and clear and adequate information also must be provided to users of the system. In addition, systems must be designed to ensure that there is appropriate human oversight, and a high level of robustness, security, and accuracy in their performance. Equally importantly, as digital journalism has by nature a global reach, providers placing AI products designed for these spheres on the EU market (regardless of the location of providers) will be subject to specific obligations. The same obligations apply no matter whether the system itself is operated outside of the EU, but its output is used in the EU. For instance, such systems must undergo an adequate risk assessment and implement mitigation measures. Concerning AI systems that are considered to pose only a limited risk, the Artificial Intelligence Act imposes transparency obligations, requiring that providers make users aware that they are interacting with a machine, while systems posing a minimal risk are not regulated by the proposed Act and may be used freely. The Commission commented that the vast majority of AI systems currently used fall into this final category.

Not specifically addressing AI but relevant to the transformation of the Union's media landscape and thus the adoption of AI-driven tools is the European Democracy Action Plan (2020). Among the objectives of the plan are "to provide sustainable funding for projects on legal and practical assistance to journalists in the EU and elsewhere" as well as "measures to support media pluralism" and "countering fake news and disinformation". The Democracy action plan goes hand in hand with the Media and Audiovisual Action Plan (('European Democracy Action Plan' n.d.), aiming to help the sector recover and make the most of the digital transformation. Actions include boosting investment in the audiovisual industry via "Media Invest", a new initiative whose target is to leverage investments of €400 million over 7 years. The Plan also includes a "News" initiative to bundle actions and support for the news media sector and a pilot investment project with foundations and other private partners, access to loans to be backed by the InvestEU guarantee, grants, and a European News Media Forum with the sector with particular attention will be paid to local media. Another chapter of the action, which may prove fundamental for the subsequent development of AI-driven tools, supports the digital transformation of journalism, to encourage European media data spaces for data sharing and innovation, fostering a European Virtual and Augmented Reality industrial coalition to help EU media benefit from these immersive technologies, and launching a VR Media Lab on projects for new ways of storytelling and interacting.

Finally, in 2020, the European Parliament's Science and Technology Options Assessment (Stoa) panel chaired by Greek MEP Eva Kaili, launched the Centre for AI (C4AI), which will produce studies, organizes public events, and acts as a platform for dialogue and information exchange on AI-relevant topics within the Parliament and beyond ('Centre for AI | Panel for the Future of Science and Technology (STOA) | European Parliament' n.d.). In particular, it provides expertise on the possibilities and limitations of AI and its implications from an ethical, legal, economic, and societal perspective.

5   Recommendations

All the initiatives described in Sect. 5 should support the positive development of digital journalism. This is good news because fostering the responsible adoption of AI in European news media is urgent and complex as in many countries the industry is still struggling with digital transformation and upskilling of the workforce. To support such transition, the following recommendations have essentially two aims: supporting responsible technological innovation and its long-term adoption and developing the appropriate ethical competencies and guidance tools for newsrooms. A direct regulatory approach on these matters is advised against, given possible consequences for freedom of the press and the democratic space.

  I. Funding. Newsrooms can be challenging environments, and developing technological solutions is only the first step in the process of innovation. As in all software based industries, after development comes implementation, testing, tweaking, and, sometimes, redesign. Structural support for R&D research in journalism automation is currently lacking in Europe and needs to become structural to develop a vibrant media innovation ecosystem. Longer-term funding would also allow to loop in ethical reviews and evaluations of the innovations introduced and make them accountable for. An example of an effective measure comes from the private sector. Over 3 years (from 2016 to 2018) Google's Digital News Innovation Fund has supported 662 digital news projects in Europe with 150 Million Euros in competitive grants. Automation driven projects have shown significant impact on newsroom. RADAR (Reporters and Data and Robots) provides subscriptions to 400 UK newsrooms to create data-driven articles using a blend of human and artificial intelligence. In 2019, RADAR articles had 95% pick-up digitally and, when it comes to print, many are front-page stories. Google's programme came to completion in 2018 and no data is available at present on the evolution of the projects it supported. A more consistent monitoring of the impact of automated systems would be extremely useful in a longer term policy approach.

  II. Media Pluralism. Specific attention should be given to supporting independent and smaller size media to ensure editorial diversity and independence. Smaller media often have a younger and more experimentation-prone workforce but, as described in Sect. 2.3, may encounter challenges in achieving financial sustainability as advertising and paywall based models require large volumes of users. Collaboration and contamination with tech startups should be encouraged to help develop a more tech-savvy European media ecosystem which, ultimately, may also provide a new talent pool to larger legacy media looking to innovate. Starting with its Audiovisual plan, the European Commission has bundled existing and new actions into a "News Initiative"('EU Media and Audiovisual Plan – The "News Initiative"', 2021), to support the news media sector. This is aimed in the right direction but more funds, and more specific calls, tailored to encourage AI use and development of new applications for the in small and medium media would find here a natural platform. Similar initiaves could be launched at National level or by private charities to match EU funds or facilitate access and submission of applications.<u>orkforce upskilling</u>. Newsrooms are a fast-paced and often understaffed environment where journalists are called to make continuous decisions about the product to which they contribute. Company management and professional bodies should encourage newsrooms to develop their guidelines in an interdisciplinary manner, involving both data experts, developers, and editorial staff to support their innovation processes involving AI solutions. A good example is what has been done by Bavarian Radio with its AI ethics guidelines (Jonas Bedford-Strohm, 2020). The aim is to continuously improve products and guidelines, through experience and learning from pilot projects and prototypes. This is done through partnershios offering practical research context for students and faculty at universities and collaboration with academia and industry to run experiments, as well as through talent acquisition to keep up with the rapid evolution of AI.

III. <u>Data and diversity culture</u>. A solid data culture within newsrooms needs to be supported and developed to help prevent algorithmic bias in the data ad reflect the diversity of society. The user experience of news media services should be designed with data sovereignty for the user in mind. Scandinavian media group Schibsted (Stembom, 2020) has moved consistently on this over the last years. Since 2019 it has a Chief Data Officer, and in 2020 appointed a Responsible data and AI specialist, Agnes Stenbom who is also a PhD candidate on AI and data ethics at the KTH Royal Institute of Technology. Use of AI varies across Schibsted's newsrooms according to their experience and size with the larger titles as Aftonbladet typically leading the way. As in many other industries, most our Shibsted's use of AI is not particularly sexy but aims at having impact both on newsroom work as on commercial operations. One example is the Machine learning and Artificial Intelligence Team development of a Natural Language Generator for Swedish to improve its internal recommendation engine but AI is applied pretty much across the board, from recommending relevant content and ads to users to helping human moderators review explicit content or predicting how many newspapers should be printed to minimize the company's environmental footprint. The company admittedly does not have an AI strategy but is working on one. At Shibsted the risks brought on by AI are perceived both on a business scale as well as at wider range, investing the social responsibility the company has as an information provider.

IV. <u>Inclusion</u>. As noted by the Council of Europe, automated filtering and sorting mechanisms can affect the exercise of an individual's right to receive information based on personal characteristics and preferences, and the use of AI-driven tools should not result in a situation in which certain parts of the population of users with particular characteristics are structurally excluded from accessing information, or where society experiences new digital divides. Such a situation would be incompatible with the positive obligation of Member States to protect and promote the right to information stemming from Article 10 ECHR.

V. <u>Codes of conduct</u>. As existing journalistic codes of conduct do not refer to the use of AI-driven tools, the media must develop journalistic ethics for the use of such tools (Helberger et al., 2020). Furthermore, it follows from Article 10 ECHR that media actors are responsible for developing professional rules regarding the risks of AI-driven tools for bias and media diversity. If journalists start using AI-driven tools without sufficiently interrogating the tools they use and without sufficient awareness of the practices and the problems that may stem from the use of AI-driven tools, including issues such as incomplete data, biased data, and faulty models, there is a risk of journalistic malpractices (Hansen et al., 2017).

VI. <u>Checklists</u>. Oaths and codes of conduct have their value but have rarely prevented misconduct. Atul Gawande once quipped that "checklists eat oaths for breakfast", and this resonates with AI ethical impacts in newsrooms. Journalist professional associations and funders need to support and advocate for checklist adoption. A good starting point would be the 13-point checklist proposed by Mike Loukides and already inspiring work in several newsrooms. A good starting point for elaborating internal guidelines and checklists, – in line with European values and, possibly, future regulation – is looking at five principles recurring across AI ethics research (Morley et al., 2020; Floridi & Cowls, 2019). Namely, AI should be: (a) beneficial to, and respectful of, people and the environment (beneficence); (b) robust and secure (non-maleficence); (c) respectful of human values (autonomy); (d) fair (justice); and (e) explainable, accountable and understandable (explicability). European policies to support the development of such tools is advisable as the additional work to the research and development process to be ethical by design translates into higher costs. As underscored by Morley and colleagues: "such overheads may directly conflict with short-term, commercial incentives. Indeed, a full ethical approach to AI design, development, deployment and use may represent a competitive disadvantage for any single 'first mover'".

6 Conclusions

From the printing press to the telegraph and the web, journalism has historically been driven and shaped by the media it uses. As news media's role in society is growingly challenged by a shift in business models driven by global digital platforms, AI presents journalists with both risks and opportunities. Risks range from abdicating accountability to giving in to excessive personalization of content and thus renouncing to the mission of helping maintain an informed society; reduction of media pluralism; abuse of readers' privacy, or copyright infringement. On the other hand, intelligent systems may assist journalists in all stages of their work, from newsgathering to its production and distribution. Moreover, if sensibly applied, they may open new revenue streams and increase the efficiency and productivity of human reporters, freeing them from the more mundane tasks and opening new fields. As in many other fields, integration of AI-driven solutions appears indispensable for any outlet willing to remain editorially competitive and financially sustainable.

Given the pivotal role of good and independent journalism as a "purveyor of information" essential to create a democratic forum for public debate, and to act as a public watchdog, as underscored by the European Court of Human Rights, we encourage the development of public guidelines and policies to support the uptake of Ai-driven applications in European newsrooms. While advising against a regulatory approach, always risky when it comes to journalism, we recommend some lines of action. These cover, quite inevitably, funding to support the adoption of AI in newsrooms, with particular attention to smaller outlets at greater risk of missing this transition; workforce training and upskilling; development of a data and diversity culture which may be also beneficial to other industries; countering excessive personalization to promote inclusion and, finally, promotion of updated journalistic codes of conduct including reference to the new issues as well as checklists to support them.

In a globalized digital information ecosystem, not integrating AI in the journalism practice is a luxury we cannot afford but how societies decide to develop, and support AI-enabled journalism may make a huge difference for democracy.

Acknowledgements We are grateful for their time and patience in sharing insights and direction, to Nikolas Diakopolous, Oliver-Elliott, Luciano Floridi, Agnes Stenbom, Dino Pedreschi, Oreste Pollicino, and Mattia Pedretti, Mariarosaria Taddeo.

Funding Guido Romeo's and Emanuela Griglié's research was supported by a fellowship funded by Google EU.

References

Angwin, J., Larson, J., Kirchner & Mattu, S. (2017). Minority Neighborhoods Pay Higher Car Insurance Premiums Than White Areas With the Same Risk. Retrieved from: https://www.propublica.org/article/minority-neighborhoods-higher-car-insurance-premiums-white-areas-same-risk

Apple News Is Excluding Local Newsrooms from Its Coveted Traffic Bump. (n.d.). *Columbia Journalism Review*. Accessed 24 Jan 2021. https://www.cjr.org/tow_center/apple-news-local-journalism.php

Artificial Intelligence (AI): New Developments and Innovations Applied to e-Commerce – Think Tank. (n.d.). Accessed 19 May 2021. https://www.europarl.europa.eu/thinktank/en/document.html?reference=IPOL_IDA(2020)648791

Automating the News — Nicholas Diakopoulos. (n.d.). Accessed 19 May 2021. https://www.hup. harvard.edu/catalog.php?isbn=9780674976986

BBC News. (2014). *Facebook emotion experiment Sparks criticism*, 30 June 2014, sec. Technology. https://www.bbc.com/news/technology-28051930

Bloomberg.Com. (2017). *A digital fact-checker fights fake News*, 12 January 2017. https://www. bloomberg.com/news/articles/2017-01-12/a-digital-fact-checker-fights-fake-news

Broussard, M., Diakopoulos, N., Guzman, A. L., Abebe, R., Dupagne, M., & Chuan, C.-H. (2019). Artificial intelligence and journalism. *Journalism & Mass Communication Quarterly, 96*(3), 673–695. https://doi.org/10.1177/1077699019859901

Brugnoli, E., Cinelli, M., Quattrociocchi, W., & Scala, A. (2019). Recursive patterns in online Echo chambers. *Scientific Reports, 9*(1), 20118. https://doi.org/10.1038/s41598-019-56191-7

Buhrmester, V., Münch, D., & Arens, M. (2019). Analysis of explainers of black box deep neural networks for computer vision: A survey. ArXiv:1911.12116 [Cs], November. http://arxiv.org/abs/1911.12116

Centre for AI | Panel for the Future of Science and Technology (STOA) | European Parliament. (n.d.). Accessed 19 May 2021. https://www.europarl.europa.eu/stoa/en/centre-for-AI

Crimemap: Polizeibericht, Brände Und Unfälle in Stuttgart – Stuttgarter Zeitung. (n.d.). Accessed 26 Dec 2020. https://www.stuttgarter-zeitung.de/crimemap

Eli Pariser on "The Filter Bubble: What the Internet Is Hiding from You". (n.d.). *Democracy Now!* Accessed 18 May 2021. http://www.democracynow.org/2011/5/27/eli_pariser_on_the_filter_bubble

EU Media and Audiovisual Plan – The "News Initiative". (2021). October 2021. II. https://digital-strategy.ec.europa.eu/en/policies/news-initiative

European Court of Human Rights – Case of Barthold v. Germany (1985) – Paragraph 59. (n.d.). Accessed 30 Dec 2020. https://www.legislationline.org/documents/id/5019

European Convention on Human Rights (n.d.) – Official Texts, Convention and Protocols. https://www.echr.coe.int/Documents/Convention_ENG.pdf

European Democracy Action Plan. (n.d.). Text. European Commission – European Commission. Accessed 16 Jan 2021. https://ec.europa.eu/commission/presscorner/detail/en/ip_20_2250

Fake News Is a Media Ecology Problem | by Luca De Biase | Medium. (n.d.). Accessed 19 May 2021. https://medium.com/@lucadebiase/fake-news-is-a-media-ecology-problem-c2e31bf17066

Floridi, L., & Chiriatti, M. (2020). GPT-3: Its nature, scope, limits, and consequences. *Minds and Machines, 30*(4), 681–694. https://doi.org/10.1007/s11023-020-09548-1

Floridi, L., & Cowls, J. (2019). A unified framework of five principles for AI in society. *Harvard Data Science Review, 1*(1). https://doi.org/10.1162/99608f92.8cd550d1

Floridi, L., & Taddeo, M. (2016). What is data ethics? *Philosophical Transactions of the Royal Society A: Mathematical, Physical and Engineering Sciences, 374*(2083), 20160360. https://doi.org/10.1098/rsta.2016.0360

GPT-3. (2020). 'A Robot Wrote This Entire Article. Are You Scared yet, Human? | GPT-3'. *The Guardian*. 8 September 2020. http://www.theguardian.com/commentisfree/2020/sep/08/robot-wrote-this-article-gpt-3

Hansen, M., Roca-Sales, M., Keegan, J. M., & King, G. (2017). *Artificial Intelligence: Practice and Implications for Journalism*. https://doi.org/10.7916/D8X92PRD

Helberger, N., Eskens, S. J, van Drunen, M. Z., Bastian, M. B., & Möller, J. E. (2020). Implications of AI-driven tools in the Media for Freedom of expression. In *Council of Europe*. https://research.vu.nl/en/publications/implications-of-ai-driven-tools-in-the-media-for-freedom-of-expre

How Trump Thrives in "News Deserts". (2018). POLITICO. 9 April 2018. https://www.politico.eu/blogs/on-media/2018/04/trump-media-news-desert-how-thrives/

ICIJ (2021). *The Panama Papers: Exposing the Rogue Offshore Finance Industry*. n.d. ICIJ (blog). Accessed 2 January 2021. https://www.icij.org/investigations/panama-papers/

Jonas Bedford-Strohm, Uli Köppen. (2020). *Ethics of artificial intelligence: Our AI ethics guidelines*. November. https://www.br.de/extra/ai-automation-lab-english/ai-ethics100.html

Journalism, Media, and Technology Trends and Predictions. (2020). Reuters Institute for the Study of Journalism. Accessed 28 December 2020. https://reutersinstitute.politics.ox.ac.uk/journalism-media-and-technology-trends-and-predictions-2020

Karp. (2021). https://www.theguardian.com/technology/2021/jan/13/google-admits-to-running-experiments-which-remove-somemedia-sites-from-its-search-results

Kobie, N. (2018). Reuters is taking a big gamble on AI-supported journalism. *Wired UK*. Accessed 19 May 2021. https://www.wired.co.uk/article/reuters-artificial-intelligence-journalism-newsroom-ai-lynx-insight

Lehman. (2016). https://www.newyorker.com/news/news-desk/how-peter-thiels-gawkerbattle-could-open-a-war-against-the-press

Mattu, J. A., Larson, J., Kirchner, L., & Surya. (n.d.). Machine Bias. ProPublica. Accessed 18 May 2021. https://www.propublica.org/article/machine-bias-risk-assessments-in-criminal-sentencing

Morley, J., Floridi, L., Kinsey, L., & Elhalal, A. (2020). From what to How: An initial review of publicly available AI ethics tools, methods and research to translate principles into practices. *Science and Engineering Ethics, 26*(4), 2141–2168. https://doi.org/10.1007/s11948-019-00165-5

Niemanlab. (2019). https://www.niemanlab.org/reading/mittmedia-homeowners-bot-boosts-digital-subscriptions-withautomated-articles/

Nyheter – Klackspark. (2020). Accessed 12 June 2021. https://klackspark.com/nyheter

Sasahara, K., Chen, W., Peng, H., Ciampaglia, G. L., Flammini, A., & Menczer, F. (2021). Social influence and unfollowing accelerate the emergence of Echo chambers. *Journal of Computational Social Science, 4*(1), 381–402. https://doi.org/10.1007/s42001-020-00084-7

Stembom. (2020). *How Schibsted Uses Artificial Intelligence*. n.d. Schibsted. Accessed 26 December 2020. https://schibsted.com/blog/how-schibsted-uses-artificial-intelligence/.

Tsamados, A., Aggarwal, N., Cowls, J., Morley, J., Roberts, H., Taddeo, M., & Floridi, L. (2021). The ethics of algorithms: Key problems and solutions. *AI & SOCIETY, 37*. https://doi.org/10.1007/s00146-021-01154-8

Universal Declaration of Human Rights. (2015). 6 October 2015. https://www.un.org/en/universal-declaration-human-rights/

Whittaker, J. P. (2019). *Tech giants, artificial intelligence, and the future of Journalism*. Published September 30, 2020 by Routledge. https://www.routledge.com/Tech-Giants-Artificial-Intelligence-and-the-Future-of-Journalism/Whittaker/p/book/9780367661090.   ISBN 9780367661090.

Yang, G.-Z., Bellingham, J., Dupont, P. E., Fischer, P., Floridi, L., Full, R., Jacobstein, N., et al. (2018). The grand challenges of science robotics. *Science Robotics, 3*(14), eaar7650. https://doi.org/10.1126/scirobotics.aar7650

Towards Equitable Health Outcomes Using Group Data Rights

Gal Wachtel

Abstract The use of Big Data and algorithmic decision-making in healthcare has been promoted over the last decade with the claim that using such methods translates into gains for marginalized populations long mis- and under-represented in biomedical research. However, a large body of works has emerged showing that these approaches disproportionately disadvantage marginalized populations as poor design and poor data can embed existing structural inequities into newly created sociotechnical systems. Within the biomedical context, these disparities risk widening existing health disparities, including disproportionate burden of acute or chronic diseases and adverse health outcomes, experienced by underprivileged populations. This chapter demonstrates how group data rights can help alleviate these potential harms by drawing on two examples of group data rights already in use by marginalized populations. By closely analyzing the potential offered by data governance approaches used for Indigenous Data Sovereignty and by Rare Disease Advocacy Organizations, this chapter shows how group data rights can both promote health equity and ultimately proposes a framework for practically implementing group data rights in a healthcare setting.

Keywords Group data · Health · Data rights · Health equity · Bias · Decision-making

1 Introduction

Over the last decade, advocates of Big Data use in healthcare have argued that increasing the scale and variability of data used for decision-making would translate into gains for marginalized populations, which have long been misrepresented and

G. Wachtel (✉)
Oxford University, Oxford, UK

Harvard University, Cambridge, MA, USA

J. Mökander, M. Ziosi (eds.), *The 2021 Yearbook of the Digital Ethics Lab*, Digital Ethics Lab Yearbook, https://doi.org/10.1007/978-3-031-09846-8_15

underrepresented in biomedical research (Zhang et al., 2017). However, a large body of works has emerged showing that Big Data- and algorithmic-based approaches for decision-making disproportionately disadvantage marginalized populations (Angwin et al., 2016; Benjamin, 2019; Eubanks, 2018; Obermeyer et al., 2019). These harms are the result of data and design practices which embed existing structural inequities into newly created sociotechnical systems (Cofone, 2018). Within the biomedical context, these disparities risk widening existing health disparities, including disproportionate burden of acute or chronic diseases and adverse health outcomes, experienced by underprivileged populations (Zhang et al., 2017). The growing literature on the limitations of algorithmic-based approaches along with the biomedical community's increased orientation towards health equity (Braveman & Gruskin, 2003), have increasingly brought to the fore the need to develop mechanisms to mitigate these harms. While attracting newfound attention, the challenge data inequities pose to marginalized communities is an old one. Indigenous Data Sovereignty and Rare Disease Advocacy Organizations are two models for data governance developed by underserved communities to alleviate the adverse outcomes of missing and misused data at the population or group level. This chapter will draw on these models to examine the potential offered by group data rights to advancing health equity and to propose a practical implementation of group data rights in a healthcare setting.

2   Health Inequities, Datafication, and Automation

Over the last 20 years, there has been a significant shift from conceptualizing health as a purely biological phenomenon to a biosocial one in which social determinants are seen as critical for understanding health outcomes (Farmer et al., 2006). This shift is reflected in the medical community's increased promotion of *equity,* defined as the absence of unfair and avoidable differences in health among population groups, over *equality,* which is defined more narrowly as equal treatment for all, as the main metric for evaluating fairness and success in health outcomes (Braveman & Gruskin, 2003). Extant research has shown that in the U.S., racial-ethnic minorities, socioeconomically disadvantaged individuals, and other underprivileged populations are disproportionately impacted by acute or chronic diseases and adverse health outcomes (Zhang et al., 2017). Moreover, there is increasing awareness that clinician bias contributes to consistent and substantial differences in the treatment of minority populations, further exacerbating health disparities (Williams & Wyatt, 2015). Given these deficiencies, the use of Big Data and algorithmic-aided systems in healthcare has been promoted as an opportunity to mitigate existing disparities (Zhang et al., 2017). However, multiple studies have shown that these systems can in fact scale and compound them (Angwin et al., 2016; Benjamin, 2019; Eubanks, 2018).

The potential for large-scale disparate impact is captured in Obermeyer et al.'s work "Dissecting racial bias in an algorithm used to manage the health of populations" (2019). In this chapter the authors show that OPTUM, a widely-used algorithm that is typical of the commercial predictive algorithms employed in the U.S. health system, exhibits significant racial bias. At a given health risk score, Black patients were found to be considerably sicker than white patients, as evidenced by signs of uncontrolled illnesses. The bias arises because the algorithm predicts health care costs rather than illness, but unequal access to care means that less money is spent caring for Black patients than for white patients (Fiscella et al., 2000). Consequently, structural racism, which is discrimination that is the result of mutually reinforcing inequitable systems that foster discriminatory beliefs, values, and distribution of resources (Krieger, 2014), is encoded into the algorithm and the care system at large. This case study is useful for analyzing the failures in the current research and regulatory landscapes that permitted the commercial deployment of a deeply flawed algorithm. In the following section I will consider how the current emphasis on personal data rights compounds existing biases, and why group data rights may offer a remediation.

3   Losing the Forest for the Trees – How Individual Data Rights Leave Marginalized Populations Vulnerable

We can think of data-driven decision-making systems as composed of three layers: data, analytical methods, and context. Each layer is situated in a different discipline with its own ethical framework and regulatory bodies (if existent). It is, therefore, appropriate to consider the safeguards provided by each discipline to mitigate unfair outcomes by the system. In this case, we will focus on data privacy, artificial intelligence and machine learning (AI/ML), and medicine.

The current legal, ethical, and analytical understanding of privacy and data protection focuses on the individual, as can be seen in the European General Data Protection Regulation which is one of the most comprehensive regulations available today (Taylor et al., 2017). Individual data rights are most useful for guarding against harms that are the result of identification since they largely focus on controlling access to personal data (Kammourieh et al., 2017). However, Big Data analytics, like predictive algorithms, are often not used to identify an individual, but rather to determine types from which inferences are drawn (Taylor et al., 2017). For example, the algorithm examined by Obermeyer et al. (2019) does not pose a unique risk to those whose data was used for its development. Instead, it poses a group-level risk to all members of the group it inadvertently discriminates against. In her work, Taylor (2017), asks whether we are able to protect groups from disparate impact only by protecting individual rights to privacy and ownership. Taylor concludes that the mechanisms offered by individual rights are insufficient. This is exemplified in the case of the Obermeyer et al. algorithm where patients are not discriminated against due to their explicit belonging to a protected class but rather due to a grouping that is a secondary outcome of race, e.g. overall lower health spending, nor is their data used for their reidentification. In sum, individual data protections are best suited to guard against individual data gathering, but are poorly equipped to guard against inferences, particularly those that can be drawn about groups, posing a grave risk for vulnerable populations.

There are also practical barriers to utilizing an individual-centered data protection framework. Legal frameworks that are intended to guarantee individuals control over their personal information are rendered ineffective when data is collected at a high frequency and often without the knowledge of the subject, since exercising control over data becomes a resource-intensive endeavor that also requires a high-level of digital literacy (Kammourieh et al., 2017). For example, American class action and anti-discrimination law, which are frameworks for pursuing legal recourse on account of disparate impact, require group members to both identify that they have been wronged and to bring their case to court (Legal Information Institute, n.d.). As a result, a significant power imbalance is created between the subject and the data gatherer.

Let us consider whether existing frameworks in medicine can better mitigate disparate impact. Unlike other fields, medicine has a rich history of holding its practitioners to a set of moral obligations and virtues through norms and professional accountability mechanisms. Although the profession's standards vary across cultures and time, and even though egregious failures of ethical negligence have occurred despite these standards, such as the Tuskegee Syphilis study in which syphilis was left untreated in hundreds of Black men for the study of the disease's progression (Gamble, 1997), there exists a common aim in medicine to promote the health and well-being of the patient which is broadly translated into regulatory frameworks that establish a fiduciary duty towards the patient (Mittelstadt, 2019). As Mittelstadt notes in his paper "Principles alone cannot guarantee ethical AI" (2019), the unique characteristics of the medical field do not translate to the current Big Data landscape, which is lacking in both professional norms that establish a fiduciary duty towards the data subject and in regulatory frameworks to hold practitioners accountable to a set of ethical obligations. These differences are particularly pronounced in the commercial development of predictive algorithms for healthcare, where developers of clinical tools are outside the purview of the ethically bound ecosystem of medicine.

Moreover, even if medicine's ethical framework mapped neatly, or could be extended to, those developing technical healthcare applications it is appropriate to question whether it provides ample protection against data-driven disparate impact. Fiduciary duties, which ground the medical ethics discourse, are set at the individual level, between the care provider and patient (Childress & Beauchamp, 1994). Similar to the argument against individual privacy rights, individual-level fiduciary duties fail to acknowledge the group-level dynamics that compound existing health inequities in Big Data methods and that are not readily apparent to the provider who, even if they were able to observe the system inputs and logic, would be unable to detect all levels of group inference that are often obscured in these systems.

In recent years several solutions have been developed in the AI/ML landscape for assessing whether algorithmic decision-making systems produce fair results. Some of these solutions focus on the algorithmic frame, while others operate at a higher level of abstraction and assess the system in which the algorithm is implemented.

Regardless of the specific assessment tool used, there remain two unresolved questions – first, who is responsible to conduct such assessments and second, who should be conducting them. One proposed solution is multi-stakeholder engagement, a participatory development process that involves key stakeholders, including marginalized populations (Katell et al., 2020; Paulus & Kent, 2020; Rajkomar et al., 2018). However, in the absence of an enforcing legal framework the implementation of multi-stakeholder processes is ad-hoc at best, significantly limiting its credibility as a mechanism for improving accountability. Furthermore, even if the framework were to become the norm, it would still fall short of promoting patient agency since data subjects do not elect who represents their interest in the multi-stakeholder process nor do they consent to their data usage.

From the discussion above it is evident that any solution aimed at mitigating data-driven disparate impact must address three issues at the group level: power imbalance, accountability, and agency. In the literature two individual data rights are acknowledged: the right to data privacy and the right to data protection. A lively debate exists about whether such rights are extensions of each other, with data protection providing the legal framework for safeguarding data privacy, or whether the two are distinct (Tzanou, 2013). The European Constitution enshrines both as a fundamental right (Tzanou, 2013) and it is useful to consider the scope and motivators of both rights in the context of group-level data rights. Privacy has been codified in myriad legal systems and "embodies a range of rights and values, such as the right to be let alone, intimacy, seclusion, personhood, and so on according to the various definitions" (Tzanou, 2013). The threat of infringing on the right to privacy via re-identification is often viewed from a reductionist lens highlighting the potential for disparate impact or discrimination, or from ownership-based theories that ground the threat in an individual's right to exclusive use of information (Mittelstadt, 2019). Data protection, on the other hand, can be viewed as stemming from the right to informational autonomy. For example, the German Constitutional Court defines the right to informational self-determination as stemming from the right to human dignity and to personality, thus guaranteeing the power of the individual to determine for himself the disclosure and use of his data. Data protection hence views data as identity constitutive and thus any alteration of it as inviolating personality. It is important to note that some, such as Floridi (2016), view damage to the integrity of information pertaining to one's identity as a violation of one's privacy if the right to privacy is legitimized by the right to personhood. In viewing both the right to privacy and protection together two central themes emerge: one of safeguarding from unwanted intervention and the other of protecting the components that constitute identity. For the remainder of this chapter, I shall be grouping these two approaches under the umbrella term – data rights. These concepts when expanded from the individual level to the group level are referred to as group data rights and have been proposed by multiple scholars as a solution to the shortcomings of individual data rights (Floridi, 2014; Taylor et al., 2017). In the following sections, I will use two case studies to explore the utility of recognizing group data rights for protecting marginalized populations and to propose a novel path for their operationalization within a healthcare setting. I will set aside, for the purposes of this essay, the question of how group data rights would map to specific legal frameworks, as this is a question that must be answered at the level of individual jurisdictions.

4 Underrepresented, Underserved and Fighting Back – Two Case Studies on Group Data Rights

Indigenous populations and rare disease patient groups are two communities that have been plagued by data inequities (Carroll et al., 2020; Merkel et al., 2016). Although the sources of their underrepresentation and misrepresentation in bodies of knowledge are different, they have both developed systems to improve their visibility and gain agency over their treatment and resources through group data control. We will be looking at Indigenous Data Sovereignty (IDS) to better understand the need for informational self-determination, while examining Disease Advocacy Groups (DAOs) as a practical model for leveraging group data for furthering group interests. Ultimately, both models will be used together to propose a new equity-promoting data governance structure.

4.1   Indigenous Data Sovereignty

Indigenous Data Sovereignty (IDS) refers to the right of indigenous nations to govern the collection, ownership, and application of their own data (Carroll et al., 2020). The 2007 United Nations Declaration on the Rights of Indigenous Peoples, the product of a decades-long process of reclamation of self-rule by indigenous communities (Carroll et al., 2020), reaffirms that IDS is integral to a nation's right to self-determination (Assembly, 2007). To date, four national-level IDS networks exist: Te Mana Raraunga – Maori Data Sovereignty Network, the United States Indigenous Data Sovereignty Network (USIDSN), the Maiamnayri Wingara Aboriginal and Torres Strait Islander Data Sovereignty Group in Australia, and the First Nations Information Governance Center (FNIGC, 2022) in Canada (Carroll et al., 2019; FNIGC).

IDS, and its practical implementation through Indigenous Data Governance, is a response to centuries of marginalization of indigenous communities by settler-colonial forces through discriminatory structures of knowledge (Carroll et al., 2020). To subjugate indigenous communities, colonial forces both weakened native knowledge forms, for example through forced residential schooling of native children as was done both in the U.S. and Canada (Smith, 2009), and used biased methods of datafication to create new structures of knowledge that furthered the colonial agenda (Carroll et al., 2020). An example of the latter is the exclusion of native populations from national census counts, such as the 1994 exclusion of First Nations people living on reservations from Canadian population surveys (FNIGC). From this legacy, IDS recognizes data as both inherent to preserving indigenous identity through the cataloging of indigenous knowledge, and as a strategic resource. Thus, the shift to IDS is inherently a process of data decolonization facilitated by replacing "external, non-Indigenous norms and priorities with Indigenous systems that define data, and inform how it is collected and used" (Carroll et al., 2019). Ultimately, IDS aims to disrupt the power imbalance created by Indigenous nations' data dependency and produce systems and findings that reflect the understandings and lived experiences of indigenous peoples (Carroll et al., 2019).

IDS is not merely a theoretical framework – several practical guidelines for its implementation have been proposed and deployed by Indigenous communities. Building upon the United Nations recognition of the right of Indigenous communities to self-determination in 2007 (Assembly, 2007), several IDS initiatives have been established in recent years. The International Indigenous Data Sovereignty Interest Group drew from existing IDS initiatives to develop 'CARE Principles for Indigenous Data Governance', a unifying framework for IDS (Carroll et al., 2020). CARE asserts that Indigenous data must: (1) facilitate Collective benefit for indigenous peoples, (2) promote Indigenous peoples Authority to control their data narratives, (3) be handled Responsibly to nurture respectful relationships with its creators, (4) should have an Ethics focus that prioritizes Indigenous Peoples' rights and wellbeing (Carroll et al., 2020). In outlining the CARE principles their developers note that a major challenge of pursuing these principles is making them compatible with the "FAIR Guiding Principles for scientific data management and stewardship" (Findable, Accessible, Interoperable, Reusable), a benchmark for data sharing (Carrol et all., 2020). CARE adds friction to accessing indigenous data, consequently risking further isolating Indigenous communities. This risk will be further explored later in this chapter.

4.2 Disease Advocacy Organizations

While IDS is useful as a framework for explicating the need for group data governance amongst marginalized populations, it is limited in its generalizability due to its grounding in the right of indigenous groups to self-determination, a legal exceptionalism that is not afforded to all marginalized communities. Next, we will consider another organizational form of group data rights that does not rely on pre-existing legal mechanisms.

Disease advocacy organizations (DAOs) are patient-centric advocacy groups which enable "individuals with a shared interest to pool their collective resources and shared knowledge of a medical condition, to work closely with clinicians and scientists" (Landy et al., 2012). DAOs emerged from the "Citizen Science" movement which aims to shift power back to the patient from the medical establishment, promoting modes of research and care that are informed by patient needs (Wiggins & Wilbanks, 2019; Banner et al., 2019). "Patient-informed care", "Citizen Science" and other models for patient engagement throughout the continuum of care emerged from the AIDS crisis in North America in the 1980s, when institutional discrimination against the queer community resulted in AIDS being largely ignored by the biomedical establishment (Epstein, 1996). As Steven Epstein details in his seminal work on AIDS activism, AIDS activist groups were able to establish themselves as "credible participants in the process of knowledge construction" thereby bringing about changes in the epistemic practices of AIDS research and biomedical research at large (Epstein, 1996). AIDS activists were appointed as full voting members of the committees of the National Institutes of Health and were named to the drug discovery boards of the Federal Drug Administration in the U.S., marking a watershed moment for patient-informed research (Epstein, 1996).

The successes of AIDS activism informed a movement that today counts tens of thousands of patient groups (in the U.S. alone) operating at different stages of the continuum of care (Smith et al., 2015). Of particular interest to this chapter are groups serving patients with rare diseases. Similar to data gaps that plague Indigenous communities, the study of rare diseases is hindered by a sparse data space, with patients of interest often living in different countries (Merkel et al., 2016). Moreover, rare disease research has been historically under-resourced and underfunded due to utilitarian allocations of health resources (Novorol, 2020). To overcome these challenges and promote investment in the study and treatment of rare diseases, rare disease DAOs have generally adopted the following course of action: (1) establish a community, (2) develop a commodity, (3) make the community an essential part of the academic enterprise, (4) advocate for the commercialization of translational research (Terry et al., 2007). This governance structure is similar to that proposed more recently by data cooperatives, which generally refer to groups that have voluntarily pooled data resources in a commonly owned enterprise, and where the stewardship of that data is a joint responsibility of the common owners (Ada Lovelace Institute, 2021).

The generation of a commodity (stage 2) is essential for the success of the DAO, as it empowers patients by transforming them from consumers to producers. In the fragmented research space of rare diseases, DAO pooled patient data, from blood samples and genome sequences to side effect logs, provides an invaluable data source. To leverage this resource, DAOs have established several patient registries and data banks (often referred to as BioBanks) with the aim of creating streamlined processes for data aggregation and collective sharing (Landy et al., 2012). For example, the U.S. Neuromuscular Disease Registry formed by the Muscular Dystrophy Association, a patient advocacy group, has been used to recruit patients into research trials for several clinically viable treatments (Huml et al., 2020). An important aspect of this approach is that such data repositories create a group data mandate, meaning that when volunteering their data patients agree to have the DAO represent their interests as part of a group to other stakeholders while maintaining the individual patient right to consent and privacy, for example by seeking renewed consent for each new research partnership (Courbier et al., 2019). The formalization of patient-centered movements in healthcare is evident in the establishment of PCORnet, the National Patient-Centered Clinical Research Network, and increased focus from the U.S. Food and Drug Administration (FDA) as well as the twenty-firstCentury Cures Act (Smith et al., 2015). This adoption of DAOs across the clinical trials enterprise by multiple stakeholders, from academia to industry, is creating a new care development landscape that is inclusive of patient voices.

5   Who Are Group Data Rights For?

With the biomedical community's reorientation towards health equity, rather than equality, as a measure of justice within the medical system it is incumbent upon its members to develop practical tools to enforce these values throughout the continuum of care. As previously described, to achieve this goal there is a need to develop frameworks that both engender accountability within the medical community and that promote group-level agency to enable group members to formally protect and advance their well-being.

The two case studies presented above demonstrate how group data rights provide a tool for protecting and promoting the interests of underserved groups. Yet, while effective for the communities they serve, these models are limited in their generalizability since they rely on restrictive definitions of who is entitled to group identity and is thus eligible for collective data rights. The proliferation of DAOs over the last three decades reflects the medical community's recognition of a disease-based group identity. However, as the biomedical world has come to recognize health as a biosocial phenomenon in which social group affiliation, such as race and ethnicity, is detrimental to health outcomes, the narrow scope of DAOs leaves marginalized communities without a mechanism for recourse. On the other hand, IDS recognizes that systemic marginalization warrants identity-based group-level protections.

However, since IDS relies on a legal framework that clearly demarcates group members and their group rights, it is difficult to expand this framework to include marginalized groups for which group affiliation is not formalized. For example, no similar legal framework recognizes the structural marginalization of Black Americans.

While imperfect on their own, IDS and DAOs offer a new perspective on collective data rights when considered together. It is useful to consider why Indigenous communities require the special protections that led to the development of IDS. The argument for Indigenous sovereignty relies on the basic recognition that Indigenous communities suffer from structural violence at the hands of the state. IDS then asserts that in order to overcome this structural violence a community must exercise control over the information that feeds the sociotechnical system that perpetuates these inequities, with group data governance as the mechanism through which this control is attained. The argument proposed by IDS is novel and significant. First, it recognizes equity as an organizing principle and the role of data in promoting equitable outcomes. From this point, it derives that marginalized groups have rights over their collective data and that exercising this right is a form of data justice essential to promoting equity. This abstraction of IDS enables us to extend its principles to include other marginalized groups that due to systemic discrimination, from colonial or other forces, experience disparate health outcomes.

6 Exercising Group Data Rights – Practical Barriers and Solutions

The recognition of group data rights requires the development of a complementary framework through which they can be exercised. Early versions of IDS advocated for Indigenous data collection and storage (Carroll et al., 2020). This approach was designed for a time in which a significant portion of data collection was survey-based and is ill-suited for the current data landscape in which data collection is frequent, large-scale, and often hidden. The recent development of CARE Principles offers value-oriented guidelines for working with Indigenous data, yet it lacks a clear structure for operationalization.

The structure of DAOs offers a possible pathway for addressing the operational deficits of IDS. As detailed above, DAOs have been normalized as a pathway for disease groups to advocate for collective benefit by leveraging group data (Landy et al., 2012). Given the medical community's familiarity and acceptance of DAOs, it is convenient and appropriate to consider their structure for organizing group data rights for health data. Having expanded our definition of which groups are eligible to assert informational self-determination to include all groups who experience health inequities, we can envision a version of DAOs that brings together individuals not based on a disease but rather by a group affiliation that increases their likelihood of experiencing disparate health outcomes, such as race, ethnicity, socioeconomic status, or geographic location. I will refer to these new DAOs as Social DAOs.

Let us test this idea by revisiting the case of the commercial algorithm assessed by Obermeyer et al. (2019). In a scenario in which Social DAOs have been normalized, we could imagine that some of the Black patients in the dataset used to develop the algorithm would be registered with a Social DAO. Then, when the commercial entity developing the algorithm attempts to gain access to patient data, they would be required to notify the DAOs with which the patients in the sample are registered. This notification is not merely procedural – the company would be required to both gain the consent of the DAOs for data usage and involve them in the algorithm development process as would be expected if one was interested in using Indigenous data. As is the case with DAOs, the introduction of the Social DAO as an intermediary generates accountability, by requiring the data consuming entity to engage with the DAO, promotes patient agency since patients consent to have the DAO advocate on their behalf, and lowers the resource and technical barriers to protecting one's right to privacy and fair treatment. In the Obermeyer et al. case this would have ideally translated to a team of experts that represent the interest of the Social DAO members contributing to the development of the algorithm and assessing its performance prior to its release. Lastly, considering the increased emphasis on fairness in AI and ML, marked by a boom in research conferences on the topic (Castelvecchi, 2020), the establishment of ethical AI boards in industry (Vincent, 2019), and increased governmental investment (Central Digital and Data Office & Office of Artificial Intelligence, 2019), Social DAOs could also function more similarly to rare disease DAOs by offering a valuable data resource, i.e. health data of under-represented groups, to entities that develop algorithmic decision-making systems for healthcare settings.

7 Limitations

While group data rights, and their operationalization through Social DAOs, offer a compelling framework for promoting health equity by enabling groups to assert informational self-determination, the model suffers from several limitations.

A central appeal of DAOs as a mechanism for realizing group data rights is that they currently exist without legal grounding. But this advantage is also a practical flaw as it is unclear whether they can be considered a legal personality in front of a court when their rights are violated. This limitation extends to any mechanism through which groups attempt to claim rights that are not already protected by law.

Another limitation of group data rights is that they require group members to be aware of their group status to claim protections. However, Big Data methods are useful particularly for extracting valuable information about passive groups, like those defined by behaviors, where group members are likely to be unaware of their group membership (Kammourieh et al., 2017). Thus, while group data rights can help active, structured groups assert their informational self-determination and protect their own interests, the pursuit of such rights must be supplemented by additional protections that recognize and address the interests of passive groups (Kammourieh et al., 2017). The challenge of awareness extends beyond knowing one's group status to technical literacy more broadly. While patients have a fundamental understanding of disease as a debilitating state the requires additional protections, knowing that group membership, even of clearly defined groups such as race and ethnicity, contributes to poor and unfair health outcomes and that these outcomes are perpetuated through data requires an advanced technical literacy. Hence, any movement to establish group data rights, be it through Social DAOs or other mechanisms, requires extensive investment in patient education so that their potential for improving health equity can be fully realized.

Lastly, as noted by Carroll et al. (2020) in their discussion of the dangers of invoking IDS, adding barriers to accessing the data of marginalized groups threatens to further isolate these populations. If data is prohibitively difficult to access and use and those developing data-driven decision-making systems are not encouraged or required to include vulnerable populations, then such entities will likely not invest in engaging these groups.

8 Conclusion

Big Data- and algorithmic-based methods are increasingly used for decision-making in healthcare, and while they have brought significant advances to this field, they also pose a grave risk to the health of minority populations. In this chapter I demonstrated the necessity and utility of establishing group data rights as a mechanism for mitigating data-driven health disparities. Group data rights recognize both that group-level dynamics are the root of existing health inequities and that group-level inferences further compound these harms, highlighting the inadequacy of individual rights to counter these.

While others have discussed the merits of group data rights more broadly, it is particularly compelling to consider their implementation in healthcare where, by providing a strong ethical norm that clearly impacts practice, the fiduciary duty helps to provide a solid grounding for ethics-based discourse and action that is not yet entrenched, although is growing, in the Big Data and algorithm development space. With these considerations in mind, I proposed Social Disease Advocacy Organizations as a novel model for implementing group data rights in a healthcare setting, drawing on Indigenous Data Sovereignty for a moral argument for the right to informational self-determination of marginalized groups and building upon existing data pooling models of Disease Advocacy Organizations as an effective and normalized operational structure. Considering alternative equity-promoting data governance structures is not only necessary, but also timely given the growing momentum around fairness and accountability in the fields of artificial intelligence and machine learning that has been increasingly permeating the broader commercial and research landscapes.

References

Ada Lovelace Institute. (2021, March 4). *Data cooperatives*. https://www.adalovelaceinstitute.org/feature/data-cooperatives/

Angwin, J., Larson, J., Mattu, S., & Kirchner, L. (2016). Machine Bias. *ProPublica*. Retrieved February 23, 2021, from https://www.propublica.org/article/machine-bias-risk-assessments-in-criminal-sentencing?token=Kc9axO_GEcQwja43fVMKIbGDdlH9IO4Z

Assembly, U. G. (2007). United Nations declaration on the rights of indigenous peoples. *UN Wash, 12*, 1–18.

Banner, D., Bains, M., Carroll, S., Kandola, D. K., Rolfe, D. E., Wong, C., & Graham, I. D. (2019). Patient and public engagement in integrated knowledge translation research: Are we there yet? *Research involvement and engagement, 5*(1), 1–14.

Childress, J. F., & Beauchamp, T. L. (1994). *Principles of biomedical ethics* (pp. 197–199). Oxford University Press.

Benjamin, R. (2019). Race after technology: Abolitionist tools for the new jim code. *Social Forces, 98*.

Braveman, P., & Gruskin, S. (2003). Defining equity in health. *Journal of Epidemiology & Community Health, 57*(4), 254–258.

Carroll, S. R., Rodriguez-Lonebear, D., & Martinez, A. (2019). *Indigenous data governance: Strategies from United States Native Nations*.

Carroll, S. R., Garba, I., Figueroa-Rodríguez, O. L., Holbrook, J., Lovett, R., Materechera, S., et al. (2020). The CARE principles for indigenous data governance. *Data Science Journal, 19*(1).

Castelvecchi, D. (2020). Prestigious AI meeting takes steps to improve ethics of research. *Nature, 589*.

Central Digital and Data Office & Office of Artificial Intelligence. (2019, October 18). *A guide to using artificial intelligence in the public sector*. GOV.UK. https://www.gov.uk/government/collections/a-guide-to-using-artificial-intelligence-in-the-public-sector#assess,-plan-and-manage-artificial-intelligence

Cofone, I. N. (2018). Algorithmic discrimination is an information problem. *Hastings Law Journal, 70*, 1389.

Courbier, S., Dimond, R., & Bros-Facer, V. (2019). Share and protect our health data: An evidence based approach to rare disease patients' perspectives on data sharing and data protection-quantitative survey and recommendations. *Orphanet Journal of Rare Diseases, 14*(1), 1–15.

Epstein, S. (1996). *Impure science: AIDS, activism, and the politics of knowledge* (Vol. 7). University of California Press.

Eubanks, V. (2018). *Automating inequality: How high-tech tools profile, police, and punish the poor.* St. Martin's Press.

Farmer, P. E., Nizeye, B., Stulac, S., & Keshavjee, S. (2006). Structural violence and clinical medicine. *PLoS Medicine, 3*(10), e449.

Fiscella, K., Franks, P., Gold, M. R., & Clancy, C. M. (2000). Inequality in quality: Addressing socioeconomic, racial, and ethnic disparities in health care. *JAMA, 283*(19), 2579–2584.

Floridi, L. (2014). Open data, data protection, and group privacy. *Philosophy & Technology, 27*(1), 1–3.

Floridi, L. (2016). On human dignity as a foundation for the right to privacy. *Philosophy & Technology, 29*(4), 307–312.

Gamble, V. N. (1997). Under the shadow of Tuskegee: African Americans and health care. *American Journal of Public Health, 87*(11), 1773–1778.

Huml, R. A., Dawson, J., Bailey, M., Nakas, N., Williams, J., Kolochavina, M., & Huml, J. R. (2020). Accelerating rare disease drug development: Lessons learned from muscular dystrophy patient advocacy groups. *Therapeutic Innovation & Regulatory Science*, 1–8.

Katell, M., Young, M., Dailey, D., Herman, B., Guetler, V., Tam, A., Bintz, C., Raz, D., & Krafft, P. (2020). Toward situated interventions for algorithmic equity: Lessons from the field. 45–55.

Kammourieh, L., Baar, T., Berens, J., Letouzé, E., Manske, J., Palmer, J., et al. (2017). Group privacy in the age of big data. In *Group privacy* (pp. 37–66). Springer.

Krieger, N. (2014). Discrimination and health inequities. *International Journal of Health Services, 44*(4), 643–710.

Landy, D. C., Brinich, M. A., Colten, M. E., Horn, E. J., Terry, S. F., & Sharp, R. R. (2012). How disease advocacy organizations participate in clinical research: A survey of genetic organizations. *Genetics in Medicine, 14*(2), 223–228.

Legal Information Institute. (n.d.). *Class action.* LII / legal information institute. Retrieved 23 March 2020, from https://www.law.cornell.edu/wex/class_action

Merkel, P. A., Manion, M., Gopal-Srivastava, R., Groft, S., Jinnah, H. A., Robertson, D., & Krischer, J. P. (2016). The partnership of patient advocacy groups and clinical investigators in the rare diseases clinical research network. *Orphanet Journal of Rare Diseases, 11*(1), 1–10.

Mittelstadt, B. (2019). Principles alone cannot guarantee ethical AI. *Nature Machine Intelligence, 1*(11), 501–507.

Novorol, C. (2020). *The challenges of combating rare diseases—And five innovations making a real difference.* Forbes. Retrieved February 24, 2021, from https://www.forbes.com/sites/clairenovorol/2020/02/28/the-challenges-of-combating-rare-diseasesand-five-innovations-making-a-real-difference/

Obermeyer, Z., Powers, B., Vogeli, C., & Mullainathan, S. (2019). Dissecting racial bias in an algorithm used to manage the health of populations. *Science, 366*(6464), 447–453.

Paulus, J. K., & Kent, D. M. (2020). Predictably unequal: Understanding and addressing concerns that algorithmic clinical prediction may increase health disparities. *Npj Digital Medicine, 3*(1), 1–8. https://doi.org/10.1038/s41746-020-0304-9

Rajkomar, A., Hardt, M., Howell, M. D., Corrado, G., & Chin, M. H. (2018). Ensuring fairness in machine learning to advance health equity. *Annals of Internal Medicine, 169*(12), 866–872. https://doi.org/10.7326/M18-1990

Smith, A. (2009). Indigenous peoples and boarding schools: A comparative study. In *Paper secretariat of the United Nations permanent forum on indigenous issues* (pp. 18–29).

Smith, S. K., Selig, W., Harker, M., Roberts, J. N., Hesterlee, S., Leventhal, D., et al. (2015). Patient engagement practices in clinical research among patient groups, industry, and academia in the United States: A survey. *PLoS One, 10*(10), e0140232.

Terry, S. F., Terry, P. F., Rauen, K. A., Uitto, J., & Bercovitch, L. G. (2007). Advocacy groups as research organizations: The PXE international example. *Nature Reviews Genetics, 8*(2), 157–164.

Taylor, L. (2017). Safety in numbers? Group privacy and big data analytics in the developing world. In *Group privacy* (pp. 13–36). Springer.

Taylor, L., Floridi, L., & Van der Sloot, B. (Eds.). (2017). *Group privacy: New challenges of data technologies* (Vol. 126). Springer.

The First Nations Information Governance Centre. (2022, July 25). Retrieved May 1, 2021, from https://fnigc.ca/

Tzanou, M. (2013). Data protection as a fundamental right next to privacy?'Reconstructing'a not so new right. *International Data Privacy Law, 3*(2), 88–99.

Vincent, J. (2019, April 3). *The problem with AI ethics*. The Verge. https://www.theverge.com/2019/4/3/18293410/ai-artificial-intelligence-ethics-boards-charters-problem-big-tech

Wiggins, A., & Wilbanks, J. (2019). The rise of citizen science in health and biomedical research. *The American Journal of Bioethics, 19*(8), 3–14.

Williams, D. R., & Wyatt, R. (2015). Racial bias in health care and health: challenges and opportunities. *Jama, 314*(6), 555–556.

Zhang, X., Pérez-Stable, E. J., Bourne, P. E., Peprah, E., Duru, O. K., Breen, N., Berrigan, D., Wood, F., Jackson, J. S., Wong, D. W. S., & Denny, J. (2017). Big data science: Opportunities and challenges to address minority health and health disparities in the 21st century. *Ethnicity & Disease, 27*(2), 95–106. https://doi.org/10.18865/ed.27.2.95

Ethical Principles for Artificial Intelligence in National Defence

Mariarosaria Taddeo, David McNeish, Alexander Blanchard, and Elizabeth Edgar

Abstract Defence agencies across the globe identify artificial intelligence (AI) as a key technology to maintain an edge over adversaries. As a result, efforts to develop or acquire AI capabilities for defence are growing on a global scale. Unfortunately, they remain unmatched by efforts to define ethical frameworks to guide the use of AI in the defence domain. This chapter provides one such framework. It identifies five principles -- justified and overridable uses; just and transparent systems and processes; human moral responsibility; meaningful human control; reliable AI systems – and related recommendations to foster ethically sound uses of AI for national defence purposes.

Keywords Artificial intelligence · Control · Defence · Digital ethics · Ethical principles · Fairness · Just war theory · Responsibility · Reliability

1 Introduction

Maintaining a technological advantage has always been pivotal to the success of national defence measures. It is even more so in mature information societies (Floridi, 2016a). This is why over the past two decades there have been growing efforts to design, develop, and deploy digital technologies for national defence. Artificial intelligence (AI), in particular, has shown to have great potential to aid

M. Taddeo (✉)
Oxford Internet Institute, University of Oxford, Oxford, UK

Alan Turing Institute, London, UK
e-mail: mariarosaria.taddeo@oii.ox.ac.uk

D. McNeish · E. Edgar
Defence Science Technology Laboratory (Dstl), Salisbury, UK

A. Blanchard
Alan Turing Institute, London, UK

J. Mökander, M. Ziosi (eds.), *The 2021 Yearbook of the Digital Ethics Lab*,
Digital Ethics Lab Yearbook, https://doi.org/10.1007/978-3-031-09846-8_16

national defence measures. Indeed, scholars, policy-makers and military experts observe that there is an on-going global race for the development of AI for defence (Taddeo & Floridi, 2018). For example, the latest national defence and innovation strategies of several governments ↓ UK,[1] US,[2] Chinese,[3] Singapore,[4] Japanese,[5] and Australian[6] ↓ explicitly mention AI capabilities, which are already deployed to improve the security of critical national infrastructures, like transport, hospitals, energy and water supply. NATO, as well, has identified AI as a key technology to maintain superiority over adversaries in its 2020 report on the future of the alliance (NATO, 2020).

The applications of AI in national defence are virtually unlimited, ranging from support to logistics and transportation systems to target recognition, combat simulation, training, and threat monitoring. As with the use in other domains, the potential of AI is coupled with serious ethical problems, ranging from possible conflict escalation, the promotion of mass surveillance measures, the spreading of misinformation, to breaches of individual rights and violation of dignity. If these problems are left undressed, the use of AI for defence purposes risks undermining the fundamental values of democratic societies and international stability (Taddeo, 2014b, 2019a, b).

This chapter offers guidance to address these problems by identifying ethical principles to inform the design, development, and use of AI for defence purposes. These principles should not be taken as an alternative to national and international laws; rather they offer guidance to the use of AI in the defence domain in ways that are coherent with existing regulations. In this sense, the proposed principles indicate what ought to be done or not to be done

> *over and above* the existing regulation, not against it, or despite its scope, or to change it, or to by-pass it (e.g. in terms of self-regulation) (Floridi, 2018, 4).

In offering these principles, the paper fills an important gap in the relevant academic and policy literature; as while numerous ethical principles and frameworks have been published which focus on civilian applications of AI (Jobin et al., 2019), very few address directly the problems inherent in the defence domain. In the rest of this chapter, Sect. 2 describes the methodology used for our analysis. Section 3 offers an analysis of the ethical problems linked to current uses of this technology in the defence domain. Section 4 focuses on the ethical principles provided by the US Defence Innovation Board (DIB). Thus far, these are the only domain-specific principles published by a defence institution. The analysis of these principles shows some key limitations of the DIB approach and paves the way to Sect. 5, which

---

[1] https://www.gov.uk/government/publications/future-force-concept-jcn-117

[2] https://media.defense.gov/2019/Feb/12/2002088963/-1/-1/1/SUMMARY-OF-DOD-AI-STRATEGY.PDF

[3] Roberts et al. (2020).

[4] https://www.csa.gov.sg/~/media/csa/documents/publications/singaporecybersecuritystrategy.pdf

[5] https://www.nisc.go.jp/eng/pdf/cs-senryaku2018-en.pdf

[6] https://www.business.gov.au/news/budget-2019-20

introduces five new ethical principles to guide the use of AI for national defence. Section 6 concludes the chapter.

2   Methodology

The first steps to identifying viable ethical principles to guide the use of AI for national defence are the definition of AI and the identification of the ethical problems that its use may pose. For the purposes of our analysis, we can abstract from specific technical aspects of AI systems (we can disregard, for example, whether the system under analysis is a statistical or a subsymbolic one) and adopt a high level of abstraction (LoA) (Floridi, 2008). Thus, we consider AI as

> a growing resource of interactive, autonomous, and self-learning agency, which can be used to perform tasks that would otherwise require human intelligence to be executed successfully (Floridi & Cowls, 2019).

The choice of the method to identify the ethical problems posed by the use of AI for national defence is not a trivial one. For example, one may think of developing a complete taxonomy of the ethical issues posed by existing uses of AI in the defence domain; but this is unfeasible and of little value: the taxonomy would be quickly outdated by the rapid developments in AI and its application to new uses. At the same time, analyses that disregard the specific domain and purpose of deployment risk defining ethical principles which are too generic to provide any concrete guidance.

Hence the choice of LoA becomes crucial to develop a correct analysis of the ethical problems and define principles able to provide actionable guidance. Given the goal of this chapter, we chose a gradient of analysis (GoA) that combines two LoAs: $LoA_{purpose}$ and LoA_{ethics}. The observables of $LoA_{purpose}$ are the purposes of deployment of AI. The observables of LoA_{ethics} are, for any given purpose, the aspects of the design, development and deployment of AI that may lead to un/ethical consequences.

The decision to focus on purposes of use rather than on the function of the technology requires clarification. It rests on two reasons. First, the dual-use nature of AI – as with any digital technology, AI is *malleable* and its original function can be easily repurposed. Hence, un/ethical implications of its uses are not necessarily defined by their design function as much as they are determined by the purpose with which these technologies are deployed. Second, within the defence domain these purposes can be clearly identified and are likely to shape both current and future uses of AI, and thus ethical principles that focus on purposes of use, rather than on the specific function, of a given technology have a better-defined scope and their guidance is more likely to stand the test of time.

The LoAs embraced for this analysis have a medium granularity. Thus, they identify problems (and inform the definition of principles in Sect. 5) that are specific to the domain but do not distinguish among specific contexts (e.g. naval or aviation)

Fig. 1 The three purposes of use of AI for national defence

of AI deployment within the defence domain and hence disregard the variation of ethical challenges that may occur between contexts. Consider, for example, the different ethical problems and related solutions for using AI in submarines or aviation operations.

Purposes of use of AI in defence span over three core categories of action by defence institutions (Fig. 1): sustainment and support, adversarial and non-kinetic, adversarial and kinetic. We shall delve into the ethical implications of each these in Sect. 3, but let us describe them briefly here. Sustainment and support uses of AI refer to all cases in which AI is deployed to support 'back-office' functions, as well as logistical distribution of resources. This category also includes uses of AI to improve the security of infrastructure and communication systems underpinning national defence. Adversarial and non-kinetic uses of AI range from uses of AI to counter cyber-attacks to active cyber defence, and offensive cyber operations with non-kinetic aims. Adversarial and kinetic uses refer to the integration of AI systems in combat operations, these range from the use of AI systems to aid the identification of targets to lethal autonomous weapons systems (LAWS).

The ethical principles for the use of AI in the defence domain that we provide in this chapter refer to sustainment and support uses and to adversarial and non-kinetic uses of AI. The ethical analysis of the use of AI for adversarial and kinetic purposes will be addressed in the second stage of our research.

3   Ethical Challenges of AI for Defence Purposes

Figure 2 below shows the minimum requirements, for each purpose of use of AI in defence that AI systems should meet to be ethically sound.

Fig. 2 A map of the ethical requirements linked to the specific purpose of the use of AI in defence

The three purposes of use of AI in the defence domain are more ethically problematic as one moves from sustainment and support uses to adversarial and kinetic uses. This is because alongside the ethical problems related to the use of AI (e.g. transparency and fairness) one also needs to consider the ethical problems related to adversarial, whether non-kinetic or kinetic, uses of this technology and its disruptive and destructive impact.

As shown in Fig. 2, each category of use has its own specific ethical requirements, but also inherits the ones from the categories on its left. For example, to be ethically sound, adversarial and non-kinetic uses of AI need to ensure some forms of meaningful control and measures to avoid escalation, while also respecting transparency and autonomy, which appear in the sustainment and support category.

Let us now consider in more details some of the key ethical problems of each purpose of use.

3.1   Sustainment and Support Uses of AI

Defence organisations already employ AI systems for different non-adversarial aspects of operations (US Army, 2017). Uses vary from applications in cybersecurity, where AI plays an ever-growing role to ensure systems robustness and resilience, to AI-based drones capturing video reconnaissance (Lysaght et al., 1988; Fraga-Lamas et al., 2016; Schubert et al., 2018).

For nations with technically advanced militaries, AI systems are likely to be fully integrated into national defence capabilities to support back-office, logistics and security tasks. For example, research estimates that the number of intelligent sensors in a military setting could reach one million per square kilometre similar to the supported connection density of the 5G network (Kott et al., 2017; International Telecommunications Union, 2017). This has been described as *the internet of battle things* (Kott et al., 2017). In these cases, AI will play a key role to ensure the robustness and resilience of the networks as well as to elaborate data and extract relevant information. All these uses pose serious ethical problems.

First, consider the use of AI to enhance system robustness. This refers to AI for software testing, which is a new area of research and development. It is defined as an

> emerging field aimed at the development of AI systems to test software, methods to test AI systems, and ultimately designing software that is capable of self-testing and self-healing.[7]

In this sense, AI can take software testing to a new level, making systems more robust (King et al., 2019). However, delegating testing to AI could lead to a complete deskilling of defence personnel deployed for verification and validation of systems and networks and a subsequent lack of control of this technology (Yang et al., 2018; Taddeo 2019a, b).

Next, let us focus on system resilience. AI is increasingly deployed for threat and anomaly detection (TAD). TAD can make use of existing security data to train for pattern recognition. As stressed by Taddeo et al. (2019) in some cases, threat scanners have access to files, emails, mobile and endpoint devices, or even traffic data on a network. Monitoring extends to users as well. AI can be used to authenticate users by monitoring behaviour and generating biometric profiles, like for example, the unique way in which a user moves her mouse around ('BehavioSec: Continuous Authentication Through Behavioral Biometrics', 2019). In this case, the risk is clear. This use of AI puts users' privacy under a sharp devaluative pressure, exposing users to extra risks should data confidentiality be breached, and creating a mass-surveillance effect (Taddeo, 2013, 2014b).

AI can extract information to support logistics and decision-making, but also for foresight analyses, internal governance and policy. These are perhaps some of the uses of AI with the greater potential to improve defence operations, as they will facilitate timely and effective management of both human and physical resources, improve risk assessment, and support decision-making processes. For example, a recent report by KPMG[8] stresses that a defence agency could have only a few minutes to decide whether a missile launch represents a threat, share the findings with allies, and decide how to respond. AI would be of great help in this scenario, for it could integrate real-time data from satellites and sensors and elaborate key information that may contribute to the decision-making process. The challenge is that these uses of AI must ensure that the systems would not perpetrate a biased decision and unduly discriminate, whilst also offering a means to maintain accountability, control and transparency.

---

[7] www.aitesting.org

[8] https://assets.kpmg/content/dam/kpmg/xx/pdf/2018/04/next-major-defense-challenge.pdf

3.2 Adversarial and Non-kinetic Uses of AI

As cyber threats escalate, so does the need for defence strategies required to meet them. The UK and the US have employed 'active' cyber defence strategies that enable computer experts to neutralise or distract viruses with decoy targets, and to break back into a hacker's system to delete data or to destroy it completely. In February 2020, the UK also established the National Cyber Force, as a joint initiative between the Ministry of Defence and GCHQ, which is tasked to target hostile foreign actors. On an international scale, NATO can now rely on sovereign cyber effects in response to cyber-attacks, as agreed at the Brussels Summit.[9] This may enable the alliance to punish (attributed) attacks and deter attackers from striking again in the future (Taddeo, 2019a).

AI will revolutionize these activities. Attacks and responses will become faster, more precise, and more disruptive. It will also expand the targeting ability of attackers, enabling them to use more complex and richer data. Enhancing current methods of attack is an obvious extension of existing technology, however, using AI within malware can change the nature and delivery of an attack. Autonomous and semi-autonomous cybersecurity systems endowed with a "playbook" of pre-determined responses to an activity, constraining the agent to known actions are already available on the market ('DarkLight Offers First of Its Kind Artificial Intelligence to Enhance Cybersecurity Defenses', 2017). Autonomous systems able to learn adversarial behaviour and generate decoys and honeypots ('Acalvio Autonomous Deception', 2019) are also being commercialised. Additionally, AI-enabled cyber weapons have already been prototyped including autonomous malware, corrupting medical imagery, and attacking autonomous vehicles (Mirsky et al., 2019; Zhuge et al., 2007). For example, IBM created a prototype autonomous malware, DeepLocker, that uses a neural network to select its targets and disguise itself until it reaches its destination ('DeepLocker: How AI Can Power a Stealthy New Breed of Malware', 2018).

As states use increasingly aggressive AI-driven strategies, opponents may respond more fiercely (Taddeo & Floridi, 2018). This may expand into an intensification of cyber-attacks and responses, which, in turn, may pose serious risks of escalation and lead to kinetic consequences (Taddeo, 2017a, b). To avoid the escalation, it is vital that uses of AI respect key principles of Just War Theory which underpins international regulations (Taddeo 2012a, b, 2014a), such as the United Nations Charter,[10] The Hague and Geneva Conventions,[11] and International Humanitarian Law,[12] and sets the parameters for both ethical and political debates on waging conflicts. It is crucial that the deployment of AI for aggressive and non-kinetic purposes respects the principles of proportionality of responses, discrimi-

---

[9] https://www.nato.int/docu/review/articles/2019/02/12/natos-role-in-cyberspace/index.html

[10] https://www.un.org/en/sections/un-charter/un-charter-full-text/

[11] https://www.loc.gov/rr/frd/Military_Law/pdf/ASubjScd-27-1_1975.pdf

[12] https://www.icrc.org/en/doc/resources/documents/misc/57jm93.htm

nates between legitimate and illegitimate targets, ensures some form of redress when mistakes are made (Taddeo 2012a, b, 2014a), and maintains responsibility and control within the chain of command. Ultimately, ethical analyses of the adversarial and non-kinetic use of AI should contribute to understanding how to apply Just War Theory in cyberspace and be used to shape the debate on the regulation of state behaviour in cyberspace (Taddeo & Floridi, 2018).

3.3 Adversarial and Kinetic Uses of AI

The use of AI for aggressive and kinetic purposes varies, ranging from automating various functions of a weapon system, to systems that follow the pre-programmed instructions of a human, to full autonomy, where the weapons system identifies, selects, and engages targets without any human input. Consider, for example, STARTLE[13] a system developed for the Royal Navy to support human decision making. It is endowed with a situational awareness software that monitors and assesses potential threats using a combination of AI techniques. Similarly, the Advanced Targeting & Lethality Automated System (ATLAS)[14] developed for the US Army support humans in identifying threats and prioritize potential targets. Ethical problems vary with the degree of autonomy of weapons systems, the level of force that they can deploy, and the nature of the possible targets, whether material or humans.

Whilst many countries have expressed their commitment to not develop or use fully autonomous weapon systems, it is still important to consider and address the ethical problems they pose in order to establish boundaries for ethical debate considering the development and use of weapons which incorporate AI but are not fully autonomous in their operation or may not target human agents.

A key challenge is to ensure that adversarial and kinetic uses of AI will be able to respect the tenets of Just War Theory, for example necessity, proportionality, and discrimination. So, for example, AI systems must be able to distinguish between combatants and non-combatants carrying a weapon or recognising the generally-accepted signs of surrender that operate in armed conflict. This may be problematic, because AI, at least in its current state of development, is insufficiently able to analyse context, in some situations its capacity to recognise who is and who is not a legitimate target could be significantly worse than that of humans (Sharkey, 2010, 2012a, b; Tamburrini, 2016).

The responsibility gap is another key ethical challenge. As mentioned in Sect. 3, whilst a responsibility gap is problematic in all the three categories of use of AI, it is particularly worrying when considering the adversarial and kinetic case, given the high stakes involved (Sparrow, 2007). This gap becomes even more pressing when

---

[13] https://www.roke.co.uk/products/startle

[14] https://breakingdefense.com/2019/03/atlas-killer-robot-no-virtual-crewman-yes/

coupled with the respect of the opponent and of her dignity. Treating opponents with respect in warfare is an important way of maintaining warfare's morality, the interpersonal relation with the opponent is considered to be key to this end. Insofar as the use of autonomous LAWS would sever this relation, they undermine the dignity of those whom they target and lead to a form of morally problematic killing (Asaro, 2012; Docherty, 2014; Sharkey, 2019; Johnson & Axinn, 2013; Sparrow, 2016; O'Connell, 2014; Ekelhof, 2019).

Finally, questions arise with respect to the impact of LAWS on international stability. On the one side, LAWS may reduce the time span of the hostilities in which states may engage and thus contribute to fostering stability. They could also be an effective deterrent against possible opponents. On the other side, LAWS may lead to unjust war and hamper international instability. This is because the use of LAWS may lower the barriers to warfare (Enemark, 2011; Brunstetter & Braun, 2013) possibly increasing the number of wars. For instance, it may be the case that the widespread use of LAWS would allow decision-makers to wage wars without the need to overcome the potential objections of military personnel (McMahan, 2013). In the same vein, asymmetric warfare that would result from one side using LAWS may lead to the weaker side resorting to insurgency and terrorist tactics more often (Sharkey, 2012a, b). Because terrorism is considered to be a form of unjust warfare (or, worse, an act of indiscriminate murder), deploying LAWS may lead to a greater incidence of immoral violence.

4 Ethical Guidelines for the Use of AI

Over the past few years several frameworks for the ethical design, development, and use of AI have been proposed (Floridi & Cowls, 2019). For example, Jobin et al. (2019) identified 84 ethical frameworks for AI in their review. Ethical guidelines can vary in a number of dimensions, e.g. by the agency putting them forward (from governments to non-governmental organisations); by the scope of their application (e.g. from guidelines for private sector, e.g. social media; to guidelines for all entities developing and using AI, e.g. the European Guidelines for Trustworthy Artificial Intelligence); and by the applications they are seeking to govern (e.g. from national defence applications to applications in the wider public sector).

Despite the wide scope covered by existing frameworks, uses of AI in the defence domain has received very little attention. Thus far, the principles defined by the US Defence Innovation Board (DIB) (2020a) for the use of AI in defence are the only exception to this lack of focus. In this section, we analyse the DIB principles to identify both their points of strength and limitations. The goal is to extract valuable lessons to learn before moving to the describe the principles that we propose in this chapter.

The DIB identifies five principles – responsible, equitable, traceable, reliable, and governable AI – they are meant to apply both to kinetic and non-kinetic uses of AI, whether adversarial or not. The following subsections analyse each principle in turn focusing on both the principles described in (DIB, 2020a) and the wider supporting report (DIB, 2020b) where the DIB describes the rational for each principle and specific recommendations for their implementation.

4.1   Responsible

The DIB principle states that

> Human beings should exercise appropriate levels of judgment and remain responsible for the development, deployment, use, and outcomes of DoD AI systems (DIB, 2020a, 8).

This principle is uncontroversial and coherent with other ethical frameworks (Department for Digital, Culture, Media, & Sport, 2018, 5; Gavaghan et al., 2019, 41; Japanese Society for Artificial Intelligence [JSAI], 2017, 3).

In the supporting document, the recommendation on how to implement this principle proposes a three-level system of responsibilities, with the first level addressing humans who control

> the design, requirements definition, development, acquisition, testing, evaluation, and training for any DoD system, including AI ones (DIB, 2020b, 27).

The second level addresses the use of AI in the conduct of hostilities (whether kinetic or not), in this case responsibilities are ascribed according to the command and control structure, insofar as commanders and operators have "appropriate information on a system's behavior, relevant training, and intelligence and situational awareness" (p. 28). The third level of responsibility refers to redressing mechanisms for actions after hostilities have ended. This level addresses both the DoD and private sector procuring AI technology. The DIB supporting documents specify that human responsibility rests on 'human appropriate judgment'.

This approach is correct only in part. There are two main limitations to it. On the one side, the definition of 'appropriate' judgment remains vague and, therefore, problematic especially when considering the problems posed by the lack of transparency and predictability of some AI systems. On the other side, the attribution of responsibility according to the three-level system risks dumping responsibilities on the first level, insofar as unintended consequences of AI systems can, in majority of the cases, be linked back to design and development issues. This may have a detrimental effect on the way actors involved in command and control may perceive their responsibilities with respect to the use of AI.

4.2   Equitable

The DIB principle prescribes that

> The DoD should take deliberate steps to avoid unintended bias in the development and deployment of combat or non-combat AI systems that would inadvertently cause harm to persons (p. 8).

This principle focuses on issues related to fairness and justice, however it avoids referring to the two concepts directly. In the supporting document, the reason given for not using the term fairness in the principle is the following:

> this principle stems from the DoD mantra that fights should not be fair, as DoD aims to create the conditions to maintain an unfair advantage over any potential adversaries, thereby increasing the likelihood of deterring conflict from the outset (DIB, 2020b, 31).

The document goes on to say that the

> DoD should have AI systems that are appropriately biased to target certain adversarial combatants more successfully and minimize any pernicious impact on civilians, non-combatants, or other individuals who should not be targeted (Defence Innovation Board [DIB] (2020b, 33).

This departure, then, is motivated by the perceived unique nature of defence AI applications. When considering fairness with respect to AI, the DIB principles centre only on the unfair impact of the use of AI on the DoD personnel, disregarding the problems that the lack of fairness may pose when deploying AI for surveillance purposes or on the cyber and kinetic battleground. This is misleading, as it may suggest that the need to seek advantage over the adversary may justify unfair, or indeed unjust, practices. This is not the case, as we distinguish between just and unjust conduct in defence and punish the latter.

There are differences between the ways in which the principle of justice is applied in civilian and non-belligerent contexts and in hostile activities. Just War Theory and International Humanitarian Law define the terms of this principle and how to respect it when conducting hostilities. These terms differ, at times radically, from the ones referring to civilian uses, but still define the space of just conduct – and hence fairness -- in defence. Ethical guidelines for AI in defence need to define principles for just uses of AI which are relevant within this domain and coherent with the principles provided by Just War Theory (Taddeo, 2014a).

4.3   Traceability

This principle addresses *indirectly* the ethical problems posed by the lack of transparency of AI. It states that

> DoD's AI engineering discipline should be sufficiently advanced such that technical experts possess an appropriate understanding of the technology, development processes, and operational methods of its AI systems, including transparent and auditable methodologies, data sources, and design procedure and documentation (DIB, 2020a, 8).

Notably, the focus of the principle is not on the transparency of the technology but on the skills of the DoD personnel and their level of understanding of AI systems, insofar as these facilitate traceability of the processes and decisions of AI systems both at development and deployment stages. As specified in the supporting document, traceability at development stage refers to the collection and sharing with appropriate stakeholders of "design methodology, relevant design documents, and data sources" (p. 34). Whereas at deployment stage, traceability includes forms of monitoring, auditing, and transparency of processes. As specified in the DIB supporting document:

> Some systems may require not just reviews of user access, but also records of use and for what purpose. This requirement can mitigate harms related to off-label use of an AI system, as well as reinforce the principle of responsibility. In short, DoD will need to rethink how it traces its AI systems, who has access to particular datasets and models, and whether those individuals are reusing them for other application areas (p. 35).

While analysis provided in the supporting document links correctly the transparency of processes to responsible uses of AI; it overlooks the relation between transparency of AI, and human responsibilities. In the event of mistakes, malfunctioning, or unintended consequences following from the use of AI, traceability of processes and decisions may compensate for the lack of transparency of this technology. Albeit useful, the approach adopted with this principle offers a remedy, not a solution to the challenges posed by the lack of transparency of AI. The DIB documents do not stress so and do not propose any suggestions to overcome the lack of transparency of AI system. This is problematic, as traceability without transparency is very limited. While it may foster responsible uses of AI, it does not shed much light on the responsibilities for mistakes and failures of this technology, nor does it offer an opportunity to identify promptly the sources of mistakes and unwanted outcomes of AI systems.

4.4   Reliable and Governable

The principle focusing on reliability of AI states that

> DoD AI systems should have an explicit, well-defined domain of use, and the safety, security, and robustness of such systems should be tested and assured across their entire life cycle within that domain of use (p. 8).

This principle resonates with one of the principles provided by Organisation for Economic Co-operation and Development (OECD), which stresses that

> AI systems must function in a robust, secure and safe way throughout their life cycles and potential risks should be continually assessed and managed.[15]

In the supporting document, the DIB (2020a, b) stresses the need for reliable AI (rather than trustworthy), whose "safety, security, and robustness [...] should be tested and assured" (DIB, 2020a, p. 8). This principle is specifically oriented at fostering verification and validation as well as to improve AI robustness. We believe

---

[15] https://www.oecd.org/going-digital/ai/principles/

that this is a crucial requirement for the use of AI in defence and one that is important to mention explicitly to reiterate the need to monitor AI systems, especially when these are deployed within a defence organisation (more on this in Sect. 5).

At the same time, the DIB supporting documents highlights the importance of human agents being able to disengage or deactivate systems that demonstrate unintended escalatory behaviour. The supporting document emphasises the need for human control given the unpredictable behaviour of some AI systems, especially those operating in complex and dynamic environments (DIB, 2020b, 39).[16]

Control is not mentioned explicitly in the principles, but it is central to the principle focusing on governable AI, which prescribes that

> DoD AI systems should be designed and engineered to fulfill their intended function while possessing the ability to detect and avoid unintended harm or disruption, and for human or automated disengagement or deactivation of deployed systems that demonstrate unintended escalatory or other behavior.

While pointing at the correct direction, insofar as it specifies the need to maintain AI under some forms of control, the principle remains vague with respect to what the desirable forms of control should be, and how this should be exerted, and what the minimum level of ethically acceptable control is.

These are important aspects to consider, especially as defence institutions increasingly deploy AI in hybrid teams, including human and artificial agents. In this scenario, a *governable* AI offers too generic a guidance to identify, for example, ethically sound forms of control over AI systems or how to attribute responsibilities for failures with respect to mis-uses or over-uses of this technology.

In this respect, a notable omission in the DIB principles is the lack of focus on human autonomy. While this is acceptable insofar as autonomy may be considered a principle attaining to personal sphere and the ability of individuals to pursue their own choices; this is also a missed opportunity. Autonomy protects individuals' ability to dissent from AI-based decisions. In this sense, as AI is increasingly embedded in the decision-making processes of defence organisations, it is important to protect the ability of human agents to contest and override AI decisions, when these should be considered mistaken or inappropriate. In this sense, the principle of autonomy enables stronger forms of control over the use of AI.

5 Five Ethical Principles for Sustainment and Support and Adversarial and Non-kinetic Uses of AI

In this section we offer five ethical principles specifically designed to address the ethical problems posed by the deployment of AI in the defence domain. The principles specified in this chapter refer to both sustainment and support uses and

---

[16] It should be noted that the High-Level Expert Group's principles also include provisions for human control, but given its focus on trustworthy AI, these are more flexible. For example, it allows that less human oversight may be exercised so long as more extensive testing and stricter governance is in place.

adversarial and non-kinetic uses of AI. They should be regarded as the first building block of a more comprehensive ethical framework addressing also the adversarial and kinetic uses of AI, which will be the focus of the second, forthcoming, part of this project.

In order to be ethically sound, sustainment and support and adversarial and non-kinetic uses of AI for national defence purposes should respect the following ethical principles:

(i) Justified and overridable uses
(ii) Just and transparent systems and processes
(iii) Human moral responsibility
(iv) Meaningful human control
(v) Reliable AI systems

5.1 Justified and Overridable Uses

The (non) adoption of AI needs to be justified to ensure that AI solutions are not being underused, thus creating opportunity costs; or overused and misused, thus creating risks. Similarly, the decision to (or not to) resort to AI should always be overridable, should it become clear that it leads to unwanted consequences.

Even when designed and deployed according to ethical principles, AI remains an ethically challenging technology. Its use may lead to great advantages for national defence. Yet, AI is not a silver bullet. This is a lesson that should be learned from the ethical governance of AI for social good. As Floridi and colleagues (Floridi et al., 2020, p. 1773) stress:

> it is important to acknowledge at the outset that there are myriad circumstances in which AI will not be the most effective way to address a particular social problem. This could be due to the existence of alternative approaches that are more efficacious or because of the unacceptable risks that the deployment of AI would introduce.

At the same time, AI can also encroach upon human rights, International Humanitarian Law or pose risks to international stability (the reader will recall the risks of the snowball effect linked to the adversarial and non-kinetic use of AI). This is why the decision to (or not to) delegate tasks to AI systems should follow a careful analysis of the ethical risks and benefits in any given context of deployment to justify it.

This principle yields different recommendations when considering sustainment and support and adversarial and non-kinetic uses. In the first case, the principle calls for an assessment of the ethical risks against the expected benefits following from the deployment of AI systems. For example, weighting the benefits of using an AI system that may speed-up a decision-making process or optimise logistic and distribution of resources against the likelihood that it may have a negative impact on jobs and human expertise; or considering the impact on human autonomy when AI is integrated in human teams (human-machine teaming).

When deciding on deploying AI for adversarial and non-kinetic purposes, for example for offensive cyber operations, it is essential to ensure that AI systems will respect the principles of necessity, humanity, distinction, and proportionality ('The UK and International Humanitarian Law 2018', n.d.). This may prove to be a complex task, as the principles of International Humanitarian Law, and the underpinning principles of Just War Theory, are geared toward kinetic forms of conflicts and therefore their implementation to the case of non-kinetic warfare may be problematic. Consider for example, proportionality and the problems of assessing the expected damage to intangible entities (e.g. data or services) against the concrete military aim to be achieved (Taddeo 2012a, b, 2014a). Satisfying this principle will require extending the scope of the fundamental tenets of Just War Theory from kinetic to non-kinetic operation. A complex but necessary, and not impossible, task.

Given the learning capability of AI and the potential lack of predictability of its outcome, even when uses of AI are justified, a constant monitoring of the ethical soundness of the solutions that they provide should be in place. Similarly, procedures to override the decision to resort to AI in a timely and effective way should be established every time an AI system is deployed.

5.2   Just and Transparent Systems and Processes

AI systems should not perpetrate any undue discrimination, nor should they lead to any breach of the principles of Just War Theory. To this end, AI defence institutions should ensure that the deployed AI systems, and the processes in which they are embedded, remain transparent (and explicable) to facilitate the identification of the origin of any breach of the principles of Just War Theory, of unintended and mistaken outcomes, the attribution of responsibilities, and guarantee the possibility of scrutinising and challenging processes and outcomes to ensure that they remain ethically sound.

Three aspects are crucial:

- establish processes for ethical auditing;
- ensure that developed and procured AI systems are deployed in ways that respect the principles of Just War Theory;
- maintain traceability for the design, development or procurement, and deployment of AI systems.

Ethical auditing should involve the entire decision-making process, and so it should focus on both human and artificial agents, to ensure that both agents respect the relevant ethical principles (Mökander & Floridi, 2021).

Transparency of AI systems and processes enables access to the relevant information. The former requires explainability, while the latter traceability. Transparency of AI follows from the effort of designing and developing explainable technologies. Thus, it is crucial that in-house and procured AI systems are designed and developed with explainability in mind. Defence agencies should consider participating actively in the 'design-develop-deploy' cycle of the AI technologies that they procure and contribute to the development phase by setting standards and offering a trusted space where these technologies could be beta-tested. To facilitate this process, procurement policies should account for an ethical scrutiny of the third parties involved.

AI systems are often designed and developed in a distributed way, models, data, training and implementation may be managed by different actors. At the same time, AI learns by experience: past deployments can impact future outcomes. This is why transparency requires traceability of sourcing and practices, to ensure that the chain of events leading to possible unwanted outcomes is not lost in the distributed and dynamic nature of design, development and deployment of AI.

5.3 Human Moral Responsibility

Humans remain the only agents morally responsible for the outcomes of AI systems deployed for defence purposes. While AI systems can be considered moral agents, insofar as they perform actions that have a moral value (Floridi & Sanders, 2004), they cannot be held morally responsible for those actions.

However, ascribing responsibilities to humans for the actions of AI systems has proved to be problematic, due to the distributed and interconnected ways in which AI is developed and the lack of transparency and predictability of its outcomes. Two approaches can be followed to enable fair processes to ascribe responsibilities:

- following the chain of command, control and communication;
- faultless, back-propagation approach.

They can be described more simply as a 'linear' and a 'radial' approach, respectively. These two approaches are complementary and serve the twin purposes of addressing unwanted consequences, mis- and overuses of AI and to foster a self-improving dynamic in the network of agents involved in the design, development, and deployment of AI for defence.

According to the linear approach, responsibility is attributed following the chain of command, control and communication. In this case, decision-makers are held responsible for the unwanted consequences of AI, whether these result from failures of AI systems, unpredictability of outcomes or bad decisions. In order to ascribe responsibility fairly, it is essential that the decision-makers have adequate information and *understanding* of the way the specific AI system works in the given context, of its robustness, of the risks that it may deliver unpredicted (and unwanted) outcomes, of the required level of meaningful control, and of the dangers that may follow if the AI systems fails to behave according to expectations. The linear approach entails a certain epistemic threshold. This means that the use of AI must be coupled with proper training of the personnel, both those who decide to deploy AI systems and those who use it, so that they understand the ways in which AI systems work, risks and benefits linked to the systems, and the ethical and legal implications of the decision to deploy AI. This approach rests on the idea that informed decision-makers choosing to use AI do so while being aware of the risks that this may imply and take responsibility for it.

The radial approach is useful to address unwanted outcomes of AI systems that do not stem from bad intentions or follow from actions that are morally neutral per se. This approach addresses unethical consequences that spur from the convergence of different, independent, morally neutral factors. In the relevant literature this has been defined as *faultless responsibility* (Floridi, 2016b). It refers to contexts in which, while it is possible to identify the causal chain of agents and actions that led to a morally good/bad outcome, it is not possible to attribute intent to perform morally good/bad actions to any of those agents individually and, therefore, all the agents are held morally responsible for that outcome insofar as they are part of the network which determined it.

This is not an entirely new approach, as it is akin to the legal concept of strict liability. According to strict liability, legal responsibility for unwanted outcomes is attributed to one or more agents for the damage caused by their actions or omissions, irrespective of the intentionality of the action and feasibility of control. When considering human-machine teaming -- the integration of AI systems in defence infrastructures, decision-making processes, and operations -- what one needs to show to attribute moral responsibility according to the radial approach is that

> some evil has occurred in the system, and that the actions in question caused such evil, but it is not necessary to show exactly whether the agents/sources of such actions were careless, or whether they did not intend to cause them (Floridi, 2016b, 8).

All the agents of the network are then held maximally responsible for the outcome of the network. As Floridi (2016b) stresses, this approach does not aim at distributing reward and punishment for the actions of a system, rather it aims at establishing a feedback mechanism that incentivises all the agents in the network to improve its outcomes ↓ if all the agents are morally responsible, they may become more cautious and careful and this may reduce the risk of unwanted outcomes. This becomes quite effective when, for example, the moral responsibility is linked to the reputation of the agents.

5.4 Meaningful Human Control

It follows from the previous principle that the deployment of AI should always envisage meaningful forms of human control. Meaningful forms of human control must be in place to limit the risks that the outcome of AI systems will not meet the original intent, to identify promptly mistakes and unintended consequences, as well as to ensure timely intervention on, or deactivation of, the systems, should this be necessary.

The concept of meaningful control has been discussed widely in the relevant literature on LAWS and indeed when considering these systems, control is a key element to consider. However, meaningful control is necessary also when considering uses of AI that may not lead to the use of force. This is because military systems must be able to function safely and effectively under a wide range of highly dynamic environments and use cases that are hard to predict or anticipate during the design phase. They must also be resilient to failure and to complex, uncertain and unpredictable events and situations where the dynamics of the military domain necessitate complex judgements regarding acceptable actions based on rules of engagement, international law and judgements over legality, proportionality and risk. Because of this the maintenance of Human Control through a combination of specification, design, training, operating procedures, and assurance processes is seen as critical in many, if not all military systems (Boardman & Butcher, 2019, 2).

Meaningful human control of AI is characterised as dynamic, multidimensional and situation dependent and it can be exercised focusing on different aspects of the human-machine team. For example, the Stockholm International Peace Research Institute and the International Committee of the Red Cross identify three main aspects of human control of weapon systems: the weapon system's parameters of use, the environment, and human-machine interaction (Boulanin et al., 2020). More aspects can also be considered. For example, Boardman and Butcher (2019) suggest that control should not just be meaningful but 'appropriate', insofar as it should be exercised in such a way to ensure that the human involvement in the decision-making process remains significant without impairing system performance.

While meaningful control can be dynamic, multidimensional and situationally-dependent; the principle that prescribes it is only effective insofar as it defines a lower threshold below which control is so minimal to become irrelevant. Hence, the principle can be implemented minimally and maximally. Minimally, the implementation of this principle requires having a human *on* the loop able to understand the functioning of the system and its implications and with the ability to 'unplug' the system timely and effectively. Maximally, the principle requires individuals in charge of AI systems to combine technical, legal and ethical training to ensure that the decision *to let the system work* is informed by all relevant dimensions and not a mere vetting of the system.

Therefore, the principle does not admit *fire and forget* uses of AI, as it considers control as an element which can be modulated with respect to a rigorous risk assessment of unintended consequences, and related negative impact on national defence and international stability. Where even lower levels of meaningful control cannot be complemented with these assessments, the use of AI systems is ethically unwarranted. It should be noted that the principle is best implemented when protocols for the attribution of responsibilities for misuses of AI and mistakes made by AI systems are in place alongside effective redressing and remedy processes. The attribution of responsibility hinges on the respect of transparency.

5.5 Reliable AI Systems

Defence organisation using AI systems must establish meaningful monitoring of the execution of the tasks delegated to AI. The monitoring should be adequate to the learning nature of the systems, and their lack of transparency, while remaining feasible in terms of resources, especially time, and hence computational feasibility.

AI has a poor shock response (robustness) and any slight alterations to inputs can degrade a model disproportionately (Rigaki & Elragal, 2017). Thus, deploying on AI for defence purposes could favour opponents (Brundage et al., 2018; Taddeo et al., 2019), if the system is not deployed according to procedure that envisage forms of monitoring and prompt intervention in case of mistakes or system degradation. This is why this principle prescribes monitoring of the systems throughout their deployment on top of having in place measures that verify and validate the systems and assess their robustness.

Monitoring may include new forms of procurement that envisage an active role of the defence institutions in the design and development process; in house design and development of AI models; use of data for system training and testing collected, curated and validated directly by the systems providers and maintained securely; mandatory forms of adversarial training with appropriate levels of refinement of AI models to test their robustness; sparring training of AI models; monitoring the output of AI systems deployed in the wild with some form of *in silico* baseline model, as suggested by (Taddeo et al., 2019).

As stressed in Sect. 2, AI systems are autonomous, self-learning agents interacting with the environment. Their behaviour depends as much on the inputs they are fed and interactions with other agents once deployed as it does on their design and training. Responsible uses of AI for defence purposes need to take into account the autonomous, dynamic, and self-learning nature of AI systems, and start envisaging forms of monitoring that span from the design to the deployment stages.

6 Conclusion

As we mentioned at the beginning of this chapter, ethical principles for the use of AI in defence do not undermine International Humanitarian Laws. Rather, they offer guidance both with respect to what can be done post-compliance as well as with respect to those uses of AI in defence which International Humanitarian Laws do not address or do not address clearly. At the same time, the principles need to be logically consistent with the broader ethical principles underpinning the wide set of uses of AI in our societies, like for example the OECD principles, and with the values shaping defence institutions and their role in democratic societies. For example, the US DIB states clearly in its documents that its principles rest on International Humanitarian Law, as well as on core values of the US Armed Forces. This consistency is important, for it ensures that despite the domain-depended differences, ethical principles shaping the uses of AI in defence remain coherent with fundamental principles of our societies. This consistency is crucial, for it will shape the trade-offs among the proposed principles that will have to be made from time to time and which will vary with the context of deployment.

Finally, we would like to conclude our analysis with a warning. These principles should not be followed as an algorithm, they do not offer a set of instructions that if followed slavishly ensure ethically sound outcomes. They offer guidelines to spur and articulate ethical considerations with respect to the uses of AI in defence. To this end, it is key that both humans making the decision to use AI and those executing these decision are able to take into account the principles offered in this chapter, along with knowledge of legal and technical aspects of AI with the aim to reconcile different principles, interests, and goals, without breaching fundamental values and rights of our societies.

Acknowledgement We are very grateful to Isaac Taylor for his work and comments on an early version of this chapter and to Rebecca Hogg and the participants of the 2020 Dstl AI Fest for their questions and comments, for they enabled us to improve several aspects of our analysis. We are responsible for any remaining mistakes.

Funding Information Mariarosaria Taddeo and Alexander Blanchard's work on this chapter has been funded by the Dstl Ethics Fellowship held at the Alan Turing Institute. The research underpinning this work was funded by the UK Defence Chief Scientific Advisor's Science and Technology Portfolio, through the Dstl Autonomy Programme. This chapter is an overview of UK Ministry of Defence (MOD) sponsored research and is released for informational purposes only. The contents of this paper should not be interpreted as representing the views of the UK MOD, nor should it be assumed that they reflect any current or future UK MOD policy. The information contained in this chapter cannot supersede any statutory or contractual requirements or liabilities and is offered without prejudice or commitment.

References

Acalvio Autonomous Deception. (2019). Acalvio. 2019. https://www.acalvio.com/

Asaro, P. (2012). On banning Autonomous weapon systems: Human rights, automation, and the dehumanization of lethal decision-making. *International Review of the Red Cross, 94*(886), 687–709. https://doi.org/10.1017/S1816383112000768

BehavioSec: Continuous Authentication Through Behavioral Biometrics. (2019). BehavioSec. 2019. https://www.behaviosec.com/

Boardman, M., & Butcher, F. (2019). *An exploration of maintaining human control in AI enabled systems and the challenges of achieving it.* STO-MP-IST-178.

Boulanin, V., Carlsson, M. P., Goussac, N., & Davidson, D. (2020). *Limits on autonomy in weapon systems: Identifying practical elements of human control.* Stockholm International Peace Research Institute and the International Committee of the Red Cross. https://www.sipri.org/publications/2020/other-publications/limits-autonomy-weapon-systems-identifying-practical-elements-human-control-0

Brundage, M., Avin, S., Clark, J., Toner, H., Eckersley, P., Garfinkel, B., & Dafoe, A., et al. (2018). The malicious use of artificial intelligence: Forecasting, prevention, and mitigation. ArXiv:1802.07228 [Cs], February. http://arxiv.org/abs/1802.07228

Brunstetter, D., & Braun, M. (2013). From jus ad bellum to jus ad vim: Recalibrating our understanding of the moral use of force. *Ethics & International Affairs, 27*(01), 87–106. https://doi.org/10.1017/S0892679412000792

DarkLight Offers First of Its Kind Artificial Intelligence to Enhance Cybersecurity Defenses. (2017). *Business Wire.* 26 July 2017. https://www.businesswire.com/news/home/20170726005117/en/DarkLight-Offers-Kind-Artificial-Intelligence-Enhance-Cybersecurity

DeepLocker: How AI Can Power a Stealthy New Breed of Malware. (2018). *Security intelligence* (blog). 8 August 2018. https://securityintelligence.com/deeplocker-how-ai-can-power-a-stealthy-new-breed-of-malware/

Department for Digital, Culture, Media & Sport. (2018). *Data Ethics Framework*. https://www. gov.uk/government/publications/data-ethics-framework/data-ethics-framework

DIB. (2020a). *AI principles: Recommendations on the ethical use of Artificial Intelligence by the Department of Defense*. https://media.defense.gov/2019/Oct/31/2002204458/-1/-1/0/DIB_AI_PRINCIPLES_PRIMARY_DOCUMENT.PDF

DIB. (2020b). *AI principles: Recommendations on the ethical use of Artificial Intelligence by the Department of Defense - supporting document*. Defence Innovation Board [DIB]. https://media. defense.gov/2019/Oct/31/2002204459/-1/-1/0/DIB_AI_PRINCIPLES_SUPPORTING_DOCUMENT.PDF

Docherty, B. (2014). *Shaking the foundations: The human rights implications of killer robots*. Human Rights Watch. https://www.hrw.org/report/2014/05/12/shaking-foundations/human-rights-implications-killer-robots

Ekelhof, M. (2019). Moving beyond semantics on Autonomous weapons: Meaningful human control in operation. *Global Policy, 10*(3), 343–348. https://doi.org/10.1111/1758-5899.12665

Enemark, C. (2011). Drones over Pakistan: Secrecy, ethics, and counterinsurgency. *Asian Security, 7*(3), 218–237. https://doi.org/10.1080/14799855.2011.615082

Floridi, L. (2008). The method of levels of abstraction. *Minds and Machines, 18*(3), 303–329. https://doi.org/10.1007/s11023-008-9113-7

Floridi, L. (2016a). Mature information societies—A matter of expectations. *Philosophy & Technology, 29*(1), 1–4. https://doi.org/10.1007/s13347-016-0214-6

Floridi, L. (2016b). Faultless responsibility: On the nature and allocation of moral responsibility for distributed moral actions. *Philosophical Transactions of the Royal Society A: Mathematical, Physical and Engineering Sciences, 374*(2083), 20160112. https://doi. org/10.1098/rsta.2016.0112

Floridi, L. (2018). Soft ethics and the governance of the digital. *Philosophy & Technology, 31*(1), 1–8. https://doi.org/10.1007/s13347-018-0303-9

Floridi, L., & Cowls, J. (2019). A unified framework of five principles for AI in Society. *Harvard Data Science Review*. https://doi.org/10.1162/99608f92.8cd550d1.

Floridi, L., Cowls, J., King, T. C., & Taddeo, M. (2020). How to design AI for social good: Seven essential factors. *Science and Engineering Ethics, 26*(3), 1771–1796. https://doi.org/10.1007/s11948-020-00213-5

Floridi, L., & Sanders, J. W. (2004). On the morality of artificial agents. *Minds and Machines, 14*(3), 349–379. https://doi.org/10.1023/B:MIND.0000035461.63578.9d

Fraga-Lamas, P., Fernández-Caramés, T. M., Suárez-Albela, M., Castedo, L., & González-López, M. (2016). A review on internet of things for defense and public safety. *Sensors (Basel, Switzerland), 16*(10). https://doi.org/10.3390/s16101644

Gavaghan, C., Knott, A., Maclaurin, J., Zerilli, J., & Liddicoat, J. (2019). Government use of artificial intelligence in New Zealand, Final report on phase 1 of the law Foundation's artificial intelligence and law in New Zealand project'. In *New Zealand Law Foundation*. https://www. cs.otago.ac.nz/research/ai/AI-Law/NZLF%20report.pdf

International Telecommunications Union. (2017). *Minimum Requirements Related to Technical Performance for IMT-2020 Radio Interface(s)*. 2017. https://www.itu.int/pub/R-REP-M.2410-2017

Japanese Society for Artificial Intelligence [JSAI]. (2017). *Ethical Guidelines*. http://ai-elsi.org/wp-content/uploads/2017/05/JSAI-Ethical-Guidelines-1.pdf

Jobin, A., Ienca, M., & Vayena, E. (2019). The global landscape of AI ethics guidelines. *Nature Machine Intelligence, 1*(9), 389–399. https://doi.org/10.1038/s42256-019-0088-2

Johnson, A. M., & Axinn, S. (2013). The morality of Autonomous robots. *Journal of Military Ethics, 12*(2), 129–141. https://doi.org/10.1080/15027570.2013.818399

King, T. M., Arbon, J., Santiago, D., Adamo, D., Chin, W., & Shanmugam, R. (2019). AI for testing today and tomorrow: Industry perspectives. In *2019 IEEE international conference on Artificial Intelligence Testing (AITest)* (pp. 81–88). IEEE. https://doi.org/10.1109/AITest.2019.000-3

Kott, A., Swami, A., & West, B. J. (2017). *The internet of Battle things*. ArXiv:1712.08980 [Cs], December. http://arxiv.org/abs/1712.08980

Lysaght, R. J., Harris, R., & Kelly, W. (1988). *Artificial intelligence for command and control*. ANALYTICS INC WILLOW GROVE PA. https://apps.dtic.mil/docs/citations/ADA229342

McMahan, J. (2013). Forward. In R. Jenkins, M. Robillard, & B. J. Strawser (Eds.), *Who should die? The ethics of killing in war* (pp. ix–xiv). Oxford University Press.

Mirsky,Y.,Mahler,T.,Shelef,I.,&Elovici,Y.(2019).*CT-GAN:Malicioustamperingof3Dmedicalimagery using deep learning*. ResearchGate. https://www.researchgate.net/publication/330357848_CT-GAN_Malicious_Tampering_of_3D_Medical_Imagery_using_Deep_Learning/figures?lo=1

Mökander, J., & Floridi, L. (2021). Ethics-based auditing to develop trustworthy AI. *Minds and Machines*, 1–5. https://doi.org/10.1007/s11023-021-09557-8

NATO. (2020). *NATO 2030: United for a new era*. Brussels. https://www.nato.int/nato_static_fl2014/assets/pdf/2020/12/pdf/201201-Reflection-Group-Final-Report-Uni.pdf

O'Connell, M. E. (2014). The American way of bombing: How legal and ethical norms change. In M. Evangelista & H. Shue (Eds.), *The American way of bombing changing ethical and legal norms, from flying fortresses to drones*. Cornel University Press.

Rigaki, M., & Elragal, A. (2017). *Adversarial deep learning against intrusion detection classifiers* (p. 14). Luleå tekniska universitet, Datavetenskap.

Roberts, H., Cowls, J., Morley, J., Taddeo, M., Wang, V., & Floridi, L. (2020). The Chinese approach to artificial intelligence: An analysis of policy, ethics, and regulation. *AI & SOCIETY, 36*. https://doi.org/10.1007/s00146-020-00992-2

Schubert, J., Brynielsson, J., Nilsson, M., & Svenmarck, P. (2018). *Artificial intelligence for decision support in command and control systems*, p. 15.

Sharkey, A. (2019). Autonomous weapons systems, killer robots and human dignity. *Ethics and Information Technology, 21*(2), 75–87. https://doi.org/10.1007/s10676-018-9494-0

Sharkey, N. (2010). Saying "no!" to lethal Autonomous targeting. *Journal of Military Ethics, 9*(4), 369–383. https://doi.org/10.1080/15027570.2010.537903

Sharkey, N. (2012a). Killing made easy: From joysticks to politics. In P. Lin, K. Abney, & G. Bekey (Eds.), *Robot ethics: The ethical and social implications of robotics* (pp. 111–128). MIT Press.

Sharkey, N. E. (2012b). The Evitability of Autonomous robot warfare. *International Review of the Red Cross, 94*(886), 787–799. https://doi.org/10.1017/S1816383112000732

Sparrow, R. (2007). Killer Robots. *Journal of Applied Philosophy, 24*(1), 62–77. https://doi.org/10.1111/j.1468-5930.2007.00346.x

Sparrow, R. (2016). Robots and respect: Assessing the case against Autonomous weapon systems. *Ethics & International Affairs, 30*(1), 93–116. https://doi.org/10.1017/S0892679415000647

Taddeo, M. (2012a). Information warfare: A philosophical perspective. *Philosophy and Technology, 25*(1), 105–120.

Taddeo, M. (2012b). An analysis for a just cyber warfare. In *Fourth international conference of cyber conflict*. NATO CCD COE and IEEE Publication.

Taddeo, M. (2013). Cyber security and individual rights, striking the right balance. *Philosophy & Technology, 26*(4), 353–356. https://doi.org/10.1007/s13347-013-0140-9

Taddeo, M. (2014a). Just information warfare. *Topoi*, 1–12. https://doi.org/10.1007/s11245-014-9245-8

Taddeo, M. (2014b). The struggle between liberties and authorities in the information age. *Science and Engineering Ethics*, 1–14. https://doi.org/10.1007/s11948-014-9586-0

Taddeo, M. (2017a). The limits of deterrence theory in cyberspace. *Philosophy & Technology, 31*. https://doi.org/10.1007/s13347-017-0290-2

Taddeo, M. (2017b). Trusting Digital Technologies Correctly. *Minds and Machines 27*(4), 565–68. https://doi.org/10.1007/s11023-017-9450-5.

Taddeo, M. (2019a). The challenges of cyber deterrence. In C. Öhman & D. Watson (Eds.), *The 2018 yearbook of the digital ethics lab* (pp. 85–103). Springer. https://doi.org/10.1007/978-3-030-17152-0_7

Taddeo, M. (2019b). Three ethical challenges of applications of artificial intelligence in cyber-security. *Minds and Machines, 29*(2), 187–191. https://doi.org/10.1007/s11023-019-09504-8

Taddeo, M., & Floridi, L. (2018). Regulate artificial intelligence to avert cyber arms race. *Nature, 556*(7701), 296–298. https://doi.org/10.1038/d41586-018-04602-6

Taddeo, M., McCutcheon, T., & Floridi, L. (2019). Trusting artificial intelligence in cyberse-curity is a double-edged sword. *Nature Machine Intelligence, 1*(12), 557–560. https://doi.org/10.1038/s42256-019-0109-1

Tamburrini, G. (2016). On banning autonomous weapons systems: From deontological to wide consequentialist reasons. In B. Nehal, S. Beck, R. Geiß, H.-Y. Liu, & C. Kreß (Eds.), *Autonomous weapons systems: Law, ethics, policy* (pp. 122–142). Cambridge University Press.

The UK and International Humanitarian Law 2018. (n.d.) Accessed 1 Nov 2020. https://www.gov.uk/government/publications/international-humanitarian-law-and-the-uk-government/uk-and-international-humanitarian-law-2018

US Army. (2017). *Robotic and autonomous systems strategy.* https://www.tradoc.army.mil/Portals/14/Documents/RAS_Strategy.pdf

Yang, G.-Z., Bellingham, J., Dupont, P. E., Fischer, P., Floridi, L., Full, R., Jacobstein, N., et al. (2018). The grand challenges of science robotics. *Science Robotics, 3*(14), eaar7650. https://doi.org/10.1126/scirobotics.aar7650

Zhuge, J., Holz, T., Han, X., Song, C., & Zou, W. (2007). Collecting Autonomous spreading mal-ware using high-interaction honeypots. In S. Qing, H. Imai, & G. Wang (Eds.), *Information and communications security* (Lecture Notes in Computer Science) (pp. 438–451). Springer.

Milton Keynes UK
Ingram Content Group UK Ltd.
UKHW020607151123
432609UK00002B/13